GLENCOE

MATH

YOUR COMMON CORE EDITION

CCSS

AUTHORS

Carter • Cuevas • Day • Malloy • Kersaint • Luchin • McClain
Molix-Bailey • Price • Reynosa • Silbey • Vielhaber • Willard

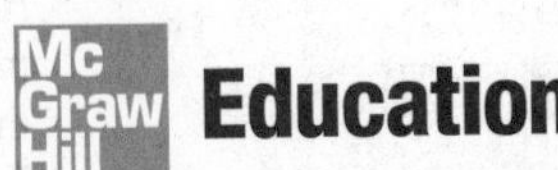

Bothell, WA • Chicago, IL • Columbus, OH • New York, NY

connectED.mcgraw-hill.com

The McGraw·Hill Companies

Education

STEM McGraw-Hill is committed to providing instructional materials in Science, Technology, Engineering, and Mathematics (STEM) that give all students a solid foundation, one that prepares them for college and careers in the 21st century.

Send all inquiries to:
McGraw-Hill Education
8787 Orion Place
Columbus, OH 43240

ISBN: 978-0-07-660553-8 (*Volume 1*)
MHID: 0-07-660553-1

Printed in the United States of America.

16 17 18 19 RMN 19 18 17 16

Our mission is to provide educational resources that enable students to become the problem solvers of the 21st century and inspire them to explore careers within Science, Technology, Engineering, and Mathematics (STEM) related fields.

CONTENTS IN BRIEF

Units organized by CCSS domain

Glencoe Math is organized into units based on groups of related standards called domains. This year, you will study and understand the five domains shown below.

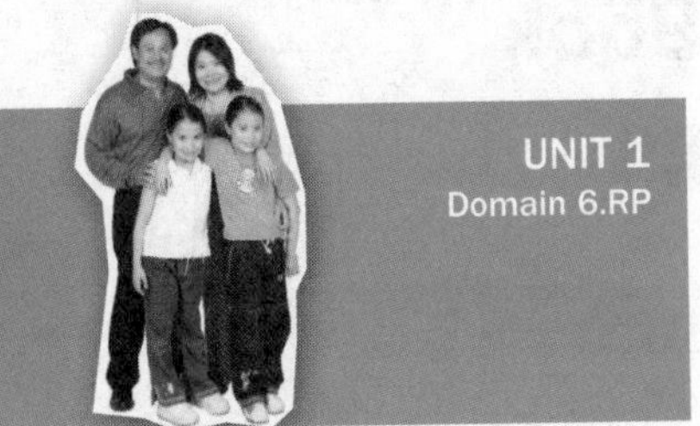

UNIT 1
Domain 6.RP

Ingram Publishing/SuperStock

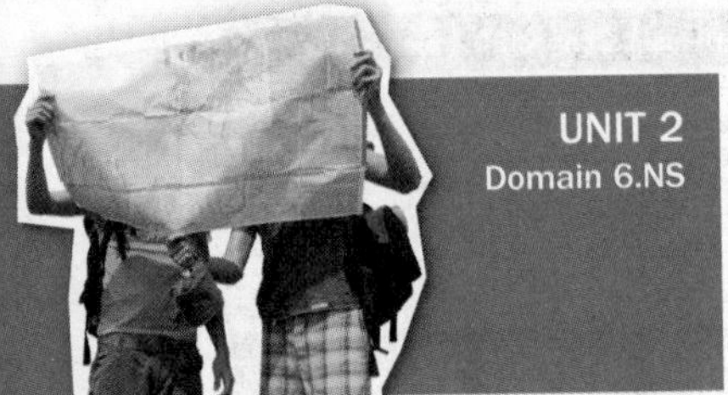

UNIT 2
Domain 6.NS

LatinStock Collection/Alamy

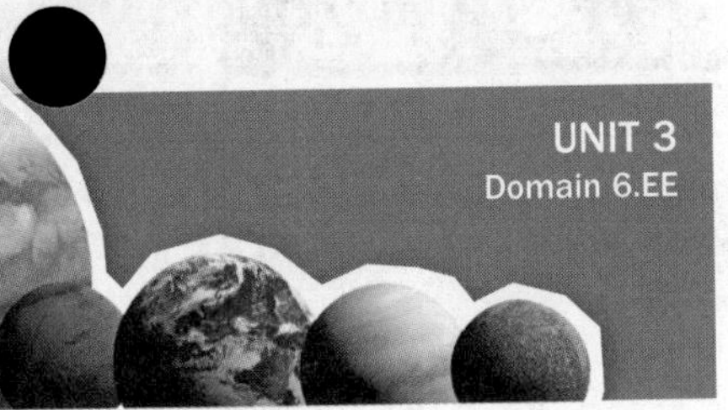

UNIT 3
Domain 6.EE

Antonio M. Rosario/Photographer's Choice/Getty Images

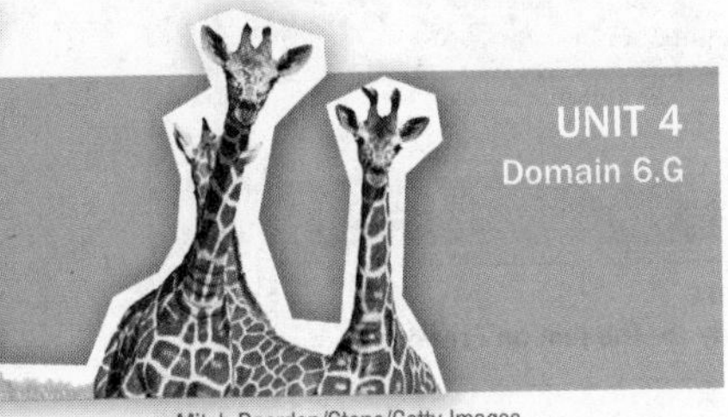

UNIT 4
Domain 6.G

Mitch Reardon/Stone/Getty Images

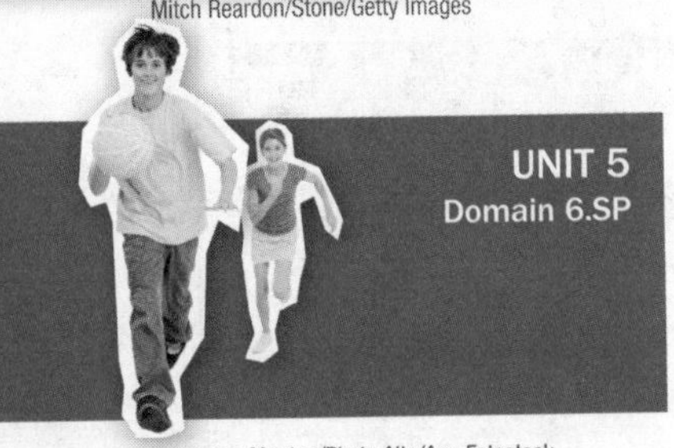

UNIT 5
Domain 6.SP

Laurence Mouton/Photo Alto/Age Fotostock

it's all at connectED.mcgraw-hill.com

Go to the Student Center for your eBook, Resources, Homework, and Messages.

Glencoe Math: Student Center

Hello, Student | Home | ConnectED | Help | Logout

GLENCOE MATH

Search | Standards

Student Center

Home | Homework | Resources

Unit 4 > Chapter 6: Transformations

Lesson 4: Dilation

Open eBook

Unit 4: Geometry > Chapter 6: Transformations > Lesson 4: Dilation

Geometry

Homework

Due: Monday, October 7, 2011

Assignment Title

You have unread teacher comments.

More

Messages

Monday, October 7, 2011

Don't forget to study for the test on Friday.

More

The McGraw-Hill Companies

Terms of Use | Privacy Policy | Technical Support | Minimum System Requirements

Copyright© The McGraw-Hill Companies, Inc.

Write your Username ______ Password ______

Get your resources online to help you in class and at home.

Vocab

Find activities for building vocabulary.

Watch

Watch animations and videos.

Tutor

See a teacher illustrate examples and problems.

Tools

Explore concepts with virtual manipulatives.

Check

Self-assess your progress.

eHelp

Get targeted homework help.

Masters

Provides practice worksheets.

GO mobile

Scan this QR code with your smart phone* or visit mheonline.com/apps.

*May require quick response code reader app.

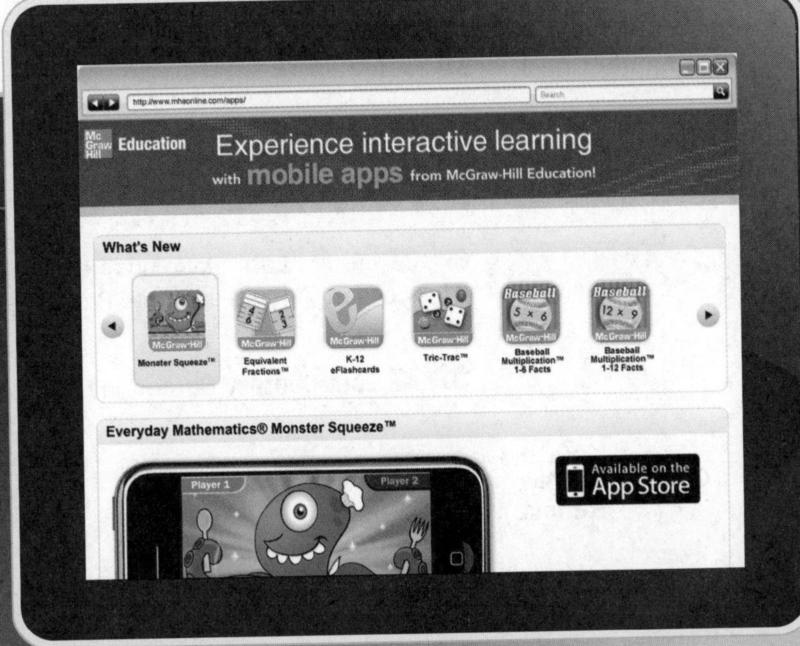

CCSS **UNIT 1** Ratios and Proportional Relationships

UNIT PROJECT PREVIEW
page 2

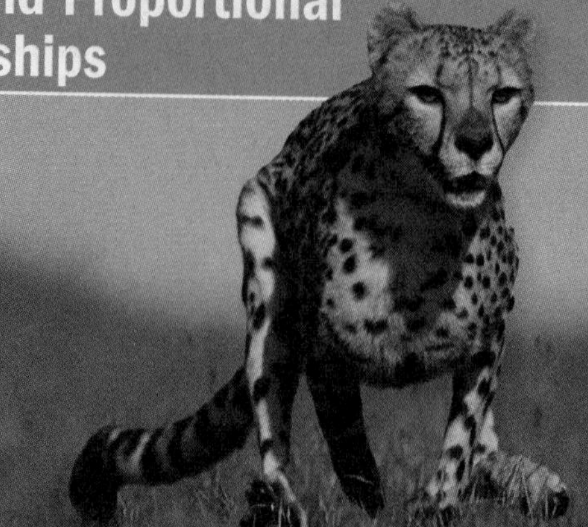

Chapter 1
Ratios and Rates

Real World
p. 31

Essential Question

HOW do you use equivalent rates in the real world?

Chapter 2
Fractions, Decimals, and Percents

Essential Question

WHEN is it better to use a fraction, a decimal or a percent?

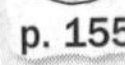

p. 155

UNIT 2 The Number System

UNIT PROJECT PREVIEW
page 172

Chapter 3
Compute with Multi-Digit Numbers

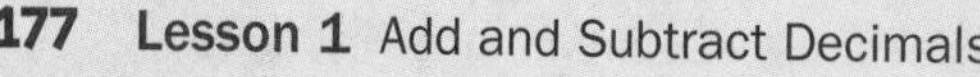

Essential Question

HOW can estimating be helpful?

p. 215

Chapter 4
Multiply and Divide Fractions

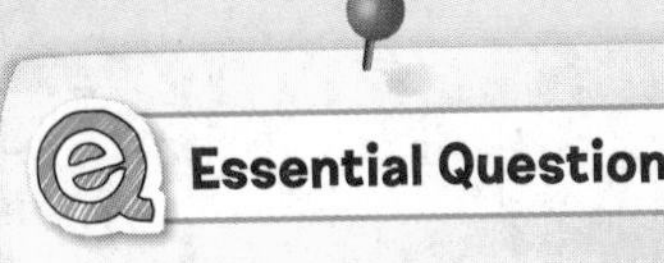

Essential Question

WHAT does it mean to multiply and divide fractions?

Chapter 5
Integers and the Coordinate Plane

Essential Question

HOW are integers and absolute value used in real-world situations?

Real World
p. 387

Get Out the Map!

UNIT 3 Expressions and Equations

UNIT PROJECT PREVIEW page 424

Chapter 6 Expressions

Essential Question

HOW is it helpful to write numbers in different ways?

p. 495

Chapter 7
Equations

Essential Question

HOW do you determine if two numbers or expressions are equal?

Chapter 8
Functions and Inequalities

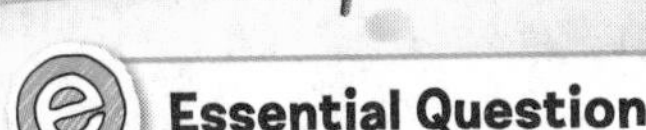

HOW are symbols, such as <, >, and =, useful?

p. 595

Chapter 9
Area

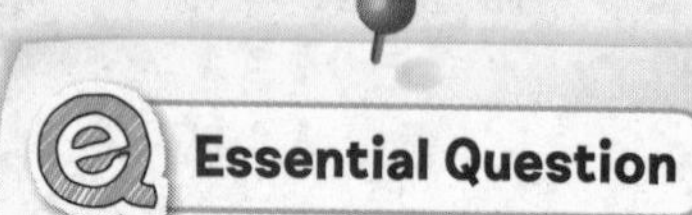

HOW does measurement help you solve problems in everyday life?

Chapter 10 Volume and Surface Area

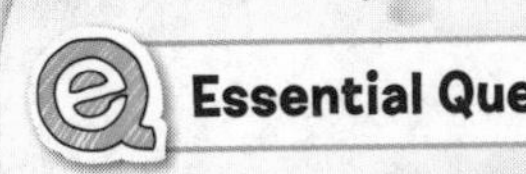

HOW is shape important when measuring a figure?

p. 783

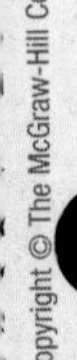

UNIT 5 Statistics and Probability

UNIT PROJECT PREVIEW page 800

Chapter 11 Statistical Measures

Essential Question

HOW are the mean, median, and mode helpful in describing data?

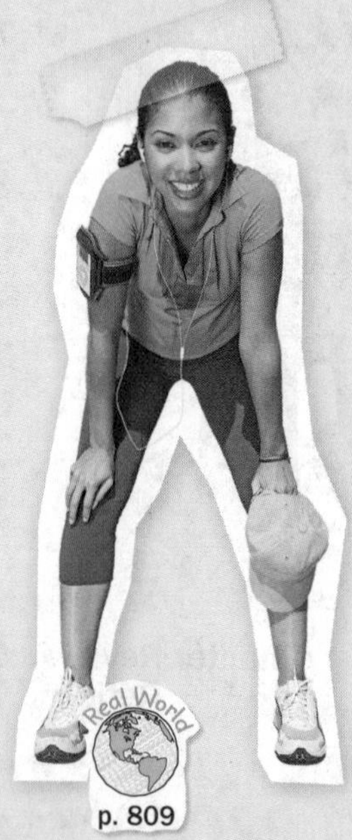

Chapter 12 Statistical Displays

p. 891

Essential Question

WHY is it important to carefully evaluate graphs?

Common Core State Standards for MATHEMATICS, Grade 6

Glencoe Math, Course 1, focuses on four critical areas: (1) using concepts of ratio and rate to solve problems; (2) understanding division of fractions; (3) using expressions and equations; and (4) understanding of statistical reasoning.

Content Standards

Domain 6.RP **Ratios and Proportional Relationships**

- Understand ratio concepts and use ratio reasoning to solve problems.

Domain 6.NS **The Number System**

- Apply and extend previous understandings of multiplication and division to divide fractions by fractions.
- Compute fluently with multi-digit numbers and find common factors and multiples.
- Apply and extend previous understandings of numbers to the system of rational numbers.

Domain 6.EE **Expressions and Equations**

- Apply and extend previous understandings of arithmetic to algebraic expressions.
- Reason about and solve one-variable equations and inequalities.
- Represent and analyze quantitative relationships between dependent and independent variables.

Domain 6.G **Geometry**

- Solve real-world and mathematical problems involving area, surface area, and volume.

Domain 6.SP **Statistics and Probability**

- Develop understanding of statistical variability.
- Summarize and describe distributions.

Mathematical Practices

1. Make sense of problems and persevere in solving them.
2. Reason abstractly and quantitatively.
3. Construct viable arguments and critique the reasoning of others.
4. Model with mathematics.
5. Use appropriate tools strategically.
6. Attend to precision.
7. Look for and make use of structure.
8. Look for and express regularity in repeated reasoning.

Ratios and Proportional Relationships

Domain 6.RP

Understand ratio concepts and use ratio reasoning to solve problems.

1. Understand the concept of a ratio and use ratio language to describe a ratio relationship between two quantities.
2. Understand the concept of a unit rate a/b associated with a ratio $a:b$ with $b \neq 0$, and use rate language in the context of a ratio relationship.
3. Use ratio and rate reasoning to solve real-world and mathematical problems, e.g., by reasoning about tables of equivalent ratios, tape diagrams, double number line diagrams, or equations.
   a. Make tables of equivalent ratios relating quantities with whole number measurements, find missing values in the tables, and plot the pairs of values on the coordinate plane. Use tables to compare ratios.
   b. Solve unit rate problems including those involving unit pricing and constant speed.
   c. Find a percent of a quantity as a rate per 100 (e.g., 30% of a quantity means 30/100 times the quantity); solve problems involving finding the whole, given a part and the percent.
   d. Use ratio reasoning to convert measurement units; manipulate and transform units appropriately when multiplying or dividing quantities.

Related Unit Project: Unit 1
People Everywhere

The Number System

Domain 6.NS

Apply and extend previous understandings of multiplication and division to divide fractions by fractions.

1. Interpret and compute quotients of fractions, and solve word problems involving division of fractions by fractions, e.g., by using visual fraction models and equations to represent the problem.

Compute fluently with multi-digit numbers and find common factors and multiples.

2. Fluently divide multi-digit numbers using the standard algorithm.
3. Fluently add, subtract, multiply, and divide multi-digit decimals using the standard algorithm for each operation.
4. Find the greatest common factor of two whole numbers less than or equal to 100 and the least common multiple of two whole numbers less than or equal to 12. Use the distributive property to express a sum of two whole numbers 1–100 with a common factor as a multiple of a sum of two whole numbers with no common factor.

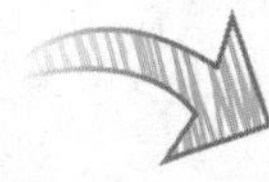

Related Unit Project: Unit 2
Get Out the Map!

Apply and extend previous understandings of numbers to the system of rational numbers.

5. Understand that positive and negative numbers are used together to describe quantities having opposite directions or values (e.g., temperature above/below zero, elevation above/below sea level, credits/debits, positive/negative electric charge); use positive and negative numbers to represent quantities in real-world contexts, explaining the meaning of 0 in each situation.

6. Understand a rational number as a point on the number line. Extend number line diagrams and coordinate axes familiar from previous grades to represent points on the line and in the plane with negative number coordinates.

- a. Recognize opposite signs of numbers as indicating locations on opposite sides of 0 on the number line; recognize that the opposite of the opposite of a number is the number itself, e.g., $-(-3) = 3$, and that 0 is its own opposite.
- b. Understand signs of numbers in ordered pairs as indicating locations in quadrants of the coordinate plane; recognize that when two ordered pairs differ only by signs, the locations of the points are related by reflections across one or both axes.
- c. Find and position integers and other rational numbers on a horizontal or vertical number line diagram; find and position pairs of integers and other rational numbers on a coordinate plane.

7. Understand ordering and absolute value of rational numbers.

- a. Interpret statements of inequality as statements about the relative position of two numbers on a number line diagram.
- b. Write, interpret, and explain statements of order for rational numbers in real-world contexts.
- c. Understand the absolute value of a rational number as its distance from 0 on the number line; interpret absolute value as magnitude for a positive or negative quantity in a real-world situation.
- d. Distinguish comparisons of absolute value from statements about order.

8. Solve real-world and mathematical problems by graphing points in all four quadrants of the coordinate plane. Include use of coordinates and absolute value to find distances between points with the same first coordinate or the same second coordinate.

Domain 6.EE

Expressions and Equations

Apply and extend previous understandings of arithmetic to algebraic expressions.

1. Write and evaluate numerical expressions involving whole-number exponents.

2. Write, read, and evaluate expressions in which letters stand for numbers.

- a. Write expressions that record operations with numbers and with letters standing for numbers.

b. Identify parts of an expression using mathematical terms (sum, term, product, factor, quotient, coefficient); view one or more parts of an expression as a single entity.

c. Evaluate expressions at specific values of their variables. Include expressions that arise from formulas used in real-world problems. Perform arithmetic operations, including those involving whole number exponents, in the conventional order when there are no parentheses to specify a particular order (Order of Operations).

3. Apply the properties of operations to generate equivalent expressions.

4. Identify when two expressions are equivalent (i.e., when the two expressions name the same number regardless of which value is substituted into them).

Related Unit Project: Unit 3
It's Out of This World

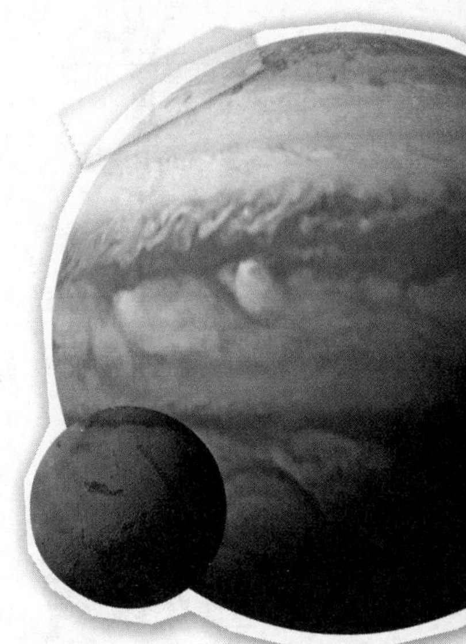

Reason about and solve one-variable equations or inequalities.

5. Understand solving an equation or inequality as a process of answering a question: which values from a specified set, if any, make the equation or inequality true? Use substitution to determine whether a given number in a specified set makes an equation or inequality true.

6. Use variables to represent numbers and write expressions when solving a real-world or mathematical problem; understand that a variable can represent an unknown number, or, depending on the purpose at hand, any number in a specified set.

7. Solve real-world and mathematical problems by writing and solving equations of the form $x + p = q$ and $px = q$ for cases in which p, q and x are all nonnegative rational numbers.

8. Write an inequality of the form $x > c$ or $x < c$ to represent a constraint or condition in a real-world or mathematical problem. Recognize that inequalities of the form $x > c$ or $x < c$ have infinitely many solutions; represent solutions of such inequalities on number line diagrams.

Represent and analyze quantitative relationships between dependent and independent variables.

9. Use variables to represent two quantities in a real-world problem that change in relationship to one another; write an equation to express one quantity, thought of as the dependent variable, in terms of the other quantity, thought of as the independent variable. Analyze the relationship between the dependent and independent variables using graphs and tables, and relate these to the equation.

Geometry

Domain 6.G

Solve real-world and mathematical problems involving area, surface area, and volume.

1. Find the area of right triangles, other triangles, special quadrilaterals, and polygons by composing into rectangles or decomposing into triangles and other shapes; apply these techniques in the context of solving real-world and mathematical problems.

Related Unit Project: Unit 4
A New Zoo

2. Find the volume of a right rectangular prism with fractional edge lengths by packing it with unit cubes of the appropriate unit fraction edge lengths, and show that the volume is the same as would be found by multiplying the edge lengths of the prism. Apply the formulas $V = l\,w\,h$ and $V = b\,h$ to find volumes of right rectangular prisms with fractional edge lengths in the context of solving real-world and mathematical problems.
3. Draw polygons in the coordinate plane given coordinates for the vertices; use coordinates to find the length of a side joining points with the same first coordinate or the same second coordinate. Apply these techniques in the context of solving real-world and mathematical problems.
4. Represent three-dimensional figures using nets made up of rectangles and triangles, and use the nets to find the surface area of these figures. Apply these techniques in the context of solving real-world and mathematical problems.

Domain 6.SP

Statistics and Probability

Develop understanding of statistical variability.

1. Recognize a statistical question as one that anticipates variability in the data related to the question and accounts for it in the answers.
2. Understand that a set of data collected to answer a statistical question has a distribution which can be described by its center, spread, and overall shape.
3. Recognize that a measure of center for a numerical data set summarizes all of its values with a single number, while a measure of variation describes how its values vary with a single number.

Related Unit Project: Unit 5
Let's Exercise

Summarize and describe distributions.

4. Display numerical data in plots on a number line, including dot plots, histograms, and box plots.
5. Summarize numerical data sets in relation to their context, such as by:
   a. Reporting the number of observations.
   b. Describing the nature of the attribute under investigation, including how it was measured and its units of measurement.
   c. Giving quantitative measures of center (median and/or mean) and variability (interquartile range and/or mean absolute deviation), as well as describing any overall pattern and any striking deviations from the overall pattern with reference to the context in which the data were gathered.
   d. Relating the choice of measures of center and variability to the shape of the data distribution and the context in which the data were gathered.

UNIT 1

Ratios and Proportional Relationships

Essential Question

HOW can you use mathematics to describe change and model real-world situations?

Chapter 1
Ratios and Rates

A ratio is a comparison of two quantities by division. In this chapter, you will explore ratio concepts and use ratio reasoning to solve rate problems.

Chapter 2
Fractions, Decimals, and Percents

Equivalent forms of fractions, decimals, and percents can be written and used to solve problems. In this chapter, you will apply these relationships to solve percent problems.

Unit Project Preview

People Everywhere You can use the U.S. Census to compare different characteristics of states' populations, such as the average number of people per square mile, the percent of youth compared to the percent of adults, and the ratio of women to men.

Choose two populations at your school, such as boys and girls, or sixth graders and seventh graders. Then choose a characteristic, such as sports played or number of siblings. Compare the characteristics of the two populations. Report your findings in a double bar graph.

At the end of Chapter 2, you'll complete a project to compare characteristics of the population of different states. Time to get in the right *state* of mind!

Comparing ______________

Kristian Sekulic/Getty Images

Chapter 1
Ratios and Rates

Essential Question

HOW do you use equivalent rates in the real world?

Common Core State Standards

Content Standards
6.RP.1, 6.RP.2, 6.RP.3, 6.RP.3a, 6.RP.3b, 6.NS.4

Mathematical Practices
1, 3, 4, 5, 6, 7, 8

Math in the Real World

Cheetahs are the fastest land animals. They can chase prey by running at speeds of 60 miles per hour.

A cheetah can only maintain top speeds for a short time. If a cheetah runs 1 mile in 60 seconds, fill in the diagram to show how far the cheetah will run in 210 seconds.

210 s	
60 s	
1 mile	

FOLDABLES® Study Organizer

1 Cut out the Foldable on page FL3 of this book.

2 Place your Foldable on page 82.

3 Use the Foldable throughout this chapter to help you learn about ratios and rates.

What Tools Do You Need?

Vocabulary

coordinate plane	origin	unit price
equivalent ratio	prime factorization	unit rate
graph	rate	*x*-axis
greatest common factor	ratio	*x*-coordinate
least common multiple	ratio table	*y*-axis
ordered pair	scaling	*y*-coordinate

Study Skill: Studying Math

New Vocabulary New vocabulary terms are clues about important concepts. Learning new vocabulary words is more than just memorizing the definition. Whenever you see a new vocabulary word, ask yourself:

- How does this fit with what I already know?
- How is this alike or different from something I learned earlier?

Definition from Text	In Your Own Words
A rate is a comparison by division of two quantities with different kinds of units.	A rate compares two amounts with different units using division.
rate	
Examples	**Nonexamples**
• 45 miles per hour • 16 books for 8 students	• 5 black cats out of 15 cats • 3 sugar cookies to 9 cookies

Organize your answers in a word map like the one shown.

Make a word map for *proper fraction*.

Definition	In Your Own Words
proper fraction	
Examples	**Nonexamples**
•	•
•	•

When Will You Use This?

Your Turn! You will solve this problem in the chapter.

Try the Quick Check below.
Or, take the Online Readiness Quiz.

CCSS Quick Review

Common Core Review 5.NBT.6, 5.NF.5b

Example 1

Find $6\overline{)348}$.

$$\begin{array}{r} 58 \\ 6\overline{)348} \\ -30 \\ \hline 48 \\ -48 \\ \hline 0 \end{array}$$

Divide each place-value position from left to right.

Since 48 – 48 = 0, there is no remainder.

Example 2

Write $\frac{40}{64}$ in simplest form.

$$\frac{40}{64} \overset{\div 8}{=} \frac{5}{8}$$

Divide the numerator and denominator by the greatest common factor (GCF), 8.

Since the GCF of 5 and 8 is 1, the fraction $\frac{5}{8}$ is in simplest form.

Quick Check

Divide Whole Numbers **Find each quotient.**

1. $3\overline{)87}$

2. $8\overline{)584}$

3. $52\overline{)312}$

Simplify Fractions **Write each fraction in simplest form.**

4. $\frac{32}{48}$ = ______

5. $\frac{7}{28}$ = ______

6. $\frac{15}{25}$ = ______

7. An airplane has flown 260 miles out of a total trip of 500 miles. What fraction, in simplest form, of the trip has been completed?

______

Which problems did you answer correctly in the Quick Check? Shade those exercise numbers below.

① ② ③ ④  ⑥ ⑦

Lesson 1

Factors and Multiples

What You'll Learn

Scan the lesson. List two headings you would use to make an outline of the lesson.

- ______________________
- ______________________

Essential Question

HOW do you use equivalent rates in the real world?

Vocabulary

greatest common factor
least common multiple

Common Core State Standards

Content Standards
6.NS.4

Mathematical Practices
1, 3, 4, 8

Vocabulary Start-Up

A *common factor* is a number that is a factor of two or more numbers. The greatest of the common factors of two or more numbers is called the **greatest common factor** (GCF).

The least number that is a multiple of two or more whole numbers is the **least common multiple** (LCM) of the numbers.

Fill in the charts below.

GCF
• stands for:
Define: • Greatest
• Common
• Factor

LCM
• stands for:
Define: • Least
• Common
• Multiple

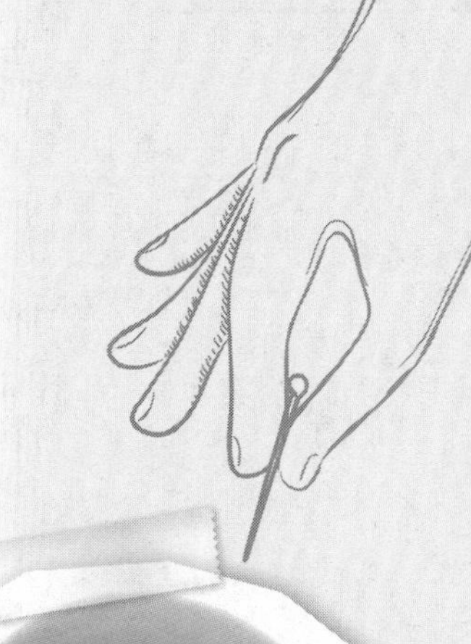

Real-World Link

Bryan is making balloon arrangements. He has 8 blue and 12 green balloons. What is the greatest amount of arrangements he can make if he wants them to be identical? ______________

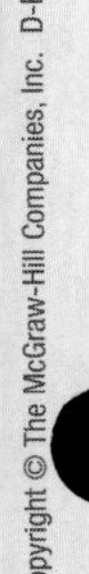

Work Zone

Prime Numbers
Remember that a prime number is a whole number that has exactly two factors, 1 and the number itself.

Find the Greatest Common Factor

You can use common factors or prime factors to find the GCF.

Example

Tutor

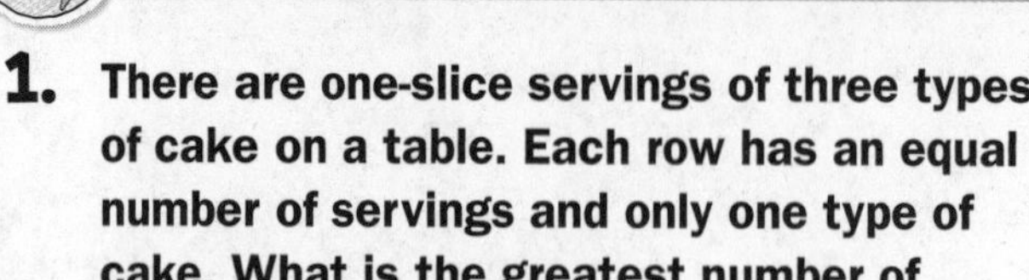

1. **There are one-slice servings of three types of cake on a table. Each row has an equal number of servings and only one type of cake. What is the greatest number of servings in each row?**

Cakes	
Type	**Number of Servings**
marble	10
red velvet	15
chocolate	20

To solve this problem, use common factors.

factors of 10: 1, 2, 5, 10

factors of 15: 1, 3, 5, 15

factors of 20: 1, 2, 4, 5, 10, 20 — The common factors are 1 and 5.

The GCF of 10, 15, and 20 is 5. So, the greatest number of pieces of cake that can be placed in each row is 5.

Got It? **Do this problem to find out.**

Show your work.

a. Lana earned \$49 on Friday, \$42 on Saturday, and \$21 on Sunday selling bracelets. She sold each bracelet for the same amount. What is the most she could have charged for each bracelet?

a. ________

Example

Tutor

2. **Find the GCF of 12 and 18.**

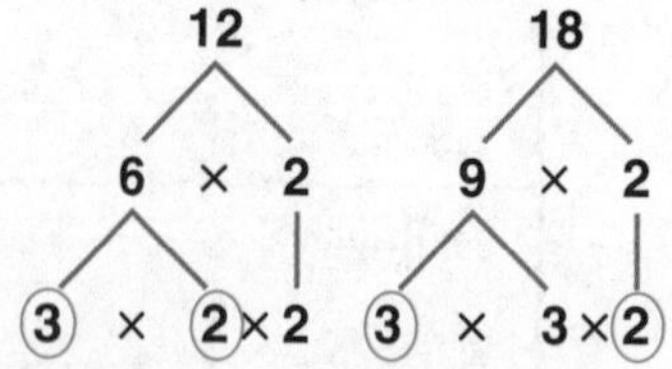

2 and 3 are the common prime factors.

So, the GCF of 12 and 18 is 2 × 3, or 6.

Got It? **Do these problems to find out.**

Find the greatest common factor of each set of numbers.

b. 12, 66 **c.** 18, 30 **d.** 32, 48

b. ________

c. ________

d. ________

Find the Least Common Multiple

You can find the least common multiple (LCM) by using a number line, making a list, or by using prime factors.

Multiples
A multiple of a number is the product of the number and any whole number (0, 1, 2, 3, ...).

Examples

3. Find the LCM of 2 and 3.

Method 1 **Use a number line.**

Put a red **X** above each multiple of 2 and a blue **X** above each multiple of 3.

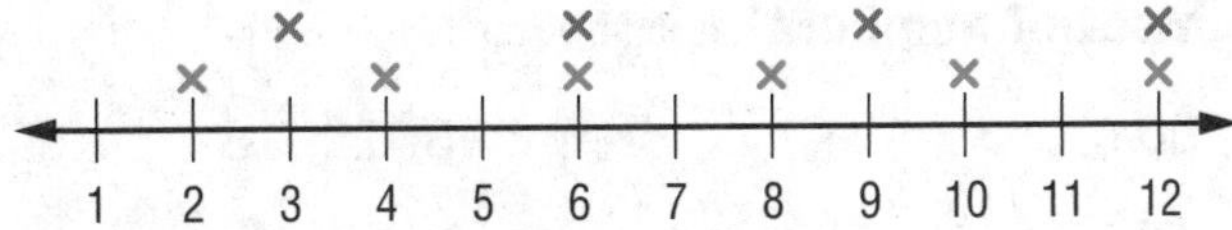

The least number with both a red and a blue X is 6.

So, 6 is the least common multiple of 2 and 3.

Method 2 **Use an organized list.**

List the nonzero multiples of 2 and 3.

multiples of 2: 2, 4, 6, 8, 10, 12,... $1 \times 2, 2 \times 2, 3 \times 2, \ldots$

multiples of 3: 3, 6, 9, 12, 15,... $1 \times 3, 2 \times 3, 3 \times 3, \ldots$

Notice that 6 and 12 are common multiples.

So, the least common multiple of 2 and 3 is 6.

4. Find the LCM of 14 and 21 using prime factorization.

Write the prime factorization of each number.

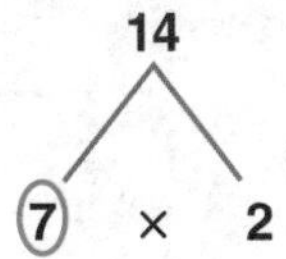

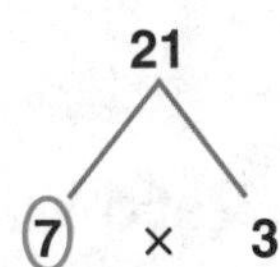

7 is the only common prime factor.

Multiply using each common prime factor only once.

So, the LCM is $7 \times 2 \times 3$ or 42.

Got It? **Do these problems to find out.**

Find the least common multiple of each set of numbers.

e. 2, 6 **f.** 4, 5, 10 **g.** 3, 5, 7

Show your work.

e. ______

f. ______

g. ______

Example

5. Ernesto has painting class every 2 weeks. Kamala has a pottery class every 5 weeks. Ernesto and Kamala met at the art building for class this week. How many weeks will it be until they see each other again?

multiples of 2: 2, 4, 6, 8, 10, 12, 14,...

multiples of 5: 5, 10, 15, 20, 25, 30,...

The least common multiple of 2 and 5 is 10. So, Ernesto and Kamala will see each other again in 10 weeks.

Guided Practice

Find the greatest common factor of each set of numbers. (Example 1 and 2)

1. 8, 32 ________

2. 24, 60 ________

3. 3, 12, 18 ________

Find the least common multiple of each set of numbers. (Examples 3 and 4)

4. 7, 9 ________

5. 6, 15 ________

6. 9, 12, 15 ________

7. The Movie House gives away a $5 coupon for every 4 movies purchased. They give away a bag of popcorn for every 3 movies purchased. How many movies would you have to purchase in all before receiving both a $5 coupon and a bag of popcorn at the same purchase? (Example 5)

____

8. **Building on the Essential Question** How does finding the greatest common factor help you to solve real-world problems? ________

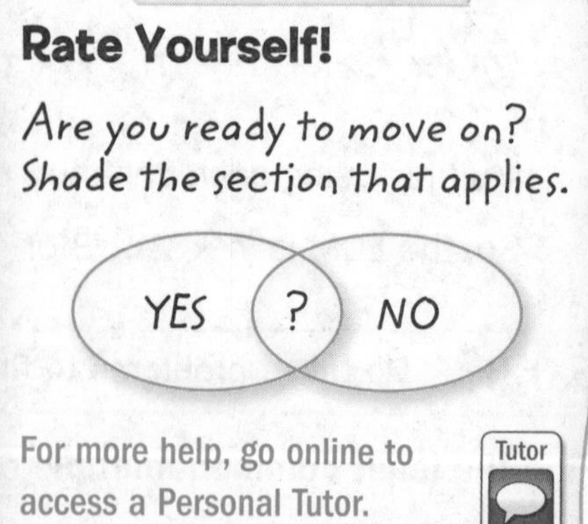

John Kelly/The Image Bank/Getty Images

Name ____________________ My Homework ____________________

Independent Practice

Go online for Step-by-Step Solutions 

Find the greatest common factor of each set of numbers. (Example 2)

1. 8, 14 ______

Show your work.

2. 21, 24, 27 ______

3. 21, 35, 49 ______

4. 12, 18, 26 ______

Find the least common multiple of each set of numbers. (Examples 3 and 4)

5. 5 and 6 ______

6. 6 and 9 ______

7 6, 12, and 15 ______

8. 3, 9, and 15 ______

9 A gardener has 27 pansies and 36 daisies. He plants an equal number of each type of flower in each row. What is the greatest possible number of pansies in each row? (Example 1)

10. Fourteen boys and 21 girls will be equally divided into groups. Find the greatest number of groups that can be created if no one is left out. (Example 1)

11. Inez waters her plants every two days. She trims them every 15 days. She did both today. When will she do both again? (Example 5)

12. CCSS **Identify Repeated Reasoning** An airport offers two shuttles that run on different schedules. If both shuttles leave the airport at 4:00 P.M., at what time will they next leave the airport together?

Shuttle Schedule	
Shuttle	**Departs**
A	every 6 minutes
B	every 9 minutes

H.O.T. Problems Higher Order Thinking

13. CCSS **Model with Mathematics** Write and solve a real-world problem that can be solved using the greatest common factor of two numbers.

14. CCSS **Identify Repeated Reasoning** How can you use number patterns to find the least common multiple of 120 and 360?

15. CCSS **Persevere with Problems** If the GCF of two numbers is 1, they are called *relatively prime.* Find three sets of relatively prime numbers.

Standardized Test Practice

16. There are 36 cans of green beans and 48 cans of corn. The display designer wants an equal number of each vegetable in each row. What is the greatest number of cans of corn that can be in each row?

Ⓐ 3 cans
Ⓑ 4 cans
Ⓒ 6 cans
Ⓓ 12 cans

Name ______ My Homework ______

Extra Practice

Find the greatest common factor of each set of numbers.

17. 15, 20 5

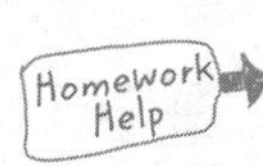

factors of 15: (1), 3, (5), 15

factors of 20: (1), 2, 4, (5), 10, 20

The common factors are 1 and 5.

The GCF is 5.

18. 30, 48, 60 ______

19. 24, 30, 42 ______

20. 24, 40, 56 ______

Find the least common multiple of each set of numbers.

21. 3 and 5 15

multiples of 3: 3, 6, 9, 12, (15), 18, 21, 24, 27, (30)

multiples of 5: 5, 10, (15), 20, 25, (30)

The common multiples are 15 and 30.

The LCM is 15.

22. 12 and 18 ______

23. 5, 10, and 15 ______

24. 9, 12, and 18 ______

25. A grocery store clerk has 16 oranges, 20 apples, and 24 pears. The clerk needs to put an equal number of apples, oranges, and pears into each basket. What is the greatest number of baskets that can be made so that no fruit is left?

______

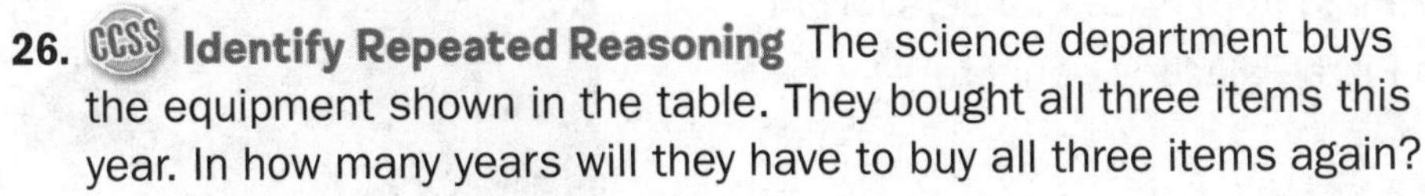

26. CCSS **Identify Repeated Reasoning** The science department buys the equipment shown in the table. They bought all three items this year. In how many years will they have to buy all three items again?

______

Item	Time Bought
Microscopes	every 5 years
Safety goggles	every 4 years
Test tubes	every 2 years

Standardized Test Practice

27. The cafeteria has 28 bottles of orange juice and 14 bottles of apple juice. An equal number of orange and apple juice bottles are displayed in each row. What is the greatest number of orange juice bottles that can be in each row?

Ⓐ 2 bottles
Ⓑ 7 bottles
Ⓒ 14 bottles
Ⓓ 18 bottles

28. Drusilla replaces the light bulb in the hall closet every 9 months and replaces the air filter every 3 months. She just replaced both items this month. How long will it be until she changes both the light bulb and the air filter?

Ⓕ 3 months
Ⓖ 9 months
Ⓗ 12 months
Ⓘ 27 months

29. **Short Response** Macy is painting a design that contains two repeating patterns. One pattern repeats every 8 inches. The other pattern repeats every 12 inches. If both patterns begin at the same place, in how many inches will they begin together again? ________

Common Core Review

Write each fraction in simplest form. 5.NF.5b

30. $\frac{9}{18}$ = ________

31. $\frac{21}{35}$ = ________

32. $\frac{36}{48}$ = ________

33. Josiah ran $\frac{4}{5}$ mile. How many tenths are equal to $\frac{4}{5}$ mile? Use bar diagrams to find the answer. 5.NF.5b

$\frac{1}{5}$	$\frac{1}{5}$	$\frac{1}{5}$	$\frac{1}{5}$	$\frac{1}{5}$

34. Pizza Palace cuts a medium pizza into 8 slices. The same size pizza at Pizza Pioneers is cut into 16 slices. Jasmine ate 4 slices of a medium pizza from Pizza Pioneers. What fraction of the pizza from Pizza Palace is equal to $\frac{4}{16}$? Explain. 5.NF.5b

Ingram Publishing/Alamy

Inquiry Lab

Ratios

HOW can you use tables to relate quantities?

CCSS **Content Standards** 6.RP.1, 6.RP.3, 6.RP.3a

Mathematical Practices 1, 3, 4

Donations Max has 3 fiction books and 6 nonfiction books to donate to the community center. He wants to package them so that there are an equal number of fiction and nonfiction books in each group. He also wants to have as many packages as possible. How many books are in each group?

What do you know? ______________________________

What do you need to find? ______________________________

Investigation 1

Step 1 Use 3 red counters to represent the fiction books. Use 6 yellow counters to represent the nonfiction books.

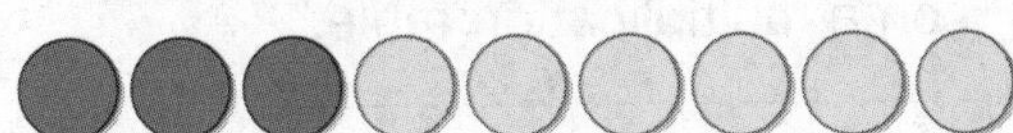

Step 2 Determine the smallest possible equal-size groups. Use mats to divide the counters into the groups.

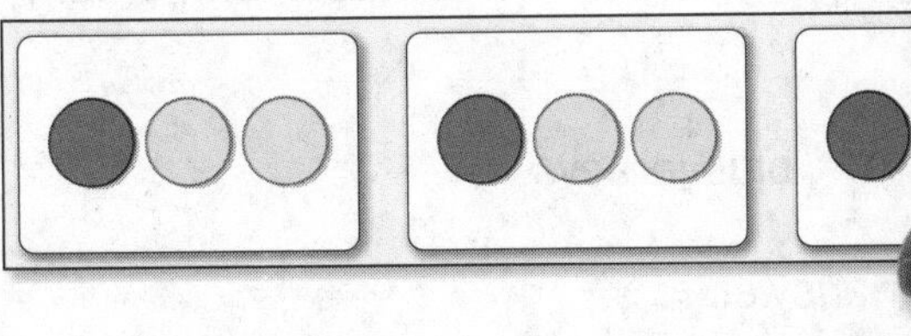

Each group has an equal number of fiction books and an equal number of nonfiction books.

Each group has ☐ fiction book and ☐ nonfiction books.

Investigation 2

Donations Maria is also collecting books. She wants to make packages that have 3 fiction books and 4 nonfiction books. She already has 9 fiction books. How many nonfiction books will she need?

Use a multiplication table to compare the numbers.

Step 1 Complete the rows for 3 and 4 on a multiplication table.

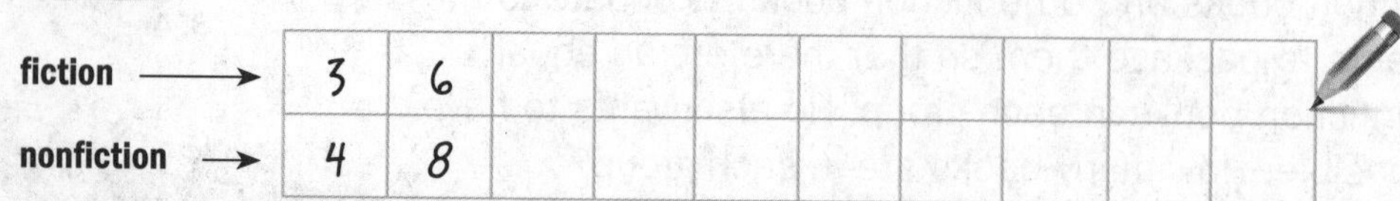

fiction →	3	6								
nonfiction →	4	8								

Step 2 Read across the top until you reach 9. Find the corresponding number in the bottom row and circle the 2 numbers.

Maria needs ☐ nonfiction books.

Investigation 3

Sports Jerseys Sanjay has 27 jerseys. Divide them into two groups so that for every 4 red jerseys, there are 5 blue jerseys.

Step 1 Complete the rows for 4 and 5 on a multiplication table.

red →	4	8								
blue →	5	10								

Step 2 Read across both rows until you find two numbers with a sum of 27.

There are ☐ red jerseys and ☐ blue jerseys.

Check Draw a picture to check your answer.

Work with a partner. Determine the number of pieces of fruit that should be put in each group. Make as many equal-size groups as possible using all the fruit. Use counters to represent the fruit.

1. 3 apples and 9 pears

2. 4 peaches and 6 oranges

3. 4 plums and 7 bananas

4. 6 apricots and 9 mangos

Work with a partner. Use a multiplication table to solve the following problems.

5. Evie wants groups of 3 notebooks and 5 pens. She already has 12 notebooks. How many pens will she need?

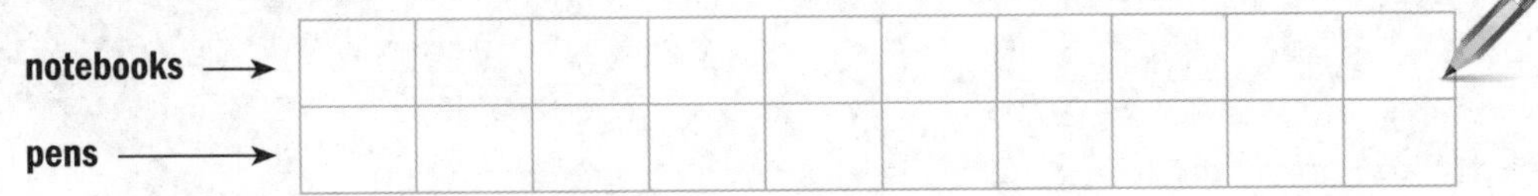

6. Louis wants groups of 6 daisies and 8 tulips for flower arrangements. He already has 24 daisies. How many tulips will he need?

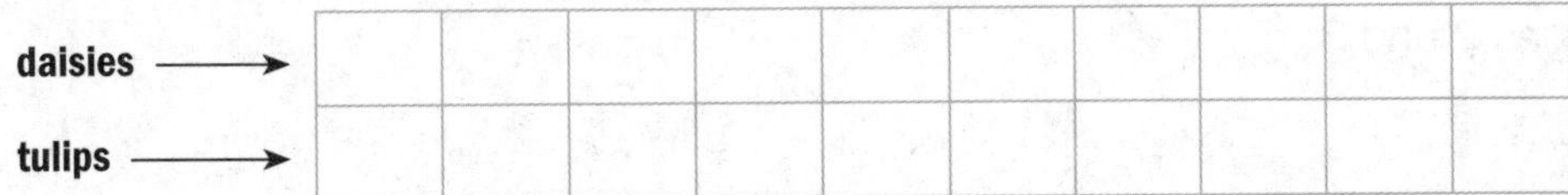

7. Selma has 77 strawberries. Divide them into two groups so that for every 4 strawberries in Group 1 there are 7 strawberries in Group 2.

Group 1 →										
Group 2 →										

Analyze

Model with Mathematics **Work with a partner. For each fraction, plot the nonzero multiples of the numerator and the denominator on separate number lines. Circle the least common multiple.**

8. 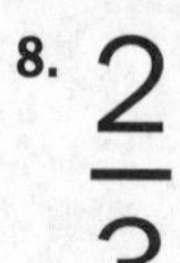

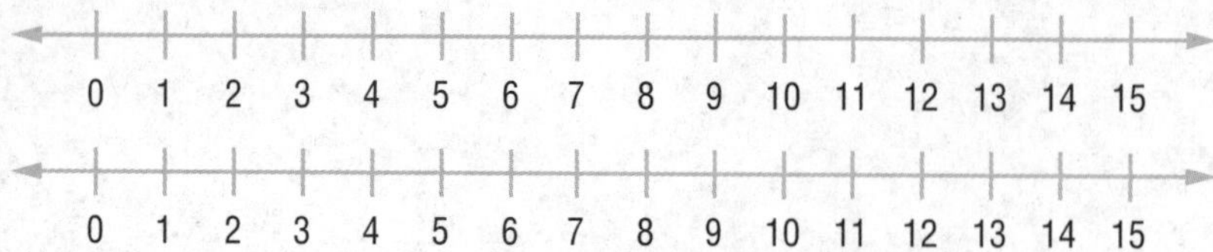

9. 

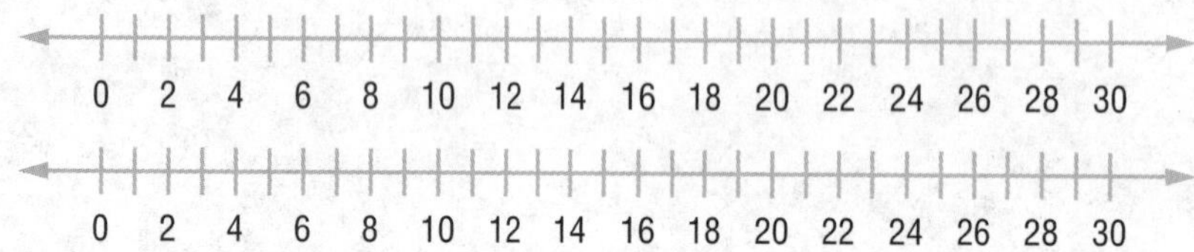

10. **Reason Inductively** How would finding the least common multiple help you when dividing items into equal groups?

Reflect

11. **Model with Mathematics** Write a word problem in which the ratio of yellow beads to blue beads is 3 to 2.

12. **Identify Repeated Reasoning** Describe the patterns used in the tables in Investigations 2 and 3.

13. **Inquiry** HOW can you use tables to relate quantities?

Lesson 2

Ratios

What You'll Learn

Scan the lesson. List two real-world scenarios in which you would use ratios.

- ______________________________

- ______________________________

Essential Question

HOW do you use equivalent rates in the real world?

Vocabulary

ratio

Common Core State Standards

Content Standards
6.RP.1, 6.RP.3

Mathematical Practices
1, 3, 4, 5

Real-World Link

Watch Tools

Dogs In her dog walking business, Mrs. DeCarbo walks 2 large dogs and 8 small dogs.

Compare the number of small dogs to large dogs. Use yellow counters to represent the large dogs. Use red counters to represent the small dogs. Draw the counters in the box.

1. 2 + ☐ = 8 There are ☐ *more* small dogs than large dogs.

2. 2 × ☐ = 8 There are ☐ *times* as many small dogs as large dogs.

3. 8 − ☐ = 2 There are ☐ *fewer* large dogs than small dogs.

4. 8 ÷ ☐ = 2 The number of large dogs is $\frac{☐}{☐}$ the number of small dogs.

Work Zone

Write a Ratio in Simplest Form

There are many different ways to compare amounts or *quantities*. A **ratio** is a comparison of two quantities by division. A ratio of 2 red paper clips to 6 blue paper clips can be written in three ways.

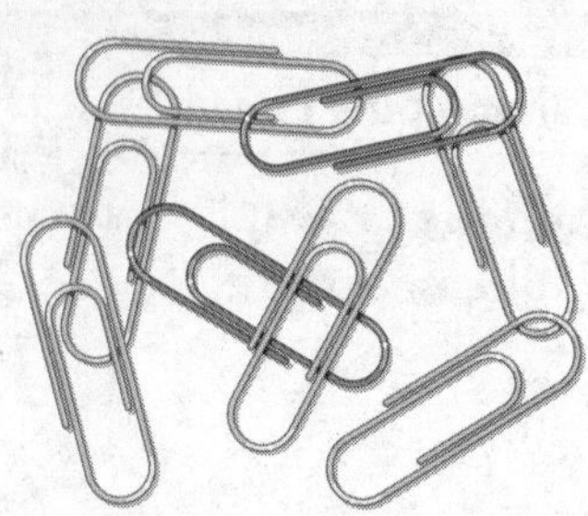

2 to 6 **2:6** $\frac{2}{6}$

As with fractions, ratios are often expressed in simplest form.

Example

Tutor

1. Write the ratio in simplest form that compares the number of red paper clips to the number of blue paper clips. Then explain its meaning.

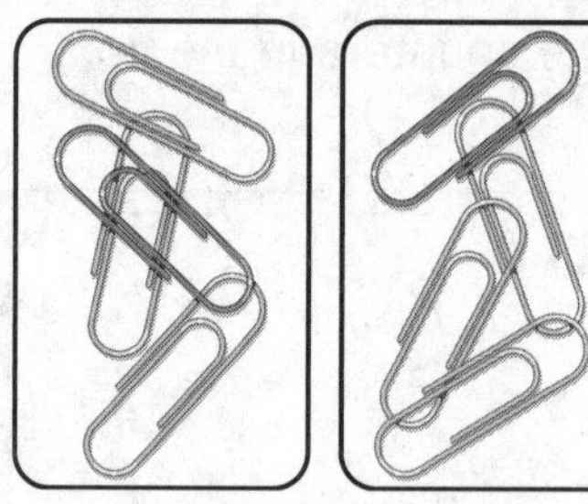

Write the ratio as a fraction. Then simplify.

red paper clips → $\frac{2}{6} = \frac{1}{3}$ ← blue paper clips (÷2 on numerator and denominator)

The GCF of 2 and 6 is 2.

The ratio of red to blue paper clips is $\frac{1}{3}$, 1 to 3, or 1:3. This means that for every 1 red paper clip there are 3 blue paper clips.

Show your work.

Got It? **Do this problem to find out.**

a. Write the ratio in simplest form that compares the number of suns to the number of moons. Then explain its meaning.

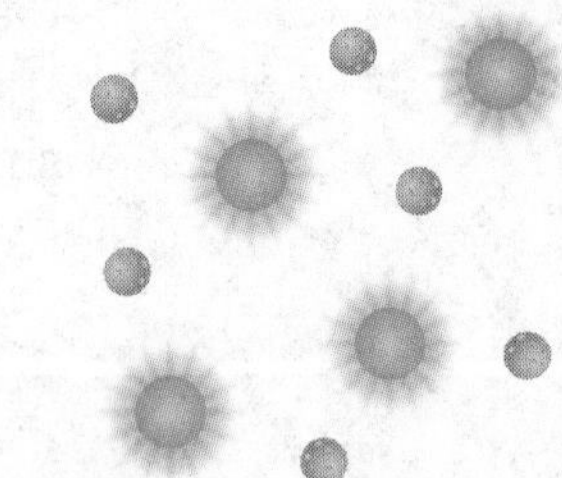

a. ______________

Use Ratios to Compare Categorical Data

Each piece of categorical data can only be assigned to one group. Bar diagrams (or tape diagrams) and frequency tables can be used to represent categorical data. Ratios can be used to compare the data.

Examples

2. Several students named their favorite flavor of gum. Write the ratio that compares the number who chose fruit to the total number of students.

Favorite Flavor of Gum	
Flavor	Number of Responses
Peppermint	9
Cinnamon	8
Fruit	3
Spearmint	1

Fruit: 3

Total: 9 + 8 + 3 + 1, or 21

fruit flavor responses → / total responses → $\frac{3}{21} = \frac{1}{7}$ (÷3 above and below) ← The GCF of 3 and 21 is 3.

The ratio is $\frac{1}{7}$, 1 to 7, or 1:7.

So, 1 out of every 7 students preferred fruit-flavored gum.

Accuracy
It is important to read the entire problem so that an accurate answer can be determined.

3. Monday's yogurt sales are recorded in the table. Write the ratio that compares the sales of strawberry yogurt to the total sales. Then explain its meaning.

Flavor	Number Sold
Peach	3
Blueberry	6
Vanilla	7
Strawberry	8

Strawberry: ☐

Total: ☐ + ☐ + ☐ + ☐, or ☐

strawberry yogurt sold → / total sold → $\frac{\square}{\square} \div \frac{\square}{\square} = \frac{\square}{\square}$ or ☐ to ☐

So, ☐ out of every ☐ yogurt cups sold were strawberry.

Got It? Do this problem to find out.

b. A pet store sold the animals listed in the table in one week. Write the ratio of cats to pets sold that week. Then explain its meaning.

Pet	Number Sold
Birds	10
Dogs	14
Cats	8

b. ______

Example

Watch Tutor

4. Katy wants to divide her 30 flowers into two groups, so that the ratio is 2 to 3.

Step 1 Use a bar diagram to show a ratio of 2 to 3.

6	6	
6	6	6

30 flowers

Step 2 There are 5 equal sections. So, each section represents 30 ÷ 5 or 6 flowers.

There are 12 flowers in one group and 18 in the other.

Guided Practice

Check

Write each ratio as a fraction in simplest form. Then explain its meaning. (Example 1)

Show your work.

1. ______

pens to pencils

2. ______

pennies:dimes

3. Last month, Ed ate 9 apples, 5 bananas, 4 peaches, and 7 oranges. Find the ratio of bananas to the total number of fruit. Then explain its meaning. (Examples 2 and 3)

______

______

4. Divide 28 cans of soda into two groups so the ratio is 3 to 4. (Example 4) ______

______

5. **Building on the Essential Question** How can you use mental math to determine if a ratio is simplified?

______

______

Independent Practice

Go online for Step-by-Step Solutions

Write each ratio as a fraction in simplest form. Then explain its meaning. (Example 1)

1. ______

Show your work.

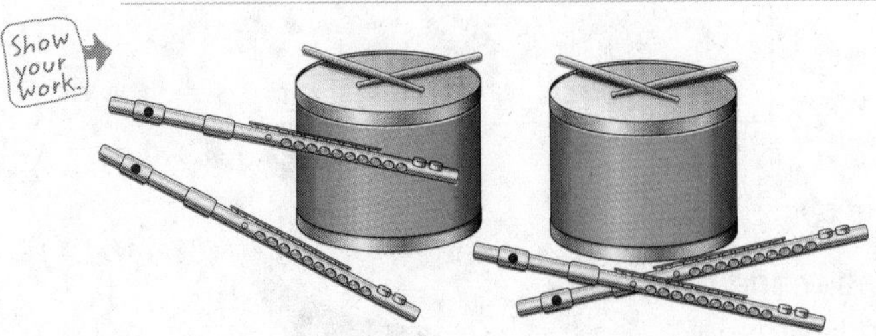

flutes:drums

2. ______

sandwiches to milk cartons

3. A class has 6 boys and 15 girls. What is the ratio of boys to girls? (Example 2) ______

4. The table shows the number of books Salvador has read. Find the ratio of mystery books to the total. Explain its meaning. (Example 3) ______

Type	Number of Books
Mystery	10
Nonfiction	7
Science Fiction	5
Western	2

5. Divide 33 photos into two groups so the ratio is 4 to 7. (Example 4) ______

6. CCSS **Model with Mathematics** Refer to the graphic novel frame below for Exercises a–b.

a. For each store, what is the ratio of the number of cans to the price? ______

b. What would be the ratio of the number of cans to the price at Super Saver and Price Busters if a coupon for $1 off the total purchase is used? ______

7 **CCSS** **Use Math Tools** The graph shows the number of appearances of hockey teams in the Stanley Cup Finals.

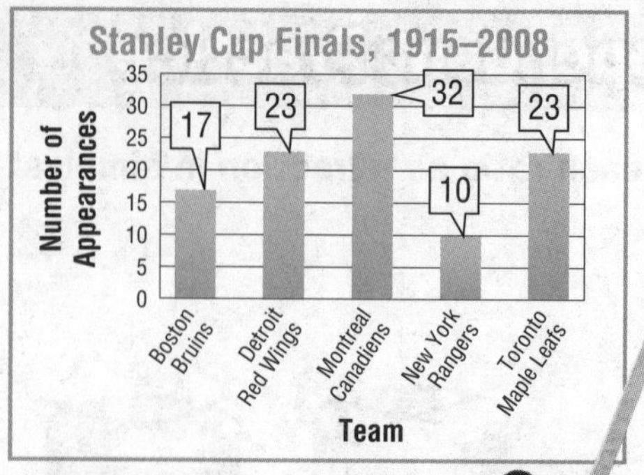

a. Write the ratio that compares the appearances made by the Rangers to the appearances made by the Canadiens in simplest form. Then explain its meaning.

b. Write the ratio that compares the appearances made by the Maple Leafs to the appearances made by the Bruins in simplest form. Then explain its meaning.

H.O.T. Problems Higher Order Thinking

8. **CCSS** **Model with Mathematics** Create three different drawings showing a number of circles and triangles in which the ratio of circles to triangles is 2:3.

9. **CCSS** **Persevere with Problems** Find the missing number in the following pattern. Explain your reasoning.

12, 24, 72, 288, ☐

Standardized Test Practice

10. The table shows how Levon spends his time at the gym. What is the ratio of the time on the treadmill to the time lifting weights?

Activity	Time (min)
Treadmill	25
Lifting weights	35

Ⓐ 2 to 3
Ⓑ 5 to 7
Ⓒ 4 to 5
Ⓓ 1 to 7

Name ______________________ My Homework ______________________

Extra Practice

Write each ratio as a fraction in simplest form. Then explain its meaning.

11. $\frac{1}{4}$; for every 1 triangle there are 4 rectangles. **12.** ______________________

Homework Help

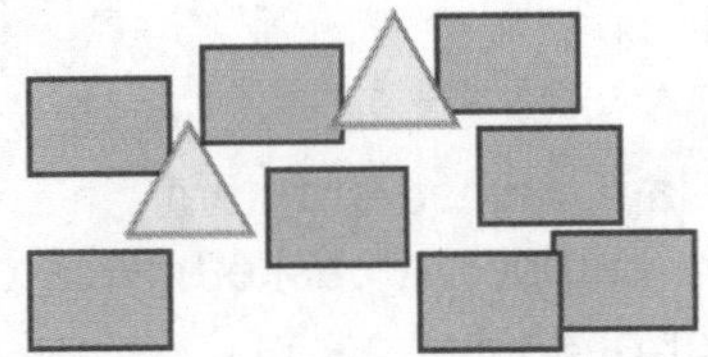

triangles to rectangles

There are 2 triangles and 8 rectangles.

The ratio is $\frac{2}{8}$. $\frac{2}{8} \div \frac{2}{2} = \frac{1}{4}$

soccer balls:footballs

13. An animal shelter has 36 kittens and 12 puppies available for adoption. What is the ratio of puppies to kittens?

14. Find the ratio of black cell phone covers sold to the total number of cell phone covers sold last week. Then explain its meaning.

Color	Number of Cell Phone Covers Sold
Green	5
Silver	6
Red	3
Black	4

15. On the first day of the food drive, Mrs. Teasley's classes brought in 6 cans of fruit, 4 cans of beans, 7 boxes of noodles, and 4 cans of soup. Find the ratio of cans of fruit to the total number of food items collected. Then explain its meaning.

16. The rise and span for a roof are shown. The pitch of a roof is the ratio of the rise to the half-span. If the rise is 8 feet and the span is 30 feet, what is the pitch in simplest form?

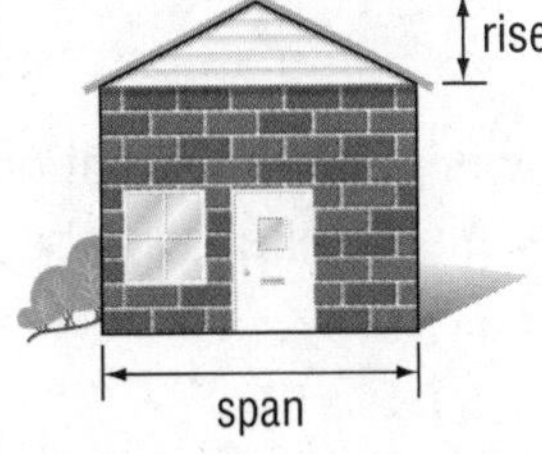

17. CCSS **Justify Conclusions** Debra found that 6 of the 24 students in her class own a cell phone. What is the ratio of students that own a cell phone to students that do not? Explain your reasoning to a classmate.

Standardized Test Practice

18. Which of the following ratios does *not* describe a relationship between the balls?

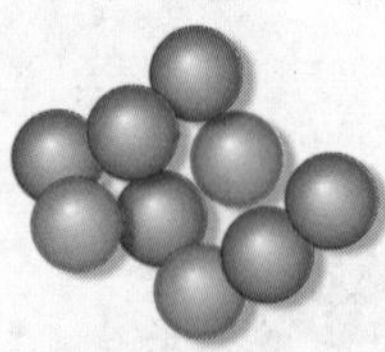

Ⓐ 3 green : 6 red
Ⓑ 3 green : 9 total
Ⓒ 1 green : 2 red
Ⓓ 1 red : 4 total

19. Of new calculators tested, 8 were defective, and 42 passed inspection. What ratio compares the number of defective calculators to the total number of new calculators?

Ⓕ 4:21
Ⓖ 4:25
Ⓗ 1:25
Ⓘ 2:13

20. **Short Response** Jaclyn counted the number of sport cards she has collected. The table shows the results.

baseball	basketball	football	soccer
45	14	20	21

Write a ratio in simplest form that compares the number of basketball cards to the number of soccer cards.

____

21. At a putt-putt course there are 50 yellow golf balls, 45 red golf balls, and 65 blue golf balls. What ratio compares the number of blue golf balls to the total number of golf balls?

Ⓐ 13:9
Ⓑ 13:32
Ⓒ 32:9
Ⓓ 16:5

Common Core Review

Find the equivalent fraction. 5.NF.5b

22. $\frac{3}{7} = \frac{\square}{21}$

23. $\frac{1}{6} = \frac{\square}{24}$

24. $\frac{4}{5} = \frac{28}{\square}$

25. The Sanchez family is going on vacation. If they drive for 3 hours at the posted speed, how many miles will they travel? 5.NBT.5

____

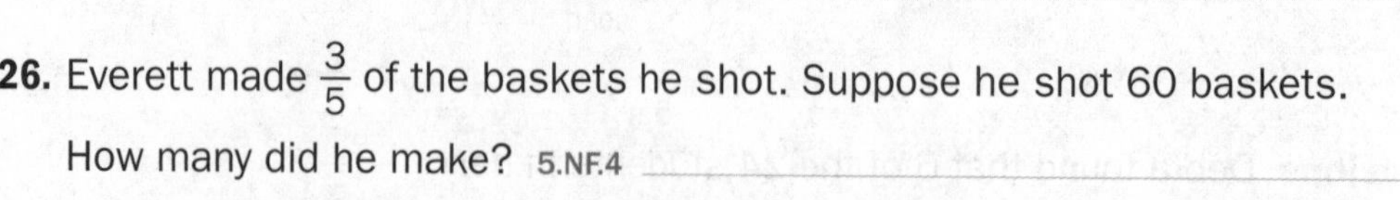

26. Everett made $\frac{3}{5}$ of the baskets he shot. Suppose he shot 60 baskets. How many did he make? 5.NF.4 ______________________

27. There are 36 students in Mrs. Keaton's sixth grade class. If $\frac{5}{12}$ of her students are girls, how many girls are in the class? 5.NF.4 ______________________

Inquiry Lab

Unit Rates

HOW can you use bar diagrams to compare quantities in real-world situations?

Content Standards
6.RP.2, 6.RP.3, 6.RP.3b

Mathematical Practices
1, 3, 8

Rollerblading Jamila and Anica were rollerblading. They skated 14 miles in 2 hours. If they skated at a constant rate, how many miles did they skate in 1 hour?

What do you know? ______________________________

____

What do you need to find? ______________________________

____

Investigation 1

Step 1 Use a bar diagram to represent 14 miles. The box is separated into two equal sections to represent 2 hours.

14 miles	
1 hour	1 hour

Step 2 Each section represents one hour. Determine the number of miles skated in one hour.

14 miles	
1 hour	1 hour
7 miles	

So, they skated ☐ miles in one hour.

Investigation 2

A package of 5 crackers contains 205 Calories. How many Calories are in one cracker?

Step 1 Draw a bar diagram to represent 205 Calories. Divide the bar diagram into 5 equal sections to represent 5 crackers.

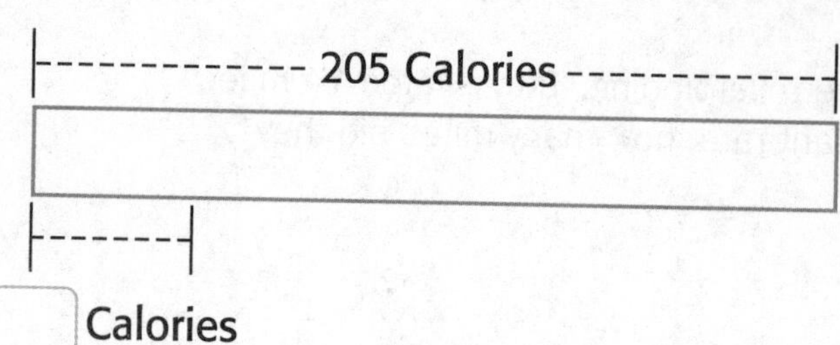

Step 2 Label the first section "1 cracker." Determine the number of Calories in 1 cracker.

So, one cracker contains ☐ Calories.

Investigation 3

A bottle of body wash costs $2.88 and contains 12 ounces. How much does it cost per ounce?

Step 1 Draw a bar diagram to represent ________. Divide the bar diagram into ☐ equal sections to represent ☐ ounces.

Step 2 Label the first section "________." Determine the cost for 1 ounce of body wash.

So, one ounce of body wash costs $☐.

Work with a partner to solve. Use a bar diagram.

1. Travis drove 129 miles in 3 hours. He drove at a constant speed. How many miles did he drive in 1 hour? ______

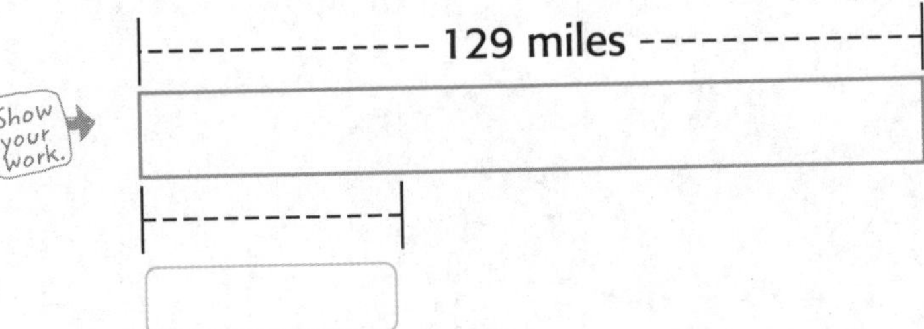

2. Six oranges cost $5.34. How much does 1 orange cost? ______

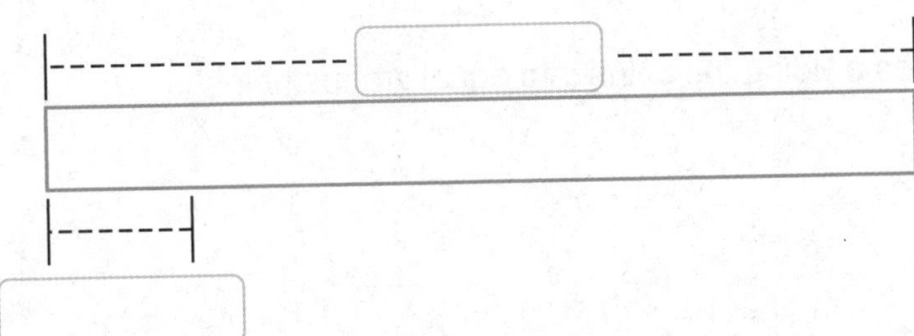

3. Doug read 231 pages in 7 hours. He read the same number of pages each hour. How many pages did he read in 1 hour? ______

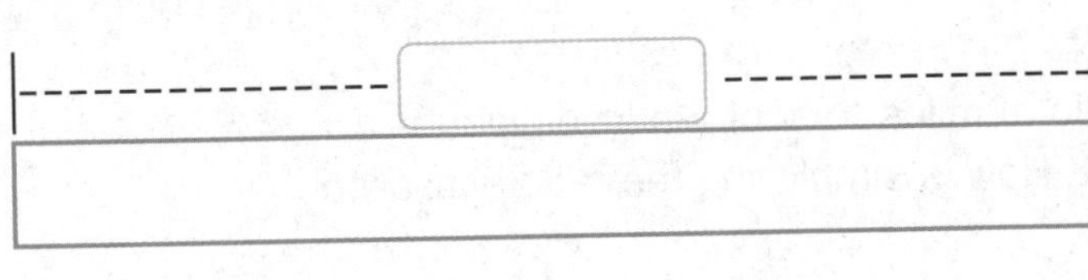

☐ pages

4. Mariah has 72 flowers in 4 vases. She put the same number of flowers in each vase. How many flowers are in 1 vase? ______

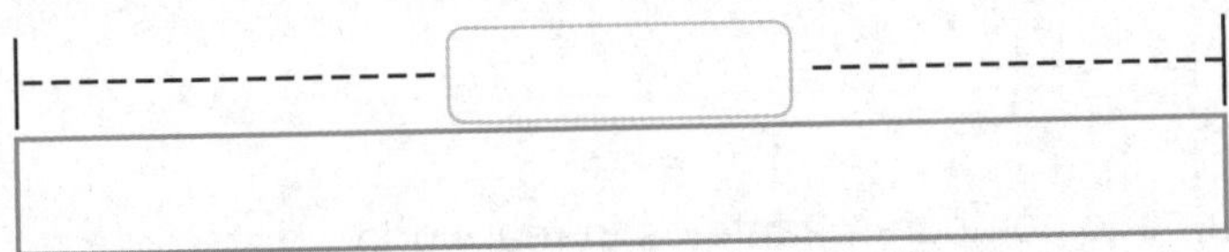

☐ flowers

Analyze

Work with a partner to complete the problem.

5. In the bakery, a container of cookies is $4.55 and contains 13 servings. The coins below equal $4.55. Divide the coins into 13 equal groups to determine the cost per serving. Circle each group. ________

6. **CCSS Reason Inductively** How does dividing the coins into equal groups help solve the problem?

____

____

Reflect

7. **CCSS Justify Conclusions** The ratio of miles to hours in Investigation 1 is 14:2, which can be reduced to 7:1. How is simplifying the ratio similar to division?

____

____

8. **CCSS Identify Repeated Reasoning** Write a rule for how to find a ratio with a denominator of 1 without using a diagram.

____

____

9. **Inquiry** HOW can you use bar diagrams to compare quantities in real-world situations? ______________________________

____

Lesson 3
Rates

Essential Question

HOW do you use equivalent rates in the real world?

Vocabulary

rate
unit rate
unit price

Common Core State Standards

Content Standards
6.RP.2, 6.RP.3, 6.RP.3b

Mathematical Practices
1, 3, 4

What You'll Learn

Scan the lesson. List two real-world scenarios in which you would use rates.

- ____________________
- ____________________

Vocabulary Start-Up

Use your glossary, which starts on page GL1, to complete the definitions of the vocabulary words in the table.

Definition	Examples
fraction: A number that represents part of a ________ or part of a ________.	$\frac{1}{2}, \frac{3}{4}, \frac{9}{12}, \frac{45}{3}$
ratio: A comparison of two ________ by ________.	2 out of 3, 2 to 3, 2:3, $\frac{2}{3}$
rate: A ________ comparing two ________ with different kinds of ________.	$\frac{36\text{ miles}}{3\text{ hours}}$ 36 miles for every 3 hours \$26 for 5 bags 19 songs in 5 minutes
unit rate: A ________ that is ________ so that it has a denominator of ________.	$\frac{12\text{ miles}}{1\text{ hour}}$, 12 miles per hour \$5.20 for 1 bag 3.8 songs in 1 minute

Real-World Link

Desiree typed a 15-character text message in 5 seconds.

1. Write the rate Desiree typed as a fraction. $\frac{\square \text{ characters}}{\square \text{ seconds}}$

2. What operation would you use to write the fraction in simplest form? ____________________

Work Zone

Find a Unit Rate

A **rate** is a ratio comparing two quantities of different kinds of units. A **unit rate** has a denominator of 1 unit when the rate is written as a fraction. To write a rate as a unit rate, divide the numerator and the denominator of the rate by the denominator.

Ratio	Rate	Unit Rate
15:5 =	$\frac{\text{15 characters}}{\text{5 seconds}}$ =	$\frac{\text{3 characters}}{\text{1 second}}$

Examples

1. Samantha picked 45 oranges in 5 minutes. Write this rate as a unit rate.

Write the rate as a fraction. Compare the number of oranges to the number of minutes. Then divide.

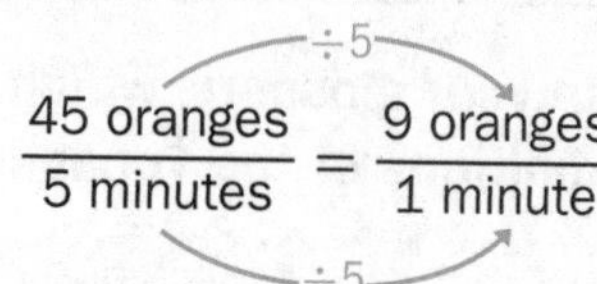

So, the unit rate is $\frac{\text{9 oranges}}{\text{1 minute}}$, or 9 oranges per minute.

2. The Australian dragonfly can travel 18 miles in 30 minutes. How far can the dragonfly travel in 1 minute?

Write the rate as a fraction. Compare the distance to the number of minutes. Then divide.

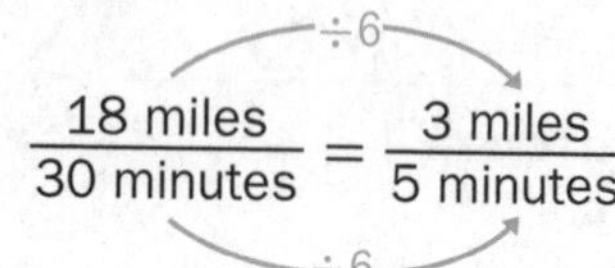

The ratio 3 to 5 cannot be simplified to a whole number rate. It can be written as $\frac{\text{3 miles}}{\text{5 minutes}}$ or as a unit rate of $\frac{3}{5}$ mile to 1 minute.

The dragonfly can travel $\frac{3}{5}$ mile every minute.

Simplifying Ratios

The lowest common factor of 3 and 5 is 1. To find the unit rate of the ratio $\frac{\text{3 miles}}{\text{5 minutes}}$, divide both the numerator and denominator by 5. So, the unit rate in fration form is $\frac{3}{5}$ mile per minute.

Got It? **Do this problem to find out.**

a. Ama downloaded 35 songs in 5 minutes. How many songs did she download per minute?

b. Jonathan is baking several loaves of bread to sell in his bakery. He used 9 cups of water and 12 cups of whole wheat flour. How much water was used per cup of flour?

a. ________

b. ________

Example

3. An adult's heart beats about 2,100 times every 30 minutes. A baby's heart beats about 2,600 times every 20 minutes. How many more beats does a baby's heart beat in 60 minutes than an adult's heart?

Step 1 Find the unit rates.

Adult: $\frac{2{,}100 \text{ beats}}{30 \text{ minutes}}$ or $\frac{70 \text{ beats}}{1 \text{ minute}}$

Baby: $\frac{2{,}600 \text{ beats}}{20 \text{ minutes}}$ or $\frac{130 \text{ beats}}{1 \text{ minute}}$

Step 2 Using the unit rate for each, determine the number of beats in 60 minutes.

Adult: $70 \times 60 = 4{,}200$ beats

Baby: $130 \times 60 = 7{,}800$ beats

Step 3 Find the difference.

$7{,}800 - 4{,}200 = 3{,}600$

So, a baby's heart beats 3,600 more times in 60 minutes than an adult's heart.

Key Phrases

Key phrases such as *per*, *in*, and *for every* are often used to describe unit rates.

Got It? **Do this problem to find out.**

c. A hummingbird's heart rate while resting is about 7,500 beats every 30 minutes. How many more beats does a hummingbird's heart beat in 60 minutes than a human baby's heart?

c. ________

Find a Unit Price

You can use what you know about unit rates to find a unit price. The **unit price** is the cost per unit. To write a price as a unit price, divide the numerator and the denominator of the rate by the denominator.

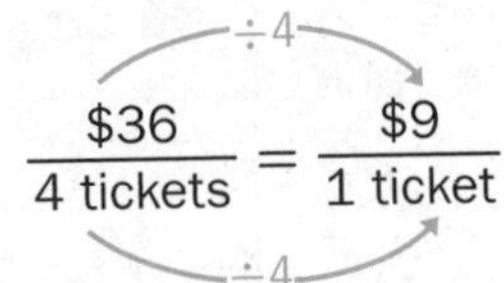

For example, it costs $36 for 4 movie tickets. So, the cost per unit, or per ticket, is $9.

Example

4. **Financial Literacy** **Four potted plants cost $88. What is the price per plant?**

Write the rate as a fraction. Compare the total cost to the number of plants. Then divide.

$$\frac{\$88}{4 \text{ plants}} = \frac{\$22}{1 \text{ plant}} \quad (\div 4)$$

So, the price per potted plant is $22.00.

Guided Practice

Write each rate as a unit rate. (Examples 1 and 2)

1. 44 points in 4 quarters = ________

2. 125 feet in 5 seconds = ________

3. 360 miles traveled on 12 gallons of gasoline = ________

4. 12 meters in 28 seconds = ________

5. Molly shot 20 baskets in 4 minutes. Nico shot 42 baskets in 6 minutes. How many more baskets did Nico shoot per minute? (Example 3) ________

6. For Carolina's birthday, her mom took her and 4 friends to a water park. Carolina's mom paid $40 for 5 student tickets. What was the price for one student ticket? (Example 4)

____

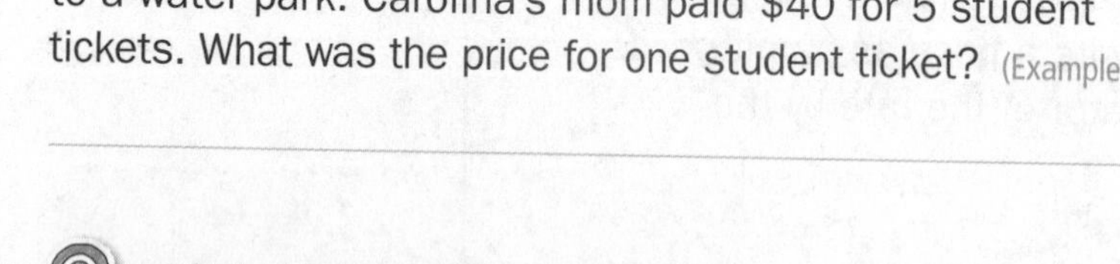

7. **Building on the Essential Question** How are rates and ratios related? ________

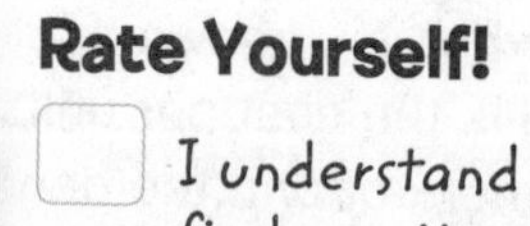

Rate Yourself!

☐ I understand how to find a unit rate.

Great! You're ready to move on!

☐ I still have some questions about rates.

No Problem! Go online to access a Personal Tutor.

Name ______________________ My Homework ______________________

Independent Practice

Go online for Step-by-Step Solutions

Write each rate as a unit rate. (Examples 1 and 2)

1. 72 ounces in 6 steaks = ________

2. 162 water bottles in 9 cases = ________

3. Marcella divided 40.8 gallons of paint among 8 containers. How much paint is in each container? (Example 1) ________

4. Central Subs made 27 sandwiches using 12 pounds of turkey. How much turkey was used per sandwich? (Example 2) ________

5. The results of a car race are shown. Determine who drove the fastest. Explain. (Example 3) ________

Drivers' Times		
Driver	Laps	Time (min)
Cutwright	35	84
Evans	42	96.6
Loza	38	102.6

6. Theo's mom bought an eight-pack of juice boxes at the store for $4. Find the unit rate for the juice boxes. (Example 4) ________

7. Joshua's cousin pledged $12 for a charity walk. If Joshua walked 3 miles, how much did his cousin pay per mile? (Example 4)

8. CCSS **Justify Conclusions** The Lovin' Lemon Company sells a 4-gallon jug of lemonade for $24. The Sweet and Sour Company sells an eight-pack of 1-quart bottles of lemonade for $16.00. Which company has a higher unit price? Explain your answer.

9. The Shanghai Maglev Train is one of the fastest trains in the world, traveling about 2,144 miles in 8 hours.

   a. How many miles does it travel in one hour? ________

   b. The distance between Columbus, Ohio, and New York City is about 560 miles. How many hours would it take the train to travel between the cities? ________

10. **Multiple Representations** The table shows the approximate population and areas of five states. *Population density* is the number of people per square unit of an area.

State	Population Estimate (as of July 2007)	Area (square miles)
California	36,500,000	163,707
Florida	18,300,000	65,758
Iowa	2,990,000	56,276
New Jersey	8,690,000	8,722
Wyoming	522,000	97,818

a. **Numbers** Find the population density of each state. Round to the nearest tenth.

b. **Graph** Make a bar graph of the five population densities.

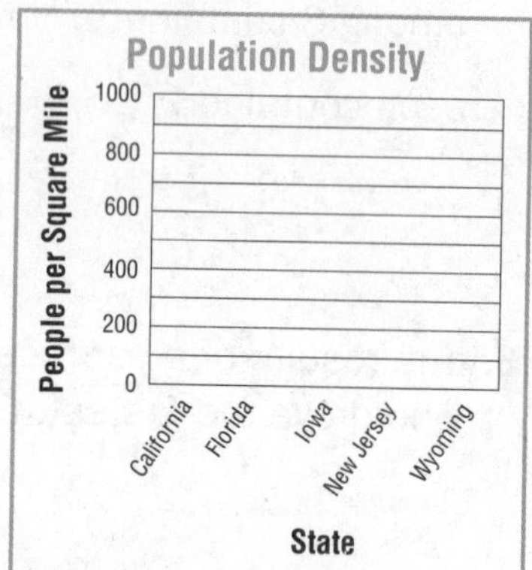

c. **Words** Connecticut has about the same population as Iowa, but its area is 4,875 square miles. Without calculating, compare Connecticut's population density to Iowa's. Justify your answer.

H.O.T. Problems Higher Order Thinking

11. **Find the Error** Julie wrote the rate \$108 in 6 weeks as a unit rate. Find her mistake and correct it.

12. **Persevere with Problems** The ratio of red jelly beans to yellow jelly beans in a dish is 3:4. If Greg eats 3 red jelly beans and 6 yellow ones, the ratio is 4:5. How many yellow jelly beans were originally in the dish?

Standardized Test Practice

13. The human heart pumps 750 gallons of blood in 9 hours. A human kidney filters 100 gallons of blood in 6 hours. How many more gallons of blood does the human heart pump than a human kidney filters during 24 hours?

Ⓐ 2,000 Ⓑ 1,600 Ⓒ 250 Ⓓ 50

Extra Practice

Write each rate as a unit rate.

14. Davis printed 24 photos in 8 minutes. How many photos did he print per minute?

3 photos per minute

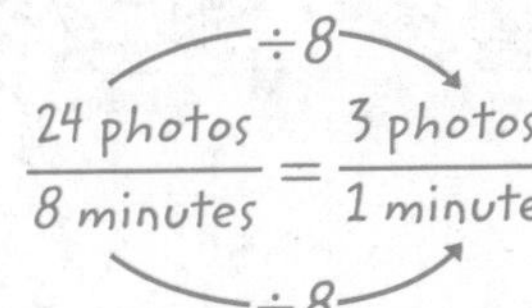
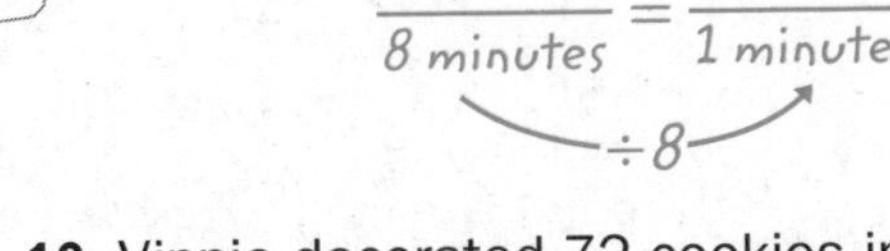

15. Carrie planted 48 tulips in 12 minutes. How many tulips did she plant per minute? ____________

16. Vinnie decorated 72 cookies in 36 minutes. How many cookies did he decorate per minute? ____________

17. Alana biked 45 miles in 3 hours. How many miles did she bike per hour? ____________

18. A Ruby Throated Hummingbird beats its wings 159 times in 3 seconds. How many times does the Ruby Throated Hummingbird beat its wings per second? ____________

19. The Reyes family bought four concert tickets for $252. What was the price per ticket? ____________

20. An adult blinks about 450 times in 30 minutes. A 12-year-old blinks about 150 times in 15 minutes. How many more times does an adult blink in 60 minutes than a 12-year-old? ____________

21. Find the number of meters each record holder ran in one second of each event. Round to the nearest tenth.

a. 200 meters, 19.30 seconds, Usain Bolt, Jamaica ____________

b. 400 meters, 43.18 seconds, Michael Johnson, USA ____________

c. 100 meters, 9.69 seconds, Usain Bolt, Jamaica ____________

22. CCSS **Justify Conclusions** The 24 students in Mr. Brown's homeroom sold 72 magazine subscriptions. The 28 students in Mrs. Garcia's homeroom sold 98 magazine subscriptions. Whose homeroom sold more magazine subscriptions per student? Explain your reasoning. ____________

Standardized Test Practice

23. Olivia printed invitations for a party. If she printed 286 invitations in 26 minutes, how many invitations did she print each minute?

Ⓐ 60
Ⓒ 11
Ⓑ 26
Ⓓ 9

24. Amy is training for a half marathon. In practice, she runs 2 miles in 15 minutes. If she continues at the same rate, how many miles will she run in 1 hour?

Ⓕ 4
Ⓗ 16
Ⓖ 8
Ⓘ 30

25. Short Response Boxes of fruit snacks are on sale at the grocery. The boxes are the same size. What is the unit rate for each kind?

26. Genevieve spent $56.25 to fill her 15-gallon tank. How much did she pay per gallon?

Ⓐ $3.25
Ⓒ $3.75
Ⓑ $3.50
Ⓓ $4.00

CCSS Common Core Review

Simplify each fraction. 5.NF.5b

27. $\frac{16}{80} = \frac{\square}{\square}$

28. $\frac{4}{10} = \frac{\square}{\square}$

29. $\frac{48}{200} = \frac{\square}{\square}$

30. Josephine wants to put a wallpaper border around the ceiling of her room. The dimensions are shown at the right. How many feet of border does she need? 5.NF.2

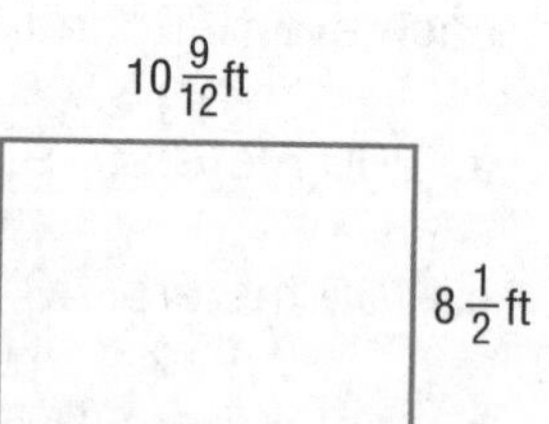

31. Miguel's grandparents live 159 miles from his house. If it takes 3 hours to drive to his grandparent's house, what is the average speed? 5.NBT.5

Lesson 4
Ratio Tables

What You'll Learn

Scan the lesson. List two things you will learn about ratio tables.

-
-

Essential Question

HOW do you use equivalent rates in the real world?

Vocabulary

ratio table
equivalent ratios
scaling

Common Core State Standards

Content Standards
6.RP.3, 6.RP.3a, 6.RP.3b

Mathematical Practices
1, 3, 4, 7, 8

Real-World Link

Tools

Refreshments A punch recipe uses one container of soda and three containers of juice to make one batch of punch.

1. Draw red counters to show the number of containers of soda and draw yellow counters to show the number of containers of juice needed to make 2 batches of punch.

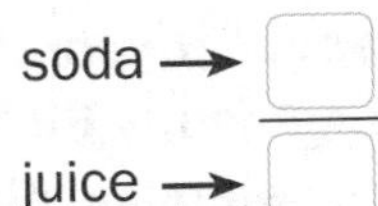

2. Draw red counters to show the number of containers of soda and draw yellow counters to show the number of containers of juice needed to make 3 batches of punch.

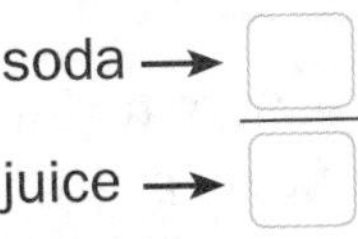

3. Find the ratio in simplest form of soda to juice needed for 1, 2, and 3 batches. What do you notice?

Work Zone

Equivalent Ratios

The quantities in the opening activity can be organized into a table. This table is called a **ratio table** because the columns are filled with pairs of numbers that have the same ratio.

Soda	1	2	3
Juice	3	6	9

The ratios $\frac{1}{3}$, $\frac{2}{6}$, and $\frac{3}{9}$ are equivalent, since each simplifies to a ratio of $\frac{1}{3}$.

Equivalent ratios express the same relationship between quantities.

Examples

1. To make yellow icing, you mix 6 drops of yellow food coloring with 1 cup of white icing. How much yellow food coloring should you mix with 5 cups of white icing to get the same shade?

Use a ratio table. Since $1 \times 5 = 5$, multiply each quantity by 5.

So, add 30 drops of yellow food coloring to 5 cups of icing.

Drops of Yellow	6	30
Cups of Icing	1	5

(×5 across each row)

2. In a recent year, Joey Chestnut won a hot dog eating contest by eating nearly 66 hot dogs in 12 minutes. If he ate at a constant rate, determine about how many hot dogs he ate every 2 minutes.

Divide each quantity by one or more common factors until you reach a quantity of 2 minutes.

So, Chestnut ate about 11 hot dogs every 2 minutes.

Hot Dogs	66	33	11
Time (min)	12	6	2

(÷2, then ÷3 across each row)

Check for Accuracy
To check your answer for Example 2, check to see if the ratio of the two new quantities is equivalent to the ratio of the original quantities.
$\frac{11}{2} \times \frac{6}{6} = \frac{66}{12}$

Got It? Do these problems to find out.

a. A patient receives 1 liter of IV fluids every 8 hours. At that rate, find how many hours it will take to receive 4 liters of IV fluids.

IV Fluids (L)	1	4
Time (h)	8	

b. To make cranberry jam, you need 12 cups of sugar for every 16 cups of cranberries. Find the amount of sugar needed for 4 cups of cranberries.

Sugar (c)	12		
Cranberries (c)	16		4

a. ______

b. ______

Use Scaling

Multiplying or dividing two related quantities by the same number is called **scaling**. Sometimes you may need to *scale back* and then *scale forward* to find an equivalent ratio.

Examples

3. Cans of corn are on sale at 10 for $4. Find the cost of 15 cans.

Cans of Corn	10		15
Cost in Dollars	4		■

There is no whole number by which you can multiply 10 to get 15. So, scale back to 5 and then scale forward to 15.

Cans of Corn	10	5	15
Cost in Dollars	4	2	6

(÷2, ×3 above and below the table)

Divide each quantity by a common factor, 2.

Then, since $5 \times 3 = 15$, multiply each quantity by 3.

So, 15 cans of corn would cost $6.

4. Joe mows lawns during his summer vacation to earn money. He took 14 hours last week to mow 8 lawns. At this rate, how many lawns could he mow in 49 hours?

Is there a whole number by which you can multiply 14 to get 49? ________

Scale back to ________, and then scale forward to ________.

Number of Hours	14	7	49
Number of Lawns	8	4	28

(÷2, ×7 above and below the table)

So, Joe can mow ________ lawns in 49 hours.

Got It? Do this problem to find out.

c. A child's height measures 105 centimeters. Estimate her height in inches.

Height (cm)	25		105
Height (in.)	10		

c. ________

Example

5. On her vacation, Leya exchanged \$50 American and received \$60 Canadian. Use a ratio table to find how many Canadian dollars she would receive for \$20 American.

Set up a ratio table. Use scaling to find the desired quantity.

÷10 ×4

Canadian Dollars	60	6	24
American Dollars	50	5	20

÷10 ×4

Divide each quantity by a common factor, 10.

Then, since $5 \times 4 = 20$, multiply each quantity by 4.

Leya would receive \$24 Canadian for \$20 American.

Guided Practice

Check

Complete each ratio table to solve each problem.

1. Santiago receives an allowance of \$7 every week. How much total does he receive after 4 weeks? (Example 1)

Allowance (\$)	7			
Number of Weeks	1			4

2. Tonya runs 8 kilometers in 60 minutes. At this rate, how long would it take her to run 2 kilometers? (Example 2)

Distance Run (km)	8		2
Time (min)	60		

3. Lamika buys 12 packs of juice boxes that are on sale and pays a total of \$48. Use a ratio table to determine how much Lamika will pay to buy 8 more packs of juice boxes at the same store. (Example 5)

Number of Juice Boxes			
Price (\$)			

4. **Building on the Essential Question** How can you determine if two ratios are equivalent?

Rate Yourself!

How well do you understand ratio tables? Circle the image that applies.

Clear

Somewhat Clear

Not So Clear

For more help, go online to access a Personal Tutor.

FOLDABLES Time to update your Foldable!

Name ______________________ My Homework ______________________

Independent Practice

Go online for Step-by-Step Solutions

Complete each ratio table to solve each problem.

1. To make 5 apple pies, you need about 2 pounds of apples. How many pounds of apples do you need to make 20 apple pies? (Example 1)

Number of Pies	5		20
Pounds of Apples	2		

____

2. Four balls of wool will make 8 knitted caps. How many balls of wool will Malcolm need if he wants to make 6 caps? (Examples 3 and 4)

Balls of Wool	4		
Number of Caps	8		6

____

3. Before leaving to visit Mexico, Levant traded 270 American dollars and received 3,000 Mexican pesos. When he returned from Mexico, he had 100 pesos left. How much will he receive when he exchanges these pesos for dollars? (Example 2)

American Dollars	270		
Mexican Pesos	3,000		100

____

4. On a bike trip across the United States, Rodney notes that he covers about 190 miles every 4 days. If he continues at this rate, use a ratio table to determine about how many miles he could bike in 6 days. (Example 5)

Miles Biked			
Days			

____

5. CCSS **Identify Repeated Reasoning** A punch recipe that serves 24 people calls for 4 liters of lemon-lime soda, 2 pints of sherbet, and 6 cups of ice.

   a. Complete a ratio table to represent this situation.

People Served	
Liters of Soda	
Pints of Sherbet	
Cups of Ice	

   b. How much of each ingredient would you need to make an identical recipe that serves 12 people? 36 people?

____

____

   c. How much of each ingredient would you need to make an identical recipe that serves 18 people? Explain your reasoning.

____

____

____

6. On a typical day, flights at a local airport arrive at a rate of 10 every 15 minutes. At this rate, how many flights would you expect to arrive in 1 hour?

Number of Flights			
Minutes			

7. CCSS **Identify Structure** Complete the graphic organizer to explain how equivalent ratios are used to find larger quantities and smaller quantities.

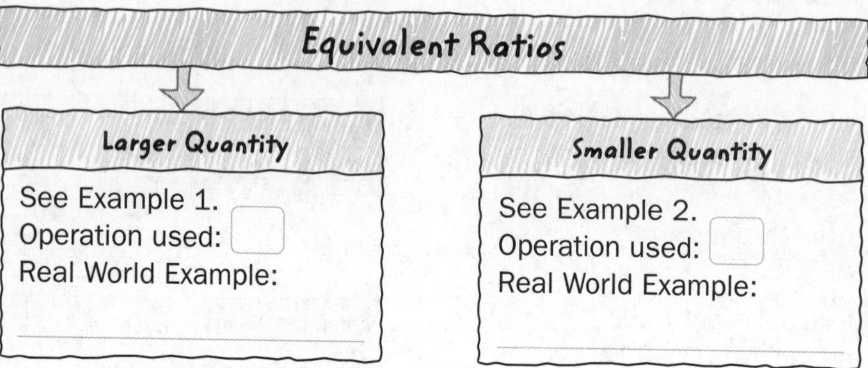

H.O.T. Problems Higher Order Thinking

8. CCSS **Persevere with Problems** Use the ratio table to determine how many people 13 subs would serve. Explain.

Number of Subs	3	5	8	13
People Served	12	20	32	

9. CCSS **Justify Conclusions** There are 10 girls and 8 boys in Mr. Augello's class. If 5 more girls and 5 more boys join the class, will the ratio of girls to boys remain the same? Justify your answer using a ratio table.

Girls			
Boys			

Standardized Test Practice

10. Leo buys 5 DVDs for $60. At this rate, how much would he pay for 3 DVDs?

Ⓐ $10 Ⓒ $36

Ⓑ $30 Ⓓ $58

Name ______________________ My Homework ______________________

Extra Practice

Complete each ratio table to solve each problem.

11. A zoo requires that 1 adult accompany every 7 students that visit the zoo. How many adults must accompany 28 students? 4 adults

Homework Help

Number of Adults	1	2	3	4
Number of Students	7	14	21	28

+1 +1 +1

+7 +7 +7

12. Valentina purchased 200 beads for $48 to make necklaces. If she needs to buy 25 more beads, how much will she pay if she is charged the same rate?

Number of Beads	200		25
Cost in Dollars	48		

____

13. If a hummingbird were to get all of its food from a feeder, then a 16-ounce nectar feeder could feed about 80 hummingbirds a day. How many hummingbirds would you expect to be able to feed with a 12-ounce feeder?

Ounces of Nectar	16		12
Number of Birds Fed	80		

____

14. When a photo is reduced or enlarged, its length to width ratio usually remains the same. Aurelia wants to enlarge a 4-inch by 6-inch photo so that it has a width of 15 inches. Use a ratio table to determine the new length of the photo.

Length (in.)	4		
Width (in.)	6		

____

15. Landon owns a hybrid SUV that can travel 400 miles on a 15-gallon tank of gas. Determine how many miles he can travel on 6 gallons.

____

16. CCSS **Justify Conclusions** A veterinarian needs to know an animal's weight in kilograms. If 20 pounds is about 9 kilograms and a dog weighs 30 pounds, use a ratio table to find the dog's weight in kilograms. Explain your reasoning.

Pounds			
Kilograms			

____

____

Standardized Test Practice

17. Jaylen is making biscuits using the recipe below.

Whole Wheat Biscuits

2 c	Whole wheat flour
4 tsp	Baking powder
$\frac{1}{2}$ tsp	Salt
2 tbsp	Shortening
1 c	Milk
1	Small egg

Makes 20 biscuits

How many cups of flour will he need to make 30 biscuits?

Ⓐ $1\frac{1}{2}$ cups Ⓒ 10 cups

Ⓑ 3 cups Ⓓ 15 cups

18. A tutor's rates are shown in the ratio table. Use the ratio table to determine how much she charges for 5 hours.

Cost ($)	30		
Number of Hours	2		5

Ⓕ $15 Ⓗ $33

Ⓖ $30 Ⓘ $75

19. Short Response Beth walks 2 blocks in 15 minutes. How many blocks would Beth walk if she walked at the same rate for an hour? Explain your reasoning.

____

____

Common Core Review

Identify each point shown on the graph. 5.G.1

20. (________, ________)

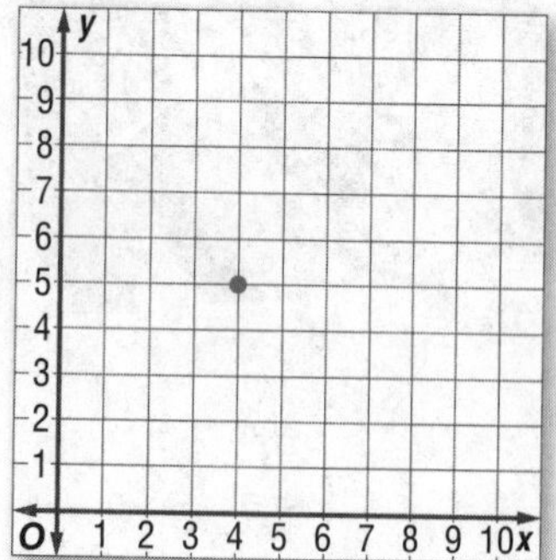

21. (________, ________)

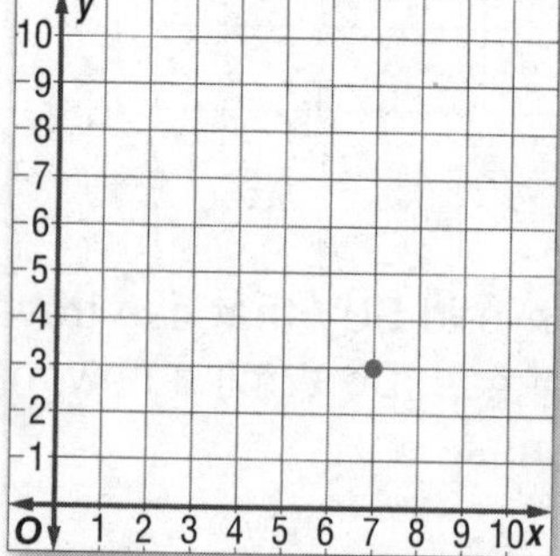

22. Liam is drawing a map. He needs to plot four points to identify four places on his map. Plot and label the following points. 5.G.2

a. the library at (3, 2)

b. the school at (6, 4)

c. the park at (8, 1)

d. Liam's house at (2, 8)

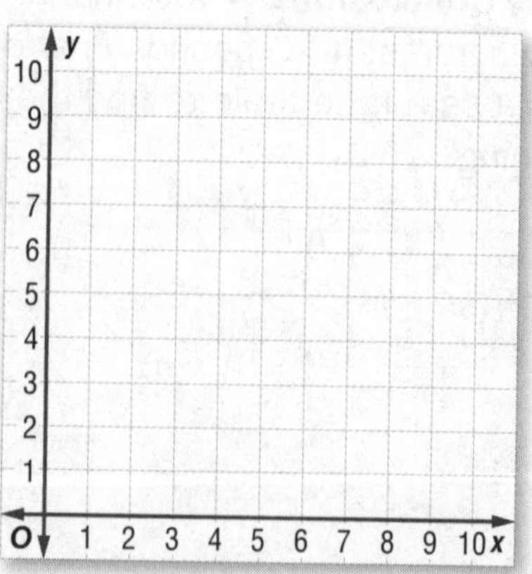

Lesson 5

Graph Ratio Tables

What You'll Learn

Scan the lesson. List two headings you would use to make an outline of the lesson.

- ______________________________
- ______________________________

Essential Question

HOW do you use equivalent rates in the real world?

Vocabulary

coordinate plane
origin
x-axis
y-axis
ordered pair
x-coordinate
y-coordinate
graph

Common Core State Standards

Content Standards
6.RP.3, 6.RP.3a

Mathematical Practices
1, 3, 4

Vocabulary Start-Up

The **coordinate plane** is formed when two perpendicular number lines intersect at their zero points. This point is called the **origin**. The horizontal number line is called the **x-axis** and the vertical number line is called the **y-axis**. An **ordered pair**, such as (2, 3), is a pair of numbers used to locate a point on the coordinate plane.

Fill in the blanks with the hightlighted words from above.

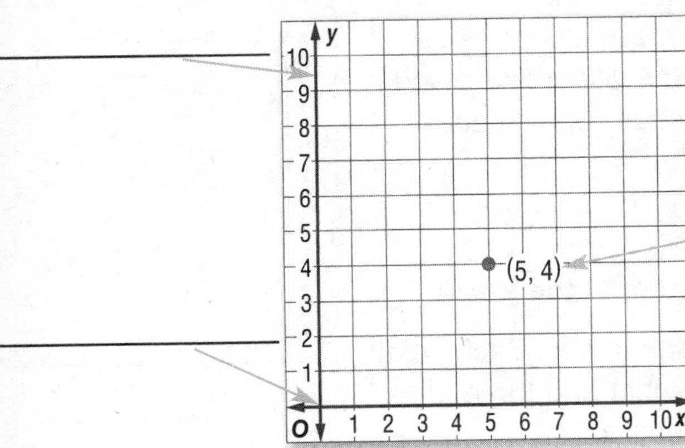

Real-World Link

In 3 minutes, a North American wood turtle can travel about 17 yards. If the *x*-axis represents minutes and the *y*-axis represents yards, write an ordered pair to represent this situation.

(___ , ___)
minutes yards

Work Zone

Graph Ordered Pairs

You can use an ordered pair to name any point on the coordinate plane. The first number in an ordered pair is the **x-coordinate**, and the second number is the **y-coordinate**.

The *x*-coordinate corresponds to a number on the *x*-axis. → **(3, 6)** ← The *y*-coordinate corresponds to a number on the *y*-axis.

You can express information in a table as a set of ordered pairs. To see patterns, **graph** the ordered pairs on the coordinate plane.

Examples

Real World

Tutor

The table shows the cost in dollars to create CDs of digital photos at a photo shop. The table also shows this information as ordered pairs (number of CDs, cost in dollars).

Cost to Create CDs		
Number of CDs, *x*	Cost in Dollars, *y*	Ordered Pair (*x*, *y*)
1	3	(1, 3)
2	6	(2, 6)
3	9	(3, 9)

1. Graph the ordered pairs.

Start at the origin. Use the *x*-coordinate and move along the *x*-axis. Then use the *y*-coordinate and move along the *y*-axis. Draw a dot at each point.

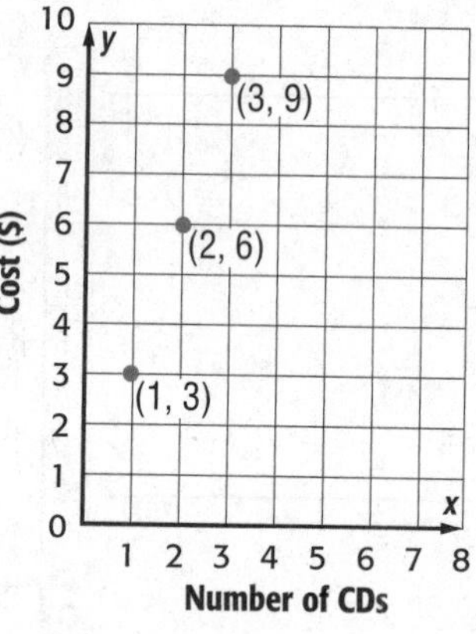

2. Describe the pattern in the graph.

The points appear in a line. Each point is one unit to the right and three units up from the previous point.

So, the cost increases by $3 for every CD created.

Got It? Do these problems to find out.

The table shows Gloria's earnings for 1, 2, and 3 hours. The table also lists this information as ordered pairs (hours, earnings).

Gloria's Earnings		
Hours, *x*	Dollars Earned, *y*	Ordered Pair (*x*, *y*)
1	5	(1, 5)
2	10	(2, 10)
3	15	(3, 15)

a. Graph the ordered pairs.

b. Describe the pattern in the graph.

Show your work.

a.

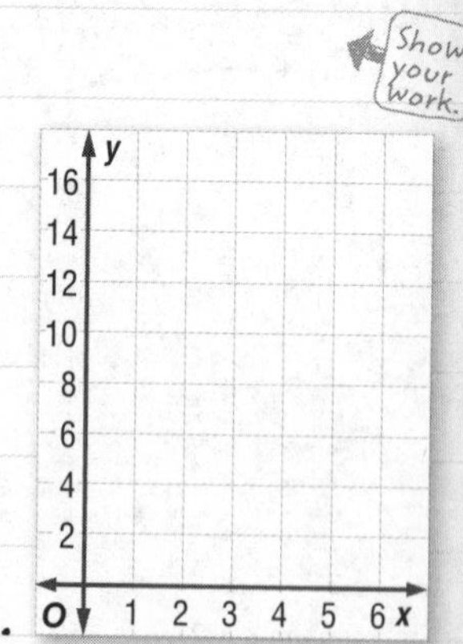

b. ____________

Compare Ratios

You can use tables and graphs to compare ratios. The greater the ratio, the steeper the line will appear.

Examples

Two friends are making scrapbooks. Renée places 4 photos on each page of her scrapbook. Gina places 6 photos on each page of her scrapbook.

3. **Make a table for each scrapbook that shows the total number of photos placed, if each book has 1, 2, 3, or 4 pages. List the information as ordered pairs (pages, photos).**

Renée's Scrapbook		
Pages, x	Photos, y	(x, y)
1	4	(1, 4)
2	8	(2, 8)
3	12	(3, 12)
4	16	(4, 16)

Gina's Scrapbook		
Pages, x	Photos, y	(x, y)
1	6	(1, 6)
2	12	(2, 12)
3	18	(3, 18)
4	24	(4, 24)

4. **Graph the ordered pairs for each friend on the same coordinate plane.**

Graph the ordered pairs for Renée's scrapbook in blue.

Graph the ordered pairs for Gina's scrapbook in red.

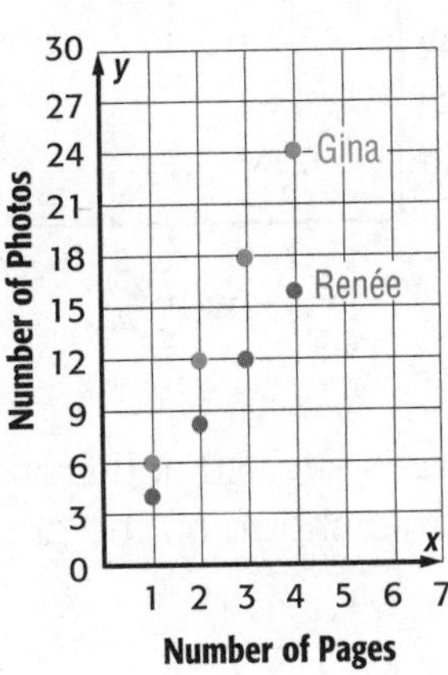

5. **How does the ratio of photos to each page compare for each person? How is this shown on the graph?**

The ratio of photos to pages for Renée's scrapbook is 4:1 while the ratio for Gina's scrapbook is 6:1. On the graph, both sets of points appear to be in a straight line, but the line for Gina is steeper than the line for Renée.

STOP and Reflect

Marta is also making a scrapbook. She places 5 photos on each page. How does the ratio of photos to each page compare for her book, Gina's book, and Renée's book?

Guided Practice

Two friends are each saving money in their bank accounts. Marcus saves $10 each week while David saves $15 each week. (Examples 1–5)

1. Make a table for each friend that shows the total amount saved for 1, 2, 3, or 4 weeks. List the information as ordered pairs (weeks,total dollars saved).

Show your work.

Marcus		
Weeks, x	Total Saved ($), y	(x, y)
1		
2		
3		
4		

David		
Weeks, x	Total Saved ($), y	(x, y)
1		
2		
3		
4		

2. Graph the ordered pairs for each friend on the same coordinate plane.

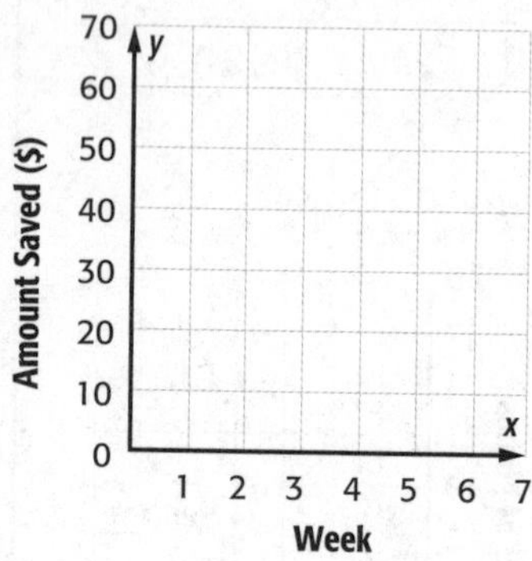

3. How do the ratios of Marcus's savings and David's savings compare? How is this shown on the graph?

4. **Building on the Essential Question** How can graphing help solve a problem involving ratios?

Independent Practice

Go online for Step-by-Step Solutions

The table shows the total time it took Samir to read 0, 1, 2, and 3 pages of the book. The table also lists this information as ordered pairs (number of pages, total minutes). (Examples 1–2)

Samir's Reading		
Number of Pages, x	Total Minutes, y	Ordered Pair (x, y)
0	0	(0, 0)
1	4	(1, 4)
2	8	(2, 8)
3	12	(3, 12)

1 Graph the ordered pairs.

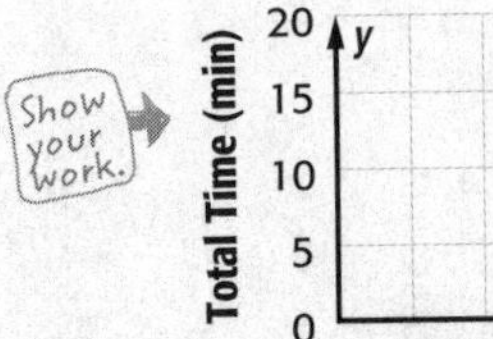

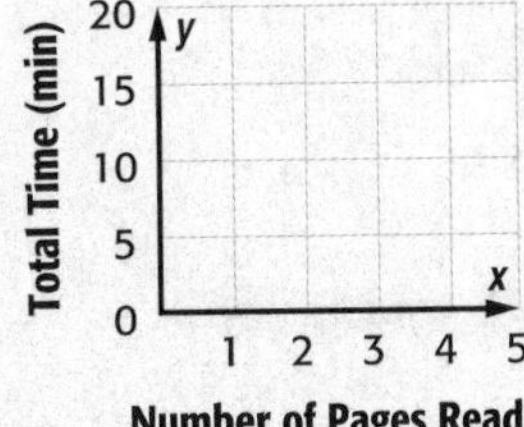

2. Describe the pattern in the graph.

____

____

Ken's Home Supply charges $5 for each foot of fencing. Wayne's Warehouse charges $6 for each foot of fencing. (Examples 3–5)

3. Make a table for each store that shows the total cost for 1, 2, 3, or 4 feet of fencing. List the information as ordered pairs (feet of fencing, total cost).

Ken's Home Supply		
Fencing (ft), x	Cost ($), y	(x, y)
1		
2		
3		
4		

Wayne's Warehouse		
Fencing (ft), x	Cost ($), y	(x, y)
1		
2		
3		
4		

4. Graph the ordered pairs for each store on the same coordinate plane.

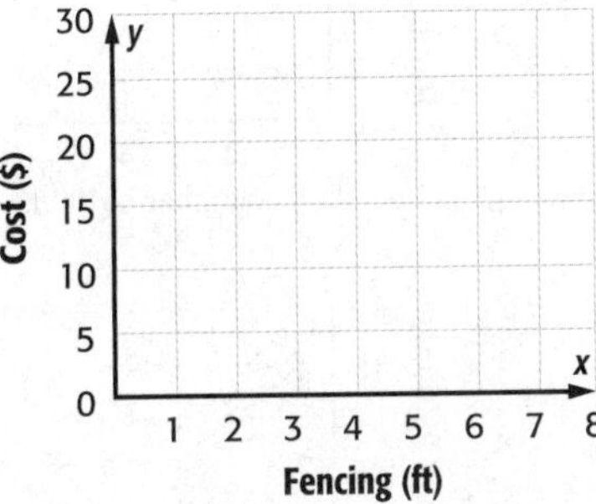

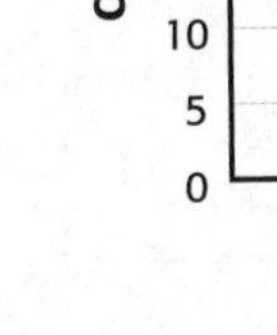

5 Using the tables and graphs, write a few sentences comparing the ratios of amount charged per foot of fencing for each store. How is this shown on the graph?

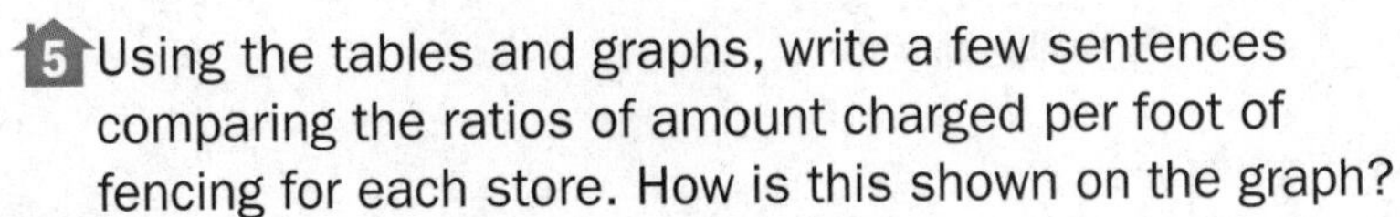

____

____

____

6.  **Justify Conclusions** Patty's Pies made 2 peach pies using 10 cups of peaches. They made 3 pies using 15 cups of peaches and 4 pies using 20 cups of peaches. Predict how many cups of peaches would be needed to make 9 peach pies. Explain.

H.O.T. Problems Higher Order Thinking

7. **Model with Mathematics** Write a real-world problem using ratios that could be represented on the coordinate plane.

8. **Persevere with Problems** Give the coordinates of the point located halfway between (2, 1) and (2, 4).

9. **Persevere with Problems** The graph below shows the cost of purchasing pencils from the school office. The graph is missing a point to indicate the cost of 12 pencils. Complete the graph by plotting the missing information. Explain your answer.

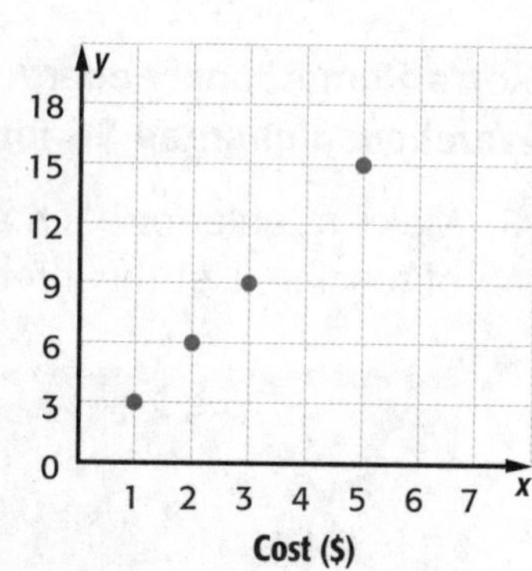

Standardized Test Practice

10. It takes an artist one hour to frame three paintings. Which graph represents this?

Ⓐ 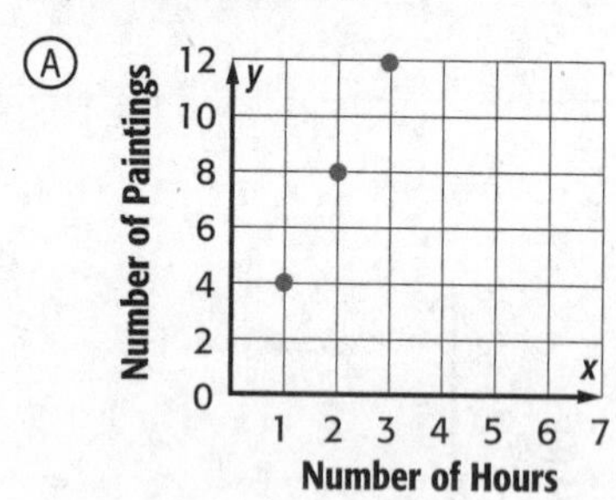

Ⓒ 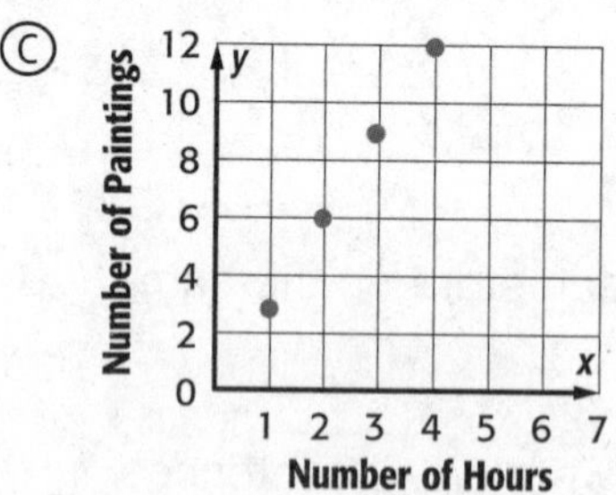

Ⓑ 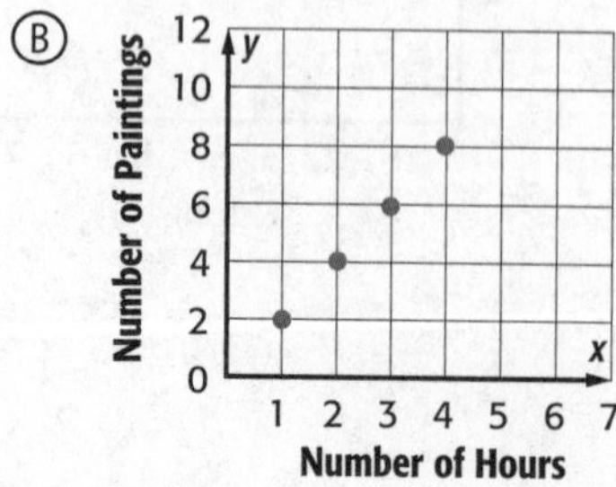

Ⓓ 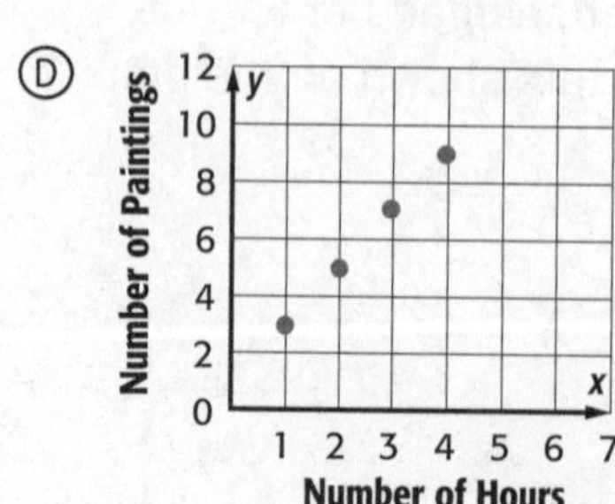

Extra Practice

The table shows the total number of miles Ariel runs for several days. The table also lists this information as ordered pairs (number of days, total miles).

11. Graph the ordered pairs.

Homework Help

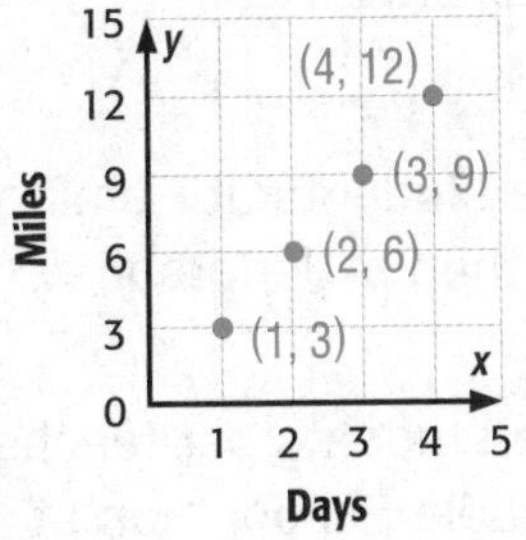

Ariel's Running Record		
Days, x	Miles, y	(x, y)
1	3	(1, 3)
2	6	(2, 6)
3	9	(3, 9)
4	12	(4, 12)

12. Describe the pattern in the graph. *The graph shows that as the number of days increases by 1, the number of miles ran increases by 3.*

There are two employees for every tiger in the tiger exhibit at a local zoo. For every elephant in the elephant exhibit, there are four employees.

13. Make a table for each animal that shows the total number of employees for 1, 2, 3, or 4 animals. List the information as ordered pairs (number of animals, number of employees).

Tiger Exhibit		
Animals, x	Employees, y	(x, y)
1		
2		
3		
4		

Elephant Exhibit		
Animals, x	Employees, y	(x, y)
1		
2		
3		
4		

14. Graph the ordered pairs for each store on the same coordinate plane.

15. CCSS **Justify Conclusions** Using the tables and graphs, write a few sentences comparing the ratios of the number of employees per animal. How is this shown on the graph?

____

____

____

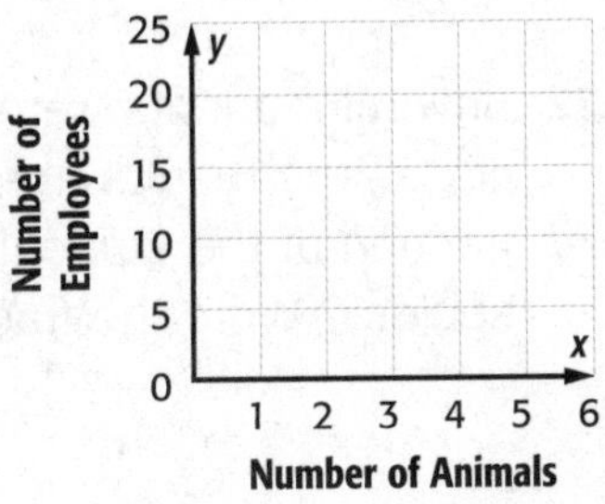

Standardized Test Practice

16. The table gives the ratio of teachers to students at Jefferson Middle School.

At Hamilton Middle School, the ratio of teachers to students is 1 to 26. Which statement correctly compares the ratio of teachers to students at the two schools?

Jefferson Middle School	
Students, x	**Teachers, y**
24	1
48	2
72	3
96	4

Ⓐ There are more students per teacher at Hamilton Middle School than at Jefferson Middle School.

Ⓑ Both schools have an equivalent ratio of students to teachers.

Ⓒ There are more students at Hamilton Middle School than at Jefferson Middle School.

Ⓓ There are more students per teacher at Jefferson Middle School than at Hamilton Middle School.

17. **Short Response** Nina earns $15 for each yard she mows. The table shows her earnings for 0, 1, 2, and 3 yards mowed. How much will Nina earn if she mows 6 yards?

____

Nina's Earnings	
Yards Mowed	**Dollars Earned ($)**
0	0
1	15
2	30
3	45

Common Core Review

Simplify each fraction. 5.NF.5b

18. $\frac{13}{78} = \frac{\square}{\square}$

19. $\frac{26}{130} = \frac{\square}{\square}$

20. $\frac{20}{240} = \frac{\square}{\square}$

21. There are 270 sixth grade students and 45 chaperones going on a field trip. How many students will be with each chaperone if the groups are divided equally? 5.NBT.5 ____________________

22. Several students were surveyed about their favorite class. The results are shown in the table. What fraction of the students chose music as their favorite subject? Write the fraction in simplest form. 5.NF.3

____

Favorite Class	
Art	26
English	19
Math	21
Music	16
Science	32

Problem-Solving Investigation
The Four-Step Plan

Content Standards
6.RP.3
Mathematical Practices
1, 3, 4

Case #1 Cabin Fever

At a summer camp, the ratio of cabins to campers is 15 to 180. An equal number of campers are staying in each cabin.

How many campers are in each cabin?

Understand What are the facts?

- You know there are 15 cabins for 180 campers.
- You need to find how many campers are in each cabin.

Plan What is your strategy to solve this problem?

Divide 180 by 15. Before you calculate, estimate.

Estimate $200 \div 20 =$ ☐

Solve How can you apply the strategy?

Use long division to find the number of campers in each cabin.

```
     ☐
15)180
  -☐
   30
  -30
    0
```

There are ☐ campers in each cabin.

Check Does the answer make sense?

Check by multiplying. Since $12 \times 15 =$ ☐, the answer is correct.

Analyze the Strategy

Justify Conclusions How many campers would be in each cabin if the ratio of cabins to campers was 15 to 225? Explain.

___

___

Case #2 Show Me the Money

The table shows Kaylee's weekly allowance.

Age	10	11	12	13
Weekly Allowance ($)	2	4	6	■

If the pattern continues, how much allowance will Kaylee earn when she is 13 years old?

Understand

Read the problem. What are you being asked to find?

I need to find ______________________________

Underline key words and values in the problem. What information do you know?

The top row shows an increase of ☐ year. The bottom row shows an increase of $☐ per year.

2 Plan

Choose an operation.

I will use ______________________________ to solve this problem.

3 Solve

Describe the pattern in the table. Then complete it using your problem-solving strategy.

____

+1 +1 +1

Age	10	11	12	13
Weekly Allowance ($)	2	4	6	

+2 +2 ☐

6 + ☐ = ☐ So, Kaylee will earn $☐ when she is 13 years old.

Check

Use information from the problem to check your answer.

Use subtraction to check your answer. ☐ − ☐ = 6

Collaborate Work with a small group to solve the following cases. Show your work on a separate piece of paper.

Case #3 Walking

Megan uses a pedometer to find how many steps she takes each school day. She took 32,410 steps over the course of 5 days.

If she took the same number of steps each day, how many did she take on Monday?

Case #4 Pools

The table shows the total amount of water in a swimming pool that is being filled.

Time (min)	5	10	15	20
Water (gal)	75	150	225	300

At this rate, how much water will be in the swimming pool after 30 minutes?

Case #5 Money

Mrs. Eddington is buying a new big-screen television. She made an initial payment of $50 and will pay $70 per month for 12 months.

How much will she spend in all for the television?

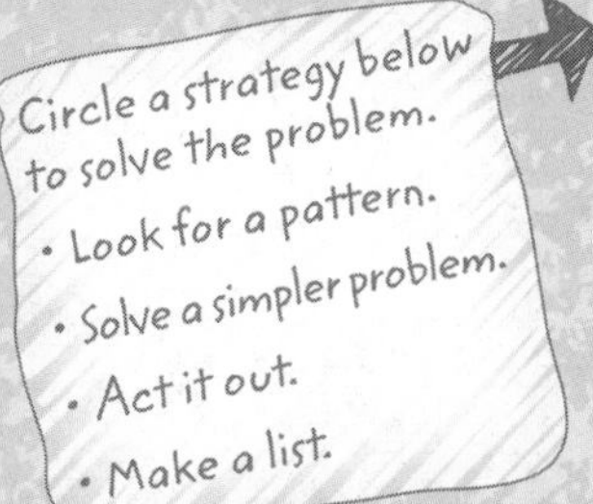

Case #6 Sports Equipment

Mrs. Dimas has $130 to buy basketballs for Edison Middle School.

How many can she buy at $15 each? Interpret the remainder.

Mid-Chapter Check

Vocabulary Check

1. Fill in the blank in the sentence below with the correct term. (Lesson 2)

A ______________ is a comparison of two quantities by division.

Skills Check and Problem Solving

Find the greatest common factor or least common multiple of each set of numbers. (Lesson 1)

2. 24 and 18

GCF = ________

3. 12 and 20

LCM = ________

4. 16 and 32

GCF = ________

5. Write *15 cookies to 40 brownies* as a ratio in simplest form. (Lesson 2) ________

6. Write 171 miles in 3 hours as a unit rate. (Lesson 3) ________

7. CCSS **Use Math Tools** The table below shows the amount in Josiah's account each week. List the information as ordered pairs and then graph the ordered pairs. Describe the pattern in the graph. (Lesson 5)

Josiah's Savings		
Week, x	**Savings ($), y**	**Ordered Pair (x, y)**
1	5	
2	10	
3	15	
4	20	
5	25	

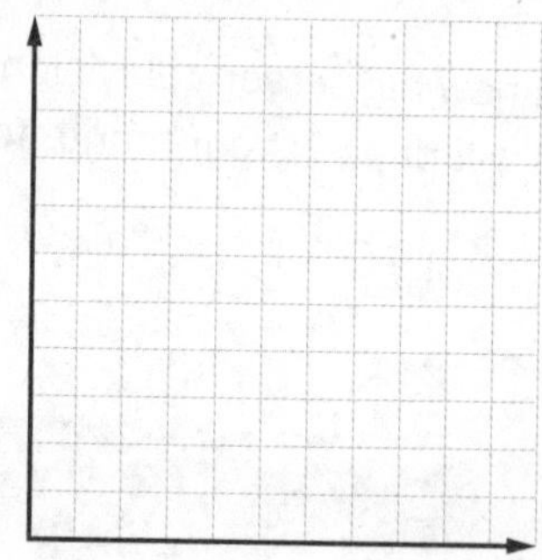

____

8. **Standardized Test Practice** The ratio of brown tiles to tan tiles is 2 to 3. If an artist needs 16 brown tiles to complete a mosaic, how many tan tiles will the artist need? (Lesson 4)

Ⓐ 8

Ⓑ 16

Ⓒ 17

Ⓓ 24

Lesson 6

Equivalent Ratios

What You'll Learn

Scan the lesson. List two things you will learn about equivalent ratios.

- ______________________________

- ______________________________

Essential Question

HOW do you use equivalent rates in the real world?

Common Core State Standards

Content Standards
6.RP.3, 6.RP.3b

Mathematical Practices
1, 3, 4, 6, 7

Real-World Link

Photography Andrea spent \$2 to make 10 prints from a photo booth. Later, she spent \$6 to make 30 prints.

Number of Prints	Cost (\$)
10	2
30	6

1. Express the relationship between the number of prints she made and the total cost for each situation as a rate in fraction form.

$\frac{\square \text{ prints}}{\square}$ and $\frac{\square \text{ prints}}{\square}$

2. Compare the relationship between the numerators of each rate in Exercise 1. Compare the relationship between the denominators of these rates.

____

____

____

3. What is the unit rate for 10 prints? ______________

4. What is the unit rate for 30 prints? ______________

5. Are the rates in Exercise 1 equivalent? Explain.

____

____

Work Zone

Use Unit Rates

There are different ways to determine if two ratios or rates are equivalent. One way is by examining unit rates. By comparing quantities as rates in simplest form, you can determine if the relationship between the two quantities stays the same.

$$\frac{10 \text{ prints}}{\$2} \overset{\div 2}{=} \frac{5 \text{ prints}}{\$1} \text{ and } \frac{30 \text{ prints}}{\$6} \overset{\div 6}{=} \frac{5 \text{ prints}}{\$1}$$

Since the rates have the same unit rate, they are equivalent ratios.

Examples

Determine if each pair of rates is equivalent. Explain your reasoning.

1. 20 miles in 5 hours; 45 miles in 9 hours

Write each rate as a fraction. Then find its unit rate.

$$\frac{20 \text{ miles}}{5 \text{ hours}} \overset{\div 5}{=} \frac{4 \text{ miles}}{1 \text{ hour}} \qquad \frac{45 \text{ miles}}{9 \text{ hours}} \overset{\div 9}{=} \frac{5 \text{ miles}}{1 \text{ hour}}$$

Since the rates do not have the same unit rate, they are not equivalent.

Unit Rates
The unit rate in Example 2, $\frac{\$7}{1 \text{ T-shirt}}$, is called the unit price since it gives the cost per unit.

2. 3 T-shirts for \$21; 5 T-shirts for \$35

$$\frac{\$21}{3 \text{ T-shirts}} \overset{\div 3}{=} \frac{\$7}{1 \text{ T-shirt}} \qquad \frac{\$35}{5 \text{ T-shirts}} \overset{\div 5}{=} \frac{\$7}{1 \text{ T-shirt}}$$

Since the rates have the same unit rate, they are equivalent.

Got It? Do these problems to find out.

Determine if each pair of rates is equivalent. Explain your reasoning.

a. 36 T-shirts in 3 boxes; 60 T-shirts in 6 boxes

b. 42 flowers in 7 vases; 54 flowers in 9 vases

a. ____________

b. ____________

Example

3. Felisa read the first 60 pages of a book in 3 days. She read the last 90 pages in 6 days. Are these reading rates equivalent? Explain your reasoning.

$$\frac{60 \text{ pages}}{3 \text{ days}} = \frac{20 \text{ pages}}{1 \text{ day}} \quad (\div 3) \qquad \frac{90 \text{ pages}}{6 \text{ days}} = \frac{15 \text{ pages}}{1 \text{ day}} \quad (\div 6)$$

Since the rates do not have the same unit rate, they are not equivalent. So, Felisa's reading rates are not equivalent.

Proportion
A proportion is an equation stating that two ratios or rates are equivalent.

Got It? Do these problems to find out.

c. Marcia made 10 bracelets for 5 friends. Jen made 12 bracelets for 4 friends. Are these rates equivalent? Explain your reasoning.

d. Club A raised \$168 by washing 42 cars. Club B raised \$152 by washing 38 cars. Are these fundraising rates equivalent? Explain your reasoning.

Show your work.

c. ____________

d. ____________

Use Equivalent Fractions

If a unit rate is not easily found, use equivalent fractions to decide whether the ratios or rates are equivalent.

Examples

Determine if the pair of ratios or rates is equivalent. Explain your reasoning.

4. 3 free throws made out of 7 attempts;
9 free throws made out of 14 attempts

Write each ratio as a fraction.

$$\frac{3 \text{ free throws}}{7 \text{ attempts}} \stackrel{?}{=} \frac{9 \text{ free throws}}{14 \text{ attempts}} \quad (\times 3 \text{ top}, \times 2 \text{ bottom})$$

The numerator and the denominator are not multiplied by the same number. So, the fractions are not equivalent.

Since the fractions are *not* equivalent, the ratios are not equivalent.

5. **Selena is comparing the cost of two packages of DVDs. A package of 6 DVDs costs $90 and a package of 3 DVDs costs $45. Are the rates equivalent? Explain your reasoning.**

$$\frac{6 \text{ DVDs}}{\$90} = \frac{3 \text{ DVDs}}{\$45} \quad (\div 2)$$

The numerator and the denominator are divided by the same number. So, the fractions are equivalent.

Since the fractions are equivalent, the ratios are equivalent.

Got It? **Do this problem to find out.**

Show your work.

e. ________

e. Mrs. Jeffries has 12 girls out of 16 students on the Student Council. The Earth Day Committee has 4 girls out of 8 students. Are the ratios equivalent? Explain your reasoning.

Guided Practice

Determine if each pair of ratios or rates is equivalent. Explain your reasoning.

1. $24 saved after 3 weeks; $52 saved after 7 weeks (Examples 1 and 2)

2. 270 Calories in 3 servings; 450 Calories in 5 servings (Examples 1 and 2)

3. Micah can do 75 push-ups in 3 minutes. Eduardo can do 130 push-ups in 5 minutes. Are these rates equivalent?

Explain. (Example 3)

4. A human adult takes about 16 breaths in 60 seconds. A puppy takes about 8 breaths in 15 seconds. Are these rates equivalent? Explain your reasoning. (Examples 4 and 5)

5. **Building on the Essential Question** How can you determine if two ratios are equivalent?

Independent Practice

Go online for Step-by-Step Solutions eHelp

Determine if each pair of ratios or rates is equivalent. Explain your reasoning. (Examples 1–2, 4–5)

1. $3 for 6 bagels; $9 for 24 bagels

2. $12 for 3 paperback books; $28 for 7 paperback books

3. 3 hours worked for $12; 9 hours worked for $36

4. 12 minutes to drive 30 laps; 48 minutes to drive 120 laps

5. Jenny is comparing the cost of two packages of socks. One package has 8 pairs of socks for $12. Another package has 3 pairs of socks for $6. Are the rates equivalent? Explain your reasoning.

6. Jade enlarged the photograph at the right to a poster. The size of the poster is 60 inches by 100 inches. Is the ratio of the poster's length and width equivalent to the ratio of the photograph's length and width? Explain your reasoning. (Example 3)

7. CCSS **Justify Conclusions** On a math test, it took Kiera 30 minutes to do 6 problems. Heath finished 18 problems in 40 minutes. Did the students work at the same rate? Explain your reasoning.

8. **Be Precise** Refer to the graphic novel frame below for Exercises a–b.

a. What is the unit price for the cans of lemonade at each of the stores?

b. From which store should Mei, Pilar, and David purchase the cans of lemonade? Explain.

H.O.T. Problems Higher Order Thinking

9. **Persevere with Problems** To verify equivalent ratios, you can use cross products. If the cross products are equal, the ratios are equivalent.

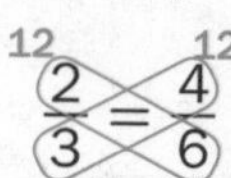 Since 12 = 12, the ratios are equivalent.

Determine whether each pair of ratios are equivalent. Explain.

a. $\frac{3}{5}, \frac{9}{15}$

b. $\frac{2}{7}, \frac{5}{21}$

10. **Identify Structure** Write two ratios that are equivalent to $\frac{5}{7}$.

Standardized Test Practice

11. The ratio of girls to boys in the junior high band is 3 to 4. Which of these shows possible numbers of the girls and boys in the band?

Ⓐ 30 girls, 44 boys
Ⓑ 27 girls, 36 boys
Ⓒ 22 girls, 28 boys
Ⓓ 36 girls, 50 boys

Extra Practice

Determine if each pair of ratios or rates is equivalent. Explain your reasoning.

12. 16 points scored in 4 games; 48 points scored in 8 games

No; $\frac{16 \text{ points}}{4 \text{ games}} = \frac{4 \text{ points}}{1 \text{ game}}$ and $\frac{48 \text{ points}}{8 \text{ games}} = \frac{6 \text{ points}}{1 \text{ game}}$; Since the unit rates are not the same, the rates are not equivalent.

13. 96 words typed in 3 minutes; 160 words typed in 5 minutes

14. 15 computers for 45 students; 45 computers for 135 students

15. 16 out of 28 students own pets; 240 out of 560 students own pets

16. 288 miles on 12 gallons of fuel; 240 miles on 10 gallons of fuel

17. Fenton is building a model of a living room. The model sofa is 16 inches long and 7 inches deep. The real sofa's dimensions are 80 inches long and 35 inches deep. Is the ratio of the model's dimensions equivalent to the ratio of the real sofa's dimensions? Explain your reasoning.

18. Store A sells 12 juice bottles for \$4 and store B sells 18 juice bottles for \$6. Are the rates equivalent? Explain your reasoning.

19. CCSS **Justify Conclusions** Rosalinda saved \$35 in 5 weeks. Her sister saved \$56 in 56 days. Are the rates at which each sister saved equivalent? Explain your reasoning.

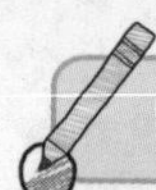

Standardized Test Practice

20. The ratio of dogs to cats at a pet store is 2 to 3. Which of these shows the possible numbers of dogs and cats in the pet store?

Ⓐ 12 dogs, 13 cats
Ⓑ 14 dogs, 21 cats
Ⓒ 5 dogs, 10 cats
Ⓓ 20 dogs, 23 cats

21. What is the cost of 8 twelve-packs of soda?

Sale!
3 12-packs for $8
2 24-packs for $10
1 18-pack for $3

Ⓕ $64
Ⓖ $40
Ⓗ $21.33
Ⓘ $2.67

22. **Short Response** What is the cost of 15 tomatoes? ______

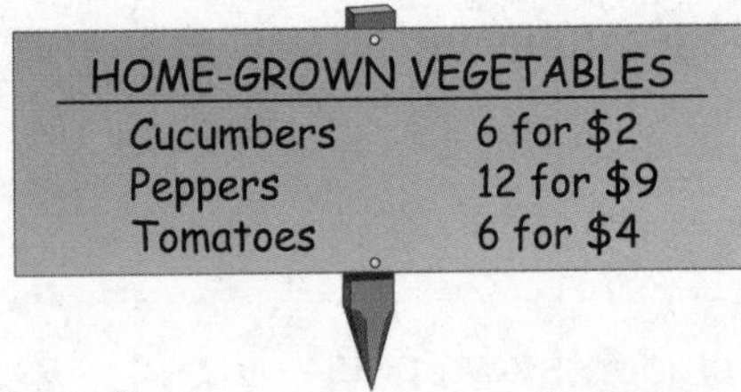

Common Core Review

Write an equivalent fraction. 5.NF.5b

23. $\frac{11}{50} = \frac{33}{\square}$

24. $\frac{4}{5} = \frac{\square}{80}$

25. $\frac{2}{9} = \frac{28}{\square}$

26. Socks are on sale 4 pairs for $5. How much would you pay for 8 pairs of socks? 5.NBT.7

______

27. Sasha bought 3 pens. Malachi bought 1 pen. How much more did Sasha spend than Malachi? 4.OA.3

______

Inquiry Lab

Ratio and Rate Problems

HOW can you use unit rates and multiplication to solve for missing measures in equivalent ratio problems?

CCSS Content Standards 6.RP.3, 6.RP.3b
Mathematical Practices 1, 3, 4, 5, 8

Racing Jill and Sammy are racing go-karts. Jill completed 6 laps in 12 minutes. If Sammy raced at the same rate, how many minutes did it take her to complete 3 laps?

What do you know? ______________________________

____

What do you need to find? ______________________________

____

Investigation 1

Step 1 Use a bar diagram to represent the number of laps Jill completed. The time to travel 6 laps is 12 minutes.

12 min

Jill's race	1 lap	1 lap	1 lap	1 lap	1 lap	1 lap

Step 2 Each section represents 1 lap. Determine the number of minutes it took Jill to complete one lap.

Jill completed each lap in 12 ÷ 6, or ☐ minutes.

Step 3 Determine the number of minutes it took Sammy to complete 3 laps.

12 min

Jill's race	1 lap	1 lap	1 lap	1 lap	1 lap	1 lap

? min

Sammy's race	1 lap	1 lap	1 lap

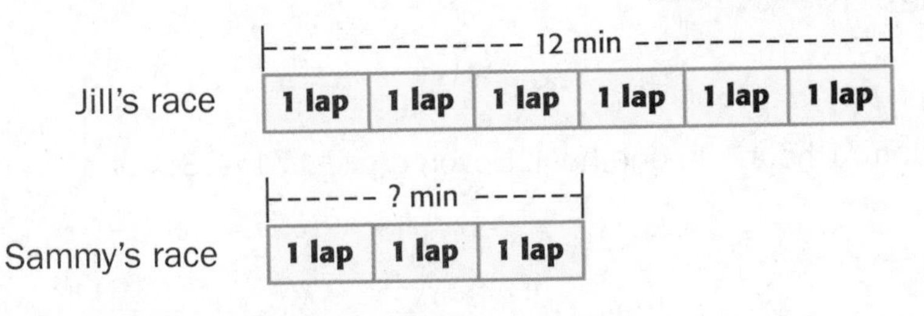

Each lap was completed in ☐ minutes.

So, Sammy's time was 3 × ☐, or ☐ minutes.

Investigation 2

Lizette and Miguel are decorating cookies for a bake sale. Lizette can decorate 4 cookies in 12 minutes. If Miguel can decorate cookies at the same rate, how many minutes will it take him to decorate 24 cookies?

Step 1 Use a bar diagram to represent the amount of time Lizette spent decorating cookies.

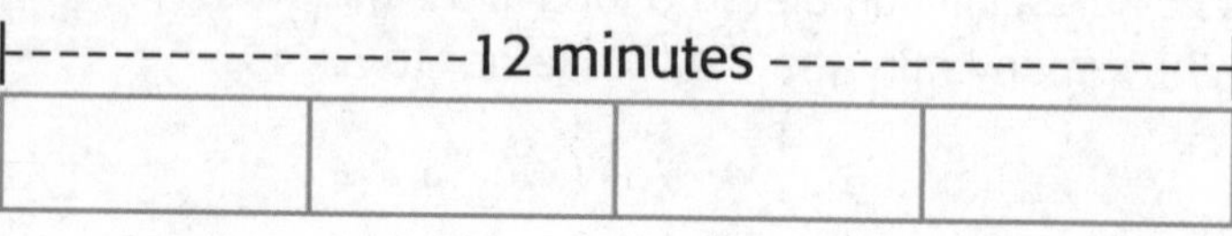

☐ minutes

Step 2 Label each section "1 cookie." Lizette decorated one cookie in $12 \div 4$, or ☐ minutes.

So, it will take Miguel $24 \times$ ☐, or ☐ minutes.

Investigation 3

Devon drives 171 miles in 3 hours. At this rate, how many miles can he drive in 7 hours?

Step 1 Use a bar diagram to represent the number of miles Devon drove.

☐ miles

Step 2 Label each section "1 hour." In one hour, Devon drove $171 \div 3$, or ☐ miles.

So, Devon will drive $7 \times$ ☐, or ☐ miles in 7 hours.

Collaborate

Work with a partner. Use a bar diagram to help solve each problem.

1. the miles traveled in 5 hours at a rate of 189 miles in 3 hours ______

Show your work.

2. the number of ice cubes in 32 glasses at a rate of 20 ice cubes in 5 glasses ______

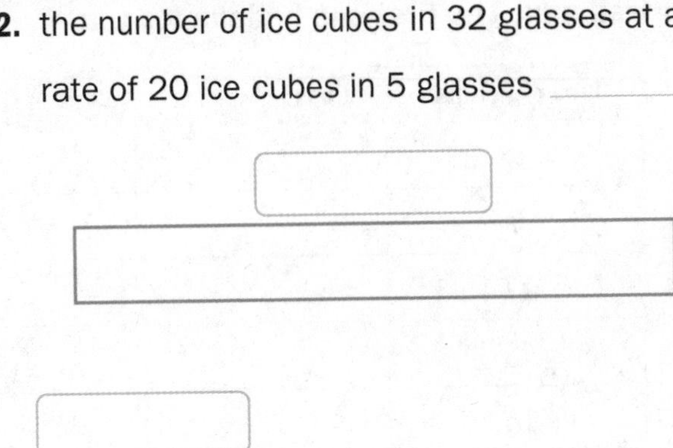

3. the cost of 5 pounds of bananas if 2 pounds cost $1.16 ______

4. the time needed to deliver 72 papers at a rate of 9 papers in 18 minutes ______

5. the number of squares in 15 quilts if 6 quilts have 288 squares ______

6. the time to run 26 miles at a rate of 12 miles in 60 minutes ______

7. the beads in 7 bracelets if 4 bracelets have 96 beads ______

8. the lemons needed for 6 pitchers of lemonade if 2 pitchers use 28 lemons ______

Analyze

Work with a partner to complete the table, using the recipe for trail mix. The first one is done for you.

	Cups of Peanuts	Cups of Raisins	Cups of Chocolate Chips	Cups of Granola
	6	4	2	8
9.	9			
10.	12			
11.	15			
12.	18			
13.	21			
14.	24			
15.	27			

Trail Mix
2 cups raisins
3 cups peanuts
1 cup chocolate chips
4 cups granola

16. CCSS **Identify Repeated Reasoning** Explain how you can use the information on the recipe card to solve for missing measures in the table.

Reflect

17. CCSS **Model with Mathematics** Lee can read 1,100 words in 5 minutes. Write and solve a word problem that uses this information.

18. Inquiry HOW can you use unit rates and multiplication to solve for missing measures in equivalent ratio problems?

Lesson 7

Ratio and Rate Problems

What You'll Learn

Scan the lesson. List two headings you would use to make an outline of the lesson.

-

-

Essential Question

HOW do you use equivalent rates in the real world?

Common Core State Standards

Content Standards
6.RP.3, 6.RP.3b

Mathematical Practices
1, 3, 4, 5, 7

Real-World Link

Games An arcade sells game tokens individually or in packages. They are having a sale on token packages, as shown below.

Number of Packages	Price ($)
1	5
2	10
3	15

1. How many token packages can you buy with $20? ☐ $25? ☐ Explain.

2. What is the unit price?

3. How much would it cost to buy 6 token packages?

4. The arcade sells individual tokens for $0.25 each. If a token package contains 25 tokens, how much would you save by buying a package of 25 tokens instead of 25 individual tokens? Explain.

Work Zone

Solve Ratio Problems

You can use bar diagrams or equations with equivalent ratios to solve ratio and rate problems.

Examples

Equivalent Ratios
Notice that the numerators of both fractions in Method 2 refer to the number of students who like gel toothpaste. The denominators of both fractions refer to the total number of students being referenced.

1. Heritage Middle School has 150 students. Two out of three students in Mrs. Mason's class prefer gel toothpaste. Use this ratio to predict how many students in the entire middle school prefer gel toothpaste.

Method 1 **Use a bar diagram.**

Step 1 Draw a bar diagram.

Step 2 Determine how many students are in each section.

150 students

50	50	50

gel | not gel

Method 2 **Use equivalent fractions.**

Write an equivalent ratio.

likes gel → $\frac{2}{3} = \frac{\blacksquare}{150}$ ← likes gel
total → ← total

$$\frac{2}{3} = \frac{100}{150}$$ (×50 numerator and denominator)

Since 3 × 50 = 150, multiply 2 by 50.

So, 100 students would prefer gel toothpaste.

2. The ratio of the number of text messages sent by Lucas to the number of text messages sent by his sister is 3 to 4. Lucas sent 18 text messages. How many text messages did his sister send?

Method 1 **Use a bar diagram.**

Step 1 Draw a bar diagram.

Step 2 Determine how many text messages are in each section.

18 texts | 24 texts

6	6	6	6	6	6	6

Lucas | Sister

Method 2 **Use equivalent fractions.**

Write an equivalent ratio.

Lucas → $\frac{3}{4} = \frac{18}{■}$ ← Lucas
his sister → ← his sister

$\frac{3}{4} = \frac{18}{24}$ (×6) Since 6 × 3 = 18, multiply 4 by 6.

So, Lucas' sister sent 24 text messages.

STOP and Reflect

What is the relationship between ratios and fractions?

Got It? **Do these problems to find out.**

a. In a survey, four out of five people preferred creamy over chunky peanut butter. There are 120 people shopping at the grocery store. Use the survey to predict how many people in the store would prefer creamy peanut butter.

b. A survey found that 12 out of every 15 people in the United States prefer eating at a restaurant over cooking at home. If 400 people selected eating at a restaurant on the survey, how many people took the survey?

a. ______

b. ______

Solve Rate Problems

You can use double number lines or equations to solve rate problems.

Example

3. The Millers drove 105 miles on 4 gallons of gas. At this rate, how many miles can they drive on 6 gallons of gas?

Draw a double number line.

105 ÷ 4 = 26.25 Find the unit rate.

26.25 × 6 = 157.5 Multiply.

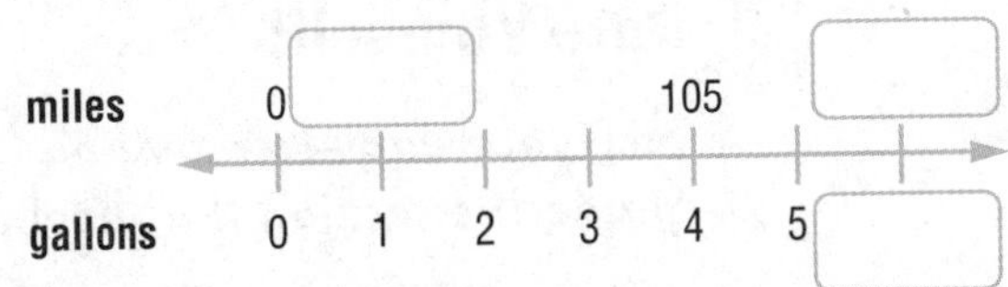

So, the Millers can drive 157.5 miles on 6 gallons of gas.

Got It? **Do this problem to find out.**

c. There are 810 Calories in 3 scoops of vanilla ice cream. How many Calories are there in 7 scoops of ice cream?

c. ______

Example

4. Jeremy drove his motorcycle 120 miles in 3 hours. At this rate, how many miles can he drive in 5 hours? At what rate did he drive his motorcycle?

$$\frac{120 \text{ miles}}{3 \text{ hours}} = \frac{\blacksquare \text{ miles}}{1 \text{ hour}} \quad (\div 3) \qquad \frac{120 \text{ miles}}{3 \text{ hours}} = \frac{40 \text{ miles}}{1 \text{ hour}}$$ Find the unit rate.

$$\frac{40}{1 \text{ hour}} \times 5 \text{ hours} = 200 \text{ miles}$$ Multiply.

So, Jeremy can drive 200 miles in 5 hours driving at a rate of 40 miles per hour.

Got It? **Do this problem to find out.**

Show your work.

d. **STEM** While resting, a human takes in about 5 liters of air in 30 seconds. At this rate, how many liters of air does he take in during 150 seconds?

d. ________

Guided Practice

1. Out of 30 students surveyed, 17 have a dog. Based on these results, predict how many of the 300 students in the school have a dog? (Example 1)

Show your work.

2. If one out of 12 students at a school share a locker, how many share a locker in a school of 456 students? (Example 2) ________

3. Sybrina jogged 2 miles in 30 minutes. At this rate, how far would she jog in 90 minutes? At what rate did she jog each hour? (Examples 3 and 4)

4. **Building on the Essential Question** How can you use diagrams and equations to solve ratio and rate problems?

Independent Practice

Go online for Step-by-Step Solutions

1. If 45 cookies will serve 15 students, how many cookies are needed for 30 students? (Examples 1 and 2)

Show your work.

2. Four students spent $12 on school lunch. At this rate, find the amount 10 students would spend on the same school lunch. (Example 3)

3. A Clydesdale drinks about 120 gallons of water every 4 days. At this rate, about how many gallons of water does a Clydesdale drink in 28 days? (Example 3)

4. **STEM** In 10 minutes, a heart can beat 700 times. At this rate, in how many minutes will a heart beat 140 times? At what rate can a heart beat? (Example 4)

5. **CCSS Make a Prediction** The table shows which school subjects are favored by a group of students. Predict the number of students out of 400 that would pick science as their favorite subject.

Favorite Subject	
Subject	**Number of Responses**
Math	6
Science	3
English	4
History	7

6. Liliana takes 4 breaths per 10 seconds during yoga. At this rate, about how many breaths would Liliana take in 2 minutes of yoga?

7. **CCSS Use Math Tools** Find a report in a newspaper or magazine, or on the Internet that uses results from a survey. Evaluate how the survey uses ratios to reach conclusions.

H.O.T. Problems Higher Order Thinking

8. **Identify Structure** One rate of an equivalent ratio is $\frac{9}{n}$. Select two other rates, one that can be solved using equivalent fractions and the other that can be solved with unit rates.

9. **Find the Error** Elisa's mom teaches at a preschool. There is 1 teacher for every 12 students at the preschool. There are 276 students at the preschool. Elisa is setting up equivalent ratios to find the number of teachers at the preschool. Find her mistake and correct it.

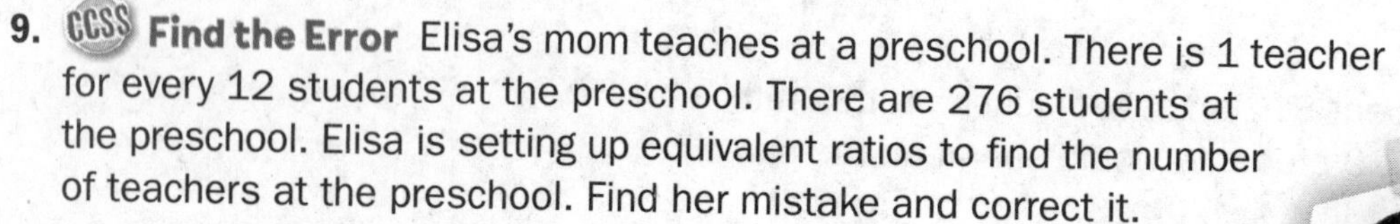

10. **Reason Inductively** Tell whether the following statement is *always*, *sometimes*, or *never* true for numbers greater than zero. Explain.

In equivalent ratios, if the numerator of the first ratio is greater than the denominator of the first ratio, then the numerator of the second ratio is greater than the denominator of the second ratio.

11. **Persevere with Problems** Suppose 25 out of 175 people said they like to play disc golf and 5 out of every 12 of the players have a personalized flying disc. At the same rates, in a group of 252 people, predict how many you would expect to have a personalized flying disc.

Standardized Test Practice

12. A car traveling at a certain speed will travel 76 feet per second. How many yards will the car travel in 120 seconds if it maintains the same speed?

Ⓐ 76 yards

Ⓑ 228 yards

Ⓒ 3,040 yards

Ⓓ 9,120 yards

3ft / 1yd

76 ft per sec

76 x 120 = 9,120 feet

Name ______________________ My Homework ______________________

Extra Practice

13. A survey reported that out of 50 teenagers, 9 said they get their news from a newspaper. At this rate, how many out of 300 teenagers would you expect to get their news from a newspaper?

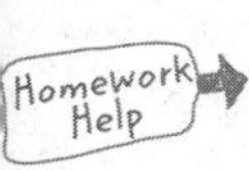

54 teenagers

$$\frac{50}{9} = \frac{300}{54} \quad (\times 6)$$

14. Nata spent $28 on 2 DVDs. At this rate, how much would 5 DVDs cost? At what rate did she spend her money?

15. If 15 baseballs weigh 75 ounces, how many baseballs weigh 15 ounces?

16. CCSS **Make a Prediction** Suppose 8 out of every 20 students are absent from school less than five days a year. Predict how many students would be absent from school less than five days a year out of 40,000 students.

17. For a store contest, 4 out of every 65 people who visit the store will receive a free DVD. If 455 people visit the store, how many DVDs were given away?

18. There were 340,000 cattle placed on feed. Write an equivalent ratio that could be used to find how many of these cattle were between 700 and 799 pounds. How many of the 340,000 cattle placed on feed were between 700 and 799 pounds?

Cattle Placed on Feed

Weight Group	Fraction of Total Cattle
Less than 600 pounds	$\frac{1}{5}$
600–699 pounds	$\frac{11}{50}$
700–799 pounds	$\frac{2}{5}$
800 pounds or more	$\frac{9}{50}$

Standardized Test Practice

19. The ratio of red poms to yellow poms on a float is 5 to 7. If there are 392 yellow poms on the float, how many red poms are there?

   Ⓐ 549 Ⓒ 280
   Ⓑ 390 Ⓓ 56

20. The ratio of green pepper plants to red pepper plants in Adeline's garden is 3 to 5. If there are 20 red pepper plants, how many green pepper plants are there?

   Ⓕ 35 Ⓗ 12
   Ⓖ 16 Ⓘ 6

21. **Short Response** Student Council sells bottled water at the cheerleading competition. At this rate, how many cases of bottled water would they sell in 3 hours? ____________

Cases Sold	3	6
Time (min)	20	40

22. At a bus station, buses depart at a rate of 3 every 10 minutes. At this rate, how many buses would you expect to depart in one hour?

   Ⓐ 6 Ⓒ 18
   Ⓑ 15 Ⓓ 30

Common Core Review

Write each fraction as a unit fraction. 5.NF.5b

23. $\frac{12}{84} =$ ________

24. $\frac{13}{143} =$ ________

25. $\frac{23}{138} =$ ________

26. Skylar gained 64 yards on 16 carries during a recent football game. Find the ratio of yards per carry. 5.NBT.5

____

27. The drama club is washing cars for a fundraiser. If the rate continues, how many cars will they wash in 4 hours? 4.OA.5

____

____

Hours	Cars Washed
1	8
2	16
3	24

28. Follow the rule to find the next three numbers in the pattern. Describe the pattern using the terms *even* and *odd*. 4.OA.5

Add 5: 1, 6, 11, ______, ______, ______ ...

____

21ST CENTURY CAREER in Chemistry

Cosmetic Chemist

Are you naturally curious and analytical? Do you like discovering new things? If so, a career as a cosmetics chemist might be a good choice for you. Cosmetics chemists spend time researching, mixing, and testing new formulas that will make cosmetic products both effective and safe. A cosmetics chemist explained, "When you're developing a product, you play with chemicals and balance ratios to get it to feel right. Basically, it's trial and error."

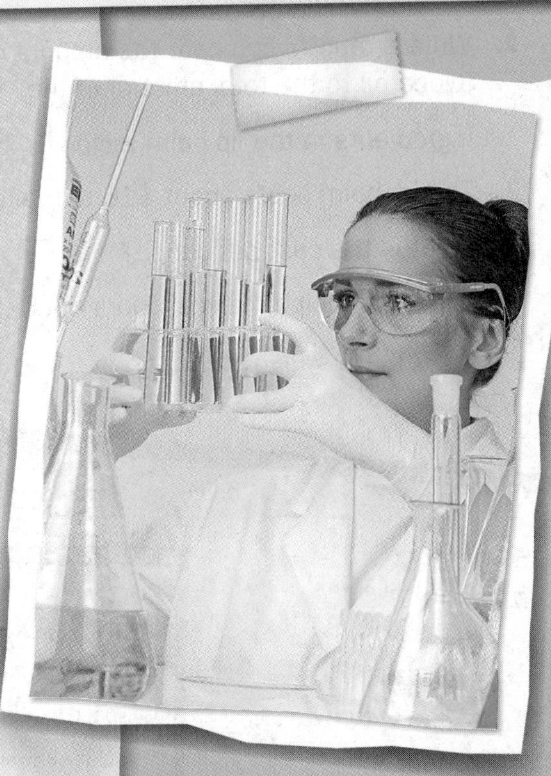

Explore college and careers at ccr.mcgraw-hill.com

Is This the Career for You?

Are you interested in a career as a cosmetics chemist? Take some of the following courses in high school.

- Algebra
- Biology
- Chemical Science
- Chemistry
- Statistics

Find out how math relates to a career in Chemisty.

Beauty is Only Science-Deep

Use the information in the recipes below to solve each problem.

1. Using the soap recipe, write a ratio comparing the amount of palm kernel oil to the amount of rose hydrosol as a fraction in simplest form. ________

2. Write a ratio to compare the amount of jojoba oil to the total amount of the ingredients in the lip balm recipe. ________

3. The lip balm costs about $16 to make. What is the cost per ounce? ________

4. The soap recipe makes 4 bars of soap. What is the weight per bar? ________

5. The lip balm recipe is increased so that 10 ounces of candelilla wax is needed. Complete the ratio table to find the amount of shea butter that is needed. ________

Candelilla wax	2				10
Shea butter	6				

6. The soap recipe is increased so that 75 grams of shea butter are needed. Complete the ratio table to find the amount of sodium hydroxide that is needed. ________

Shea butter	30		75
Sodium hydroxide	42		

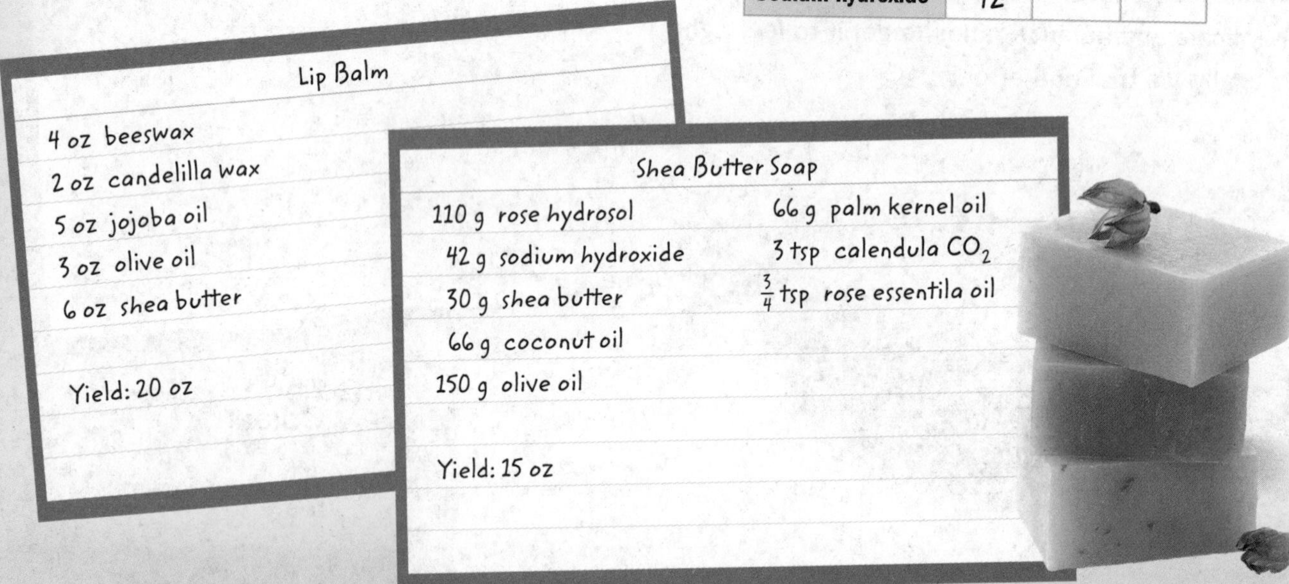

Lip Balm

4 oz beeswax
2 oz candelilla wax
5 oz jojoba oil
3 oz olive oil
6 oz shea butter

Yield: 20 oz

Shea Butter Soap

110 g rose hydrosol
42 g sodium hydroxide
30 g shea butter
66 g coconut oil
150 g olive oil
66 g palm kernel oil
3 tsp calendula CO_2
$\frac{3}{4}$ tsp rose essentila oil

Yield: 15 oz

Career Project

It's time to update your career portfolio! There are many different types of jobs in cosmetic chemistry. Research one of these jobs and write a two- or three-sentence job description.

____
____
____
____
____
____

List other careers that someone with an interest in chemistry could pursue.

- ________
- ________
- ________
- ________
- ________

Chapter Review

Vocabulary Check

Complete the crossword puzzle using the vocabulary list at the beginning of the chapter.

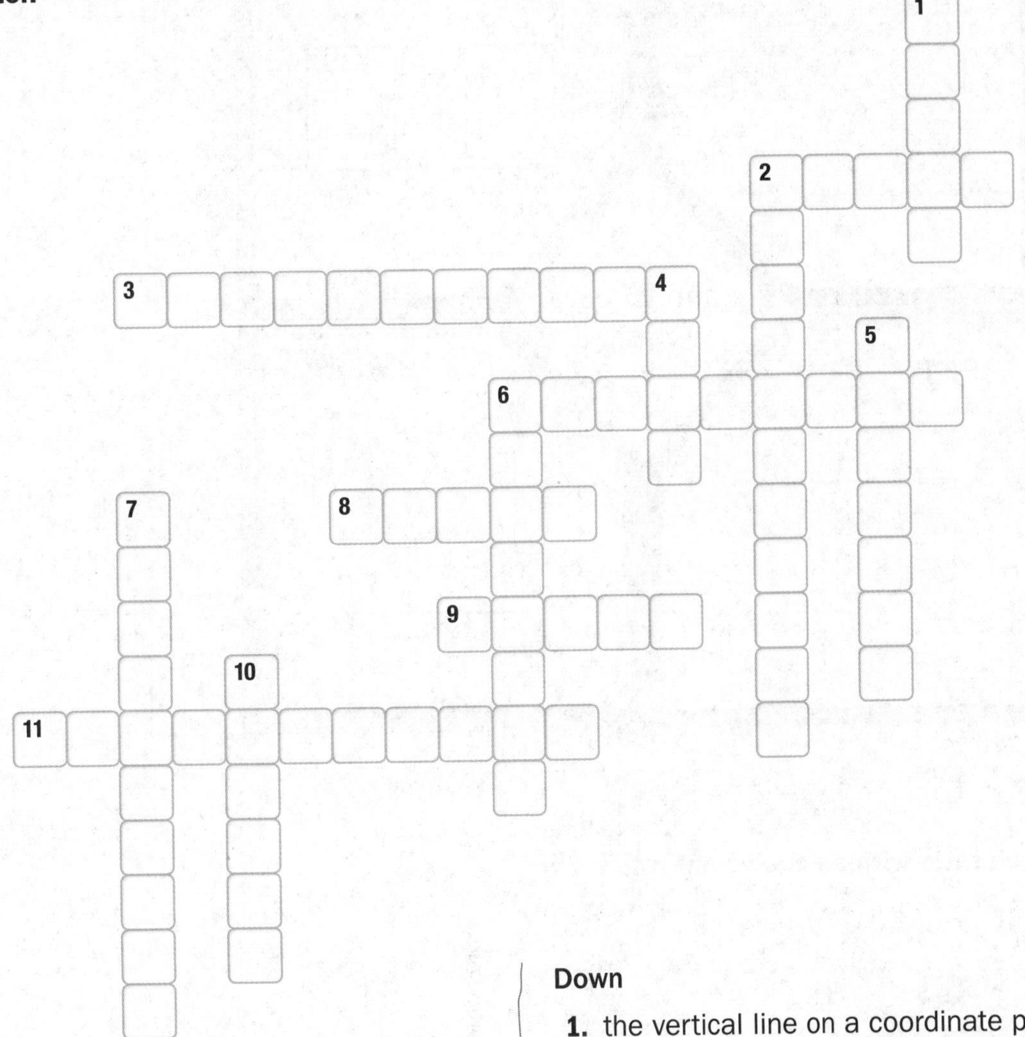

Across

2. the horizontal line on a coordinate plane
3. used to locate a point on the coordinate plane
6. the cost per unit
8. a comparison of two quantities by division
9. to place a dot a the point named by an ordered pair
11. the second number of an ordered pair

Down

1. the vertical line on a coordinate plane
2. the first number of an ordered pair
4. a ratio comparing two quantities with different kinds of units
5. multiply or divide two quantities by the same number
6. a rate simplified so that it has a denominator of 1
7. columns filled with pairs of numbers that have the same ratio
10. (0, 0)

Key Concept Check

Use Your FOLDABLES

Use your Foldable to help review the chapter.

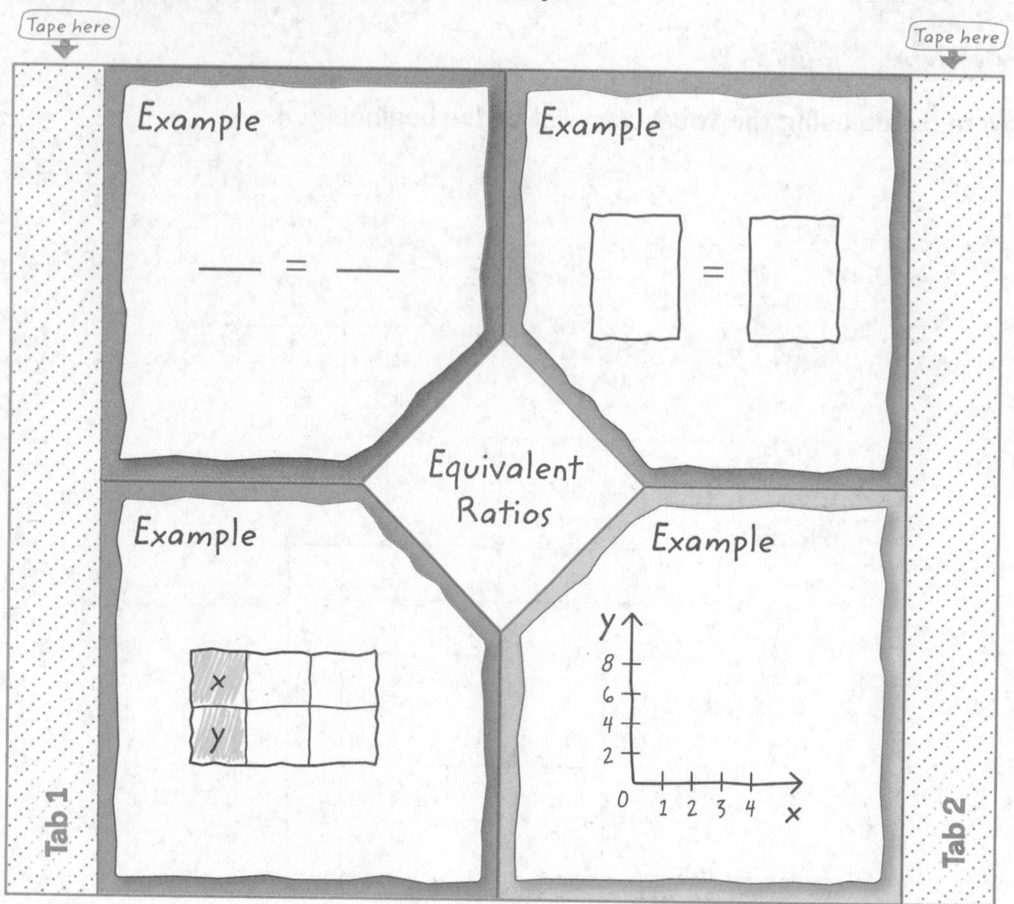

Got it?

Match each ratio with an equivalent ratio.

1. 65:390

2. $\frac{64}{256}$

3. 156:390

4. $\frac{204}{306}$

5. 56:84

6. $\frac{87}{174}$

a. $\frac{2}{5}$

b. $\frac{2}{3}$

c. $\frac{1}{3}$

d. $\frac{1}{6}$

e. $\frac{1}{4}$

f. $\frac{1}{2}$

Problem Solving

1. Amos has 12 action, 15 comedy, and 9 drama DVDs. Find the ratio of action DVDs to the total number of DVDs. Then explain its meaning. (Lesson 2)

2. A basketball player signs a contract that pays him \$16 million over 4 years. What is his average pay per year? (Lesson 3)

3. In a parking lot, 3 out of 8 vehicles were trucks. If there were 128 vehicles, complete the ratio table to find the number of trucks. (Lesson 4)

Number of Trucks	3	
Number of Vehicles	8	128

4. Isabelle bought 12 wallet-sized photos for \$36. Use a ratio table to determine how much she will pay for 5 more photos. (Lesson 4)

Number of Photos	12		5
Price (\$)	36		

5. CCSS **Justify Conclusions** The temperature rose 4°F every 90 minutes before noon and rose 2°F every 45 minutes after noon. Are these rates equivalent? Explain your reasoning. (Lesson 6)

6. CCSS **Justify Conclusions** Stacey made 8 necklaces in 48 minutes. Nick made 4 necklaces in 24 minutes. Is the rate at which they made necklaces equivalent? Explain your reasoning. (Lesson 6)

7. In the sixth grade, 12 out of 27 students have a dog. If there are 162 students, how many would have a dog? (Lesson 7)

Reflect

Answering the Essential Question

Use what you learned about ratios and rates to complete the graphic organizer.

Essential Question

HOW do you use equivalent rates in the real world?

Ratio

What is it?

Examples

Non-examples

Rate

What is it?

Examples

Non-examples

How are rates and ratios the same?

How are rates and ratios different?

Answer the Essential Question. HOW do you use equivalent rates in the real world?

Chapter 2
Fractions, Decimals, and Percents

Essential Question

WHEN is it better to use a fraction, a decimal, or a percent?

Common Core State Standards

Content Standards
6.RP.3, 6.RP.3c

Mathematical Practices
1, 2, 3, 4, 5, 6, 7

Math in the Real World

Outer Space Due to the pull of gravity, an astronaut who weighs 180 pounds on Earth would weigh $\frac{1}{6}$ of that on the moon.

Write the astronaut's weight on the moon in the box below.

FOLDABLES® Study Organizer

Cut out the Foldable on page FL5 of this book.

Place your Foldable on page 166.

Use the Foldable throughout this chapter to help you learn about fractions, decimals, and percents.

What Tools Do You Need?

Vocabulary

least common denominator
percent
percent proportion
proportion
rational number

Study Skill: Reading Math

Everyday Meaning The key to understanding word problems is to understand the meaning of the mathematical terms in the problem.

You will use the terms *factor* and *multiple* in this chapter. Here are two sentences that show their everyday meanings.

- Weather was a *factor* in their decision to postpone the picnic.
- The star quarterback won *multiple* post-season awards.

The table shows how the everyday meaning is connected to the mathematical meaning.

Term	Everyday Meaning	Mathematical Meaning	Connection
Factor	something that actively contributes to a decision or result	2 and 3 are factors of 6.	A factor helps to make a decision. In mathematics, factors "make up" a product.
Multiple	consisting of more than one or shared by many	The multiples of 2 are 0, 2, 4, 6, ...	Multiple means many. In mathematics, a number has infinitely many multiples.

Practice Make a list of other words that have the prefixes *fact-* or *multi-*. Determine what the words in each list have in common.

Word	Meaning	Connection

When Will You Use This?

Your Turn! You will solve this problem in the chapter.

Try the Quick Check below.
Or, take the Online Readiness Quiz.

CCSS Quick Review

Common Core Review 6.NS.4

Example 1

Find the GCF of 30 and 54.

First, make an organized list of the factors for each number. Then circle the common factors.

30: (1), (2), (3), 5, (6), 10, 15, 30
54: (1), (2), (3), (6), 9, 18, 27, 54

So, the greatest common factor, or GCF, is 6.

Example 2

Find the LCM of 15 and 40.

Write the prime factorization

$15 = 3 \times 5$
$40 = 2 \times 2 \times 2 \times 5$

Find the product of the prime factors. Use the common prime factor, 5, only once.

The least common multiple, or LCM, is $2 \times 2 \times 2 \times 3 \times 5$ or 120.

Quick Check

Greatest Common Factor **Find the GCF of each set of numbers.**

1. 32 and 52 ______

2. 48 and 60 ______

3. 18, 54, and 72 ______

Least Common Multiple **Find the LCM for each set of numbers.**

4. 5 and 7 ______

5. 12 and 30 ______

6. 6, 2, 22 ______

7. The front gear of a bicycle has 54 teeth. The back gear has 18 teeth. How many complete rotations must the smaller gear make for both gears to be aligned in the original starting positions?

______

Which problems did you answer correctly in the Quick Check? Shade those exercise numbers below.

1 2 3 4 5 6 7

Lesson 1

Decimals and Fractions

What You'll Learn

Scan the lesson. List two real-world scenarios in which you would use decimals written as fractions.

- ______________________________
- ______________________________

Essential Question

WHEN is it better to use a fraction, a decimal, or a percent?

Vocabulary

rational number

Common Core State Standards

Content Standards
Preparation for 6.RP.3c

Mathematical Practices
1, 3, 4, 5

Real-World Link

Music The instruments below show the part of students in the school orchestra that play each type of instrument.

Brass 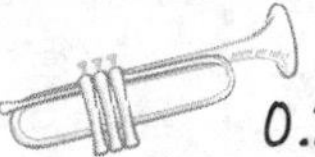 0.25

1. Write 0.25 in word form: ______________

2. Write 0.25 as a fraction: $\frac{\square}{100}$

Percussion  0.15

3. Write 0.15 in word form: ______________

4. Write 0.15 as a fraction: $\frac{\square}{100}$

Strings  0.31

5. Write 0.31 in word form: ______________

6. Write 0.31 as a fraction: $\frac{\square}{100}$

Woodwind 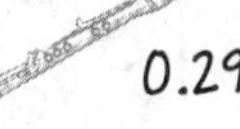 0.29

7. Write 0.29 in word form: ______________

8. Write 0.29 as a fraction: $\frac{\square}{100}$

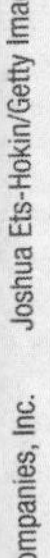

Work Zone

Write Decimals as Fractions and Mixed Numbers

Decimals like 0.25, 0.15, 0.31, and 0.29 can be written as fractions with denominators of 10, 100, 1,000, and so on. Any number that can be written as a fraction is a **rational number**.

Decimals like 3.25, 26.82, and 125.54 can be written as mixed numbers in simplest form.

Examples

Write each decimal as a fraction in simplest form.

1. **0.6**

The place-value chart shows that the place value of the last decimal place is tenths.

$0.6 = \frac{6}{10}$ Say six tenths.

$= \frac{\overset{3}{\cancel{6}}}{\underset{5}{\cancel{10}}}$ Simplify. Divide the numerator and denominator by the GCF, 2.

$= \frac{3}{5}$

1,000	100	10	1	0.1	0.01	0.001
thousands	hundreds	tens	ones	tenths	hundredths	thousandths
0	0	0	0.	6	0	0

2. **0.45**

$0.45 = \frac{45}{100}$ *Say forty-five hundredths.*

$= \frac{\overset{9}{\cancel{45}}}{\underset{20}{\cancel{100}}}$ Simplify.

$= \frac{9}{20}$

1,000	100	10	1	0.1	0.01	0.001
thousands	hundreds	tens	ones	tenths	hundredths	thousandths
0	0	0	0.	4	5	0

3. **0.375**

$0.375 = \frac{375}{1,000}$ *Say three hundred seventy-five thousandths.*

$= \frac{\overset{3}{\cancel{375}}}{\underset{8}{\cancel{1,000}}}$ Simplify.

$= \frac{3}{8}$

1,000	100	10	1	0.1	0.01	0.001
thousands	hundreds	tens	ones	tenths	hundredths	thousandths
0	0	0	0.	3	7	5

Mental Math

Here are some commonly used decimal-fraction equivalencies:

$0.1 = \frac{1}{10}$ $0.2 = \frac{1}{5}$

$0.25 = \frac{1}{4}$ $0.5 = \frac{1}{2}$

$0.75 = \frac{3}{4}$

It is helpful to memorize these.

Got It? Do these problems to find out.

a. 0.8 **b.** 0.28 **c.** 0.125

Show your work.

a. ________

b. ________

c. ________

Example

4. The average length of a conch shell is 9.85 inches. Express 9.85 as a mixed number in simplest form.

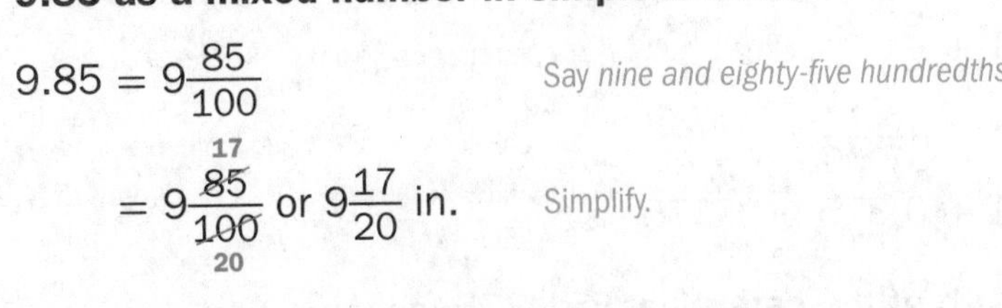

$9.85 = 9\frac{85}{100}$ Say *nine and eighty-five hundredths.*

$= 9\frac{\overset{17}{\cancel{85}}}{\underset{20}{\cancel{100}}}$ or $9\frac{17}{20}$ in. Simplify.

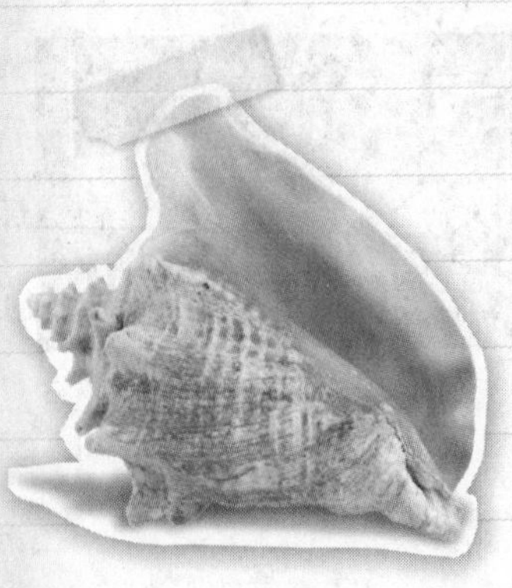

Got It? Do this problem to find out.

d. It takes approximately 4.65 quarts of milk to make a pound of cheese. Express this amount as a mixed number in simplest form.

d. __________

Write Fractions and Mixed Number as Decimals

For fractions with denominators that are *factors* of 10, 100, or 1,000, you can write equivalent fractions with these denominators.

Example

5. Write $\frac{9}{12}$ as a decimal.

Method 1 **Write an equivalent fraction.**

$\frac{9}{12} = \frac{3}{4}$ (÷ 3) $\frac{3}{4} = \frac{75}{100}$ (× 25) Simplify $\frac{9}{12}$. Then multiply the numerator and denominator of $\frac{3}{4}$ by 25.

$= 0.75$ Read 0.75 as *seventy-five hundredths.*

Method 2 **Divide the numerator by the denominator.**

$\frac{9}{12}$ →

$$\begin{array}{r} 0.75 \\ 12\overline{)9.00} \\ -\ 84 \\ \hline 60 \\ -\ 60 \\ \hline 0 \end{array}$$

To divide 9 by 12, place a decimal point after 9 and annex as many zeros as necessary to complete the division.

Got It? Do these problems to find out.

e. $\frac{3}{5}$ **f.** $\frac{14}{25}$ **g.** $\frac{102}{250}$

e. __________

f. __________

g. __________

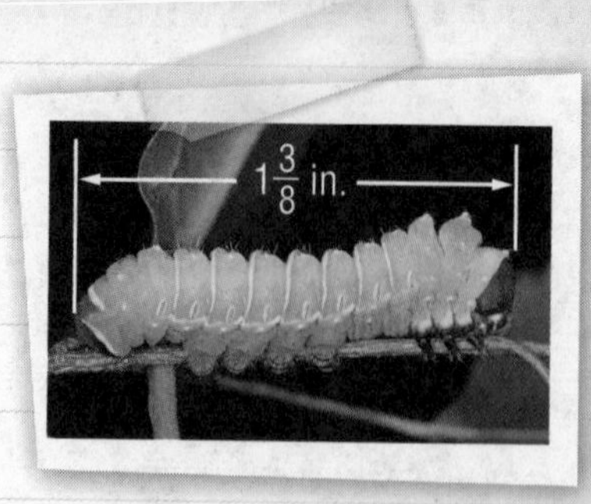

Example

6. A caterpillar can have as many as 4,000 muscles, compared to humans, who have about 600. Write the length of the caterpillar as a decimal.

$1\frac{3}{8} = 1 + \frac{3}{8}$ — Definition of a mixed number

$= 1 + \frac{375}{1{,}000}$ — Multiply the numerator and the denominator by 125.

$= 1 + 0.375$ or 1.375 — Read 1.375 as *one and three hundred seventy-five thousandths.*

The length of the caterpillar is 1.375 inches.

Guided Practice

Write each decimal as a fraction or mixed number in simplest form. (Examples 1–4)

1. 0.4 = ______

2. 0.64 = ______

3. 2.75 = ______

Write each fraction or mixed number as a decimal. (Examples 5 and 6)

4. $\frac{27}{75}$ = ______

5. $\frac{7}{2}$ = ______

6. $3\frac{1}{5}$ = ______

7. Mr. Ravenhead's car averages 23.75 miles per gallon of gasoline. Express this amount as a mixed number in simplest form. (Example 4) ______

8. **STEM** The Siberian tiger can grow up to $10\frac{4}{5}$ feet long. Express this length as a decimal. (Example 6) ______

9. **Building on the Essential Question** What is the relationship between fractions and decimals?

Rate Yourself!

Are you ready to move on? Shade the section that applies.

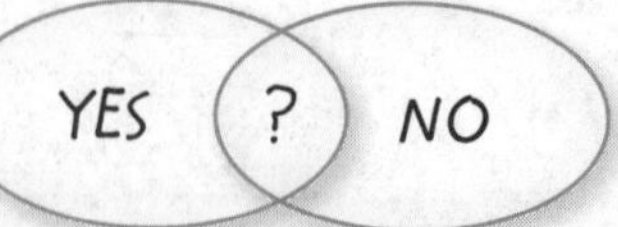

Creatas/Alamy

Name ______________________ My Homework ______________

Independent Practice

Go online for Step-by-Step Solutions eHelp

Write each decimal as a fraction in simplest form. (Examples 1–3)

1. 0.5 = ______

2. 0.7 = ______

3. 0.33 = ______

4. 0.875 = ______

Write each fraction or mixed number as a decimal. (Examples 5 and 6)

5. $\frac{77}{200}$ = ______

6. $\frac{1}{20}$ = ______

7. $\frac{12}{75}$ = ______

8. $8\frac{21}{40}$ = ______

9. **STEM** Mercury orbits the Sun in $87\frac{24}{25}$ Earth days. Venus orbits the Sun in $224\frac{7}{10}$ Earth days, and Mars orbits the Sun in $686\frac{49}{50}$ days. Write each mixed number as a decimal. (Example 6)

______

10. **STEM** Last week, a share of stock gained a total of 1.64 points. Express this gain as a fraction in simplest form. (Example 4)

______

11. CCSS **Use Math Tools** The table shows the ingredients in an Italian sandwich.

Ingredient	Amount (lb)
meat	0.35
vegetables	0.15
secret sauce	0.05
bread	0.05

a. What fraction of a pound is each ingredient?

______

b. How much more meat is in the sandwich than vegetables? Write the amount as a fraction in simplest form.

______

c. What is the total weight of the Italian sandwich? Write the amount as a fraction in simplest form. ______

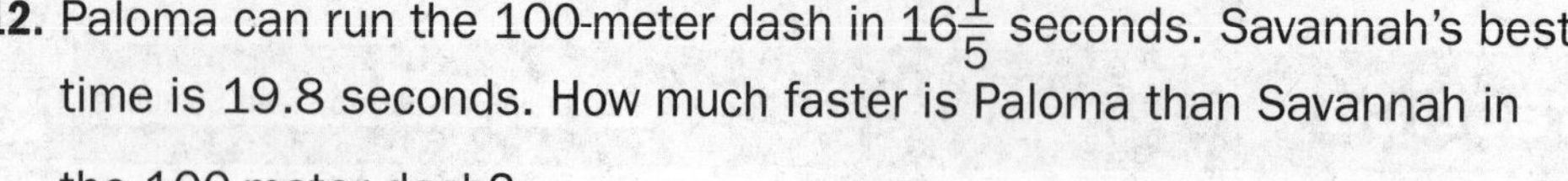

12. Paloma can run the 100-meter dash in $16\frac{1}{5}$ seconds. Savannah's best time is 19.8 seconds. How much faster is Paloma than Savannah in the 100-meter dash? ______

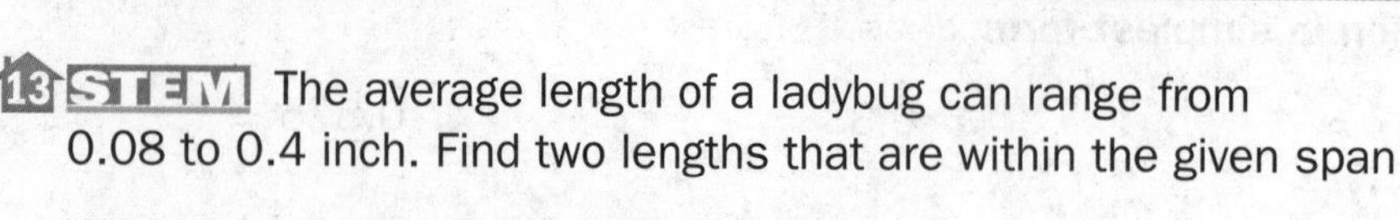

13 **STEM** The average length of a ladybug can range from 0.08 to 0.4 inch. Find two lengths that are within the given span. Write them as fractions in simplest form. ______

H.O.T. Problems Higher Order Thinking

14. **CCSS Find the Error** Mei is writing 4.28 as a mixed number. Find her mistake and correct it.

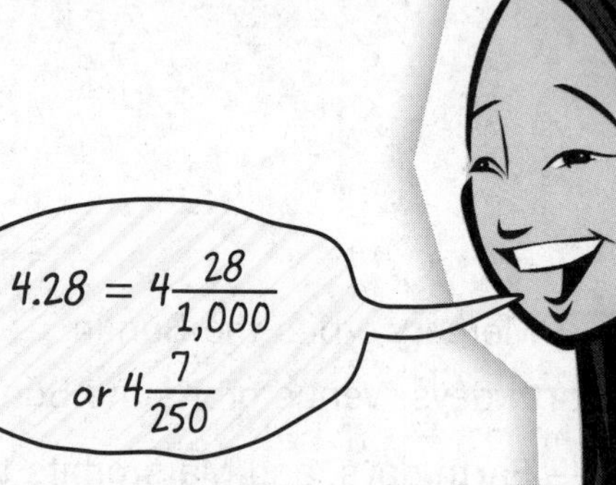

15. **CCSS Persevere with Problems** Decide whether the following statement is *always, sometimes,* or *never* true. Explain your reasoning.

Any decimal that ends with a digit in the thousandths place can be written as a fraction with a denominator that is divisible by both 2 and 5.

16. **CCSS Reason Inductively** Write a fraction with a decimal value between $\frac{1}{2}$ and $\frac{3}{4}$. Write both the fraction and the equivalent decimal.

Standardized Test Practice

17. Which of the following statements is *not* true?

Ⓐ $0.6 = \frac{3}{5}$

Ⓑ $0.125 = \frac{1}{8}$

Ⓒ $2.015 = 2\frac{1}{100}$

Ⓓ $10.38 = 10\frac{19}{50}$

Name ______________________ My Homework ______________________

Extra Practice

Write each decimal as a fraction or mixed number in simplest form.

18. 0.3 = $\frac{3}{10}$

0.3 is three tenths.

Homework Help

19. 0.65 = ______

20. 0.425 = ______

21. 9.35 = ______

Write each fraction or mixed number as a decimal.

22. $\frac{19}{25}$ = ______

23. $\frac{311}{500}$ = ______

24. $\frac{5}{8}$ = ______

25. $14\frac{3}{5}$ = ______

26. Evita lives 0.85 mile from her school. Write this distance as a fraction in simplest form.

______

27. Rancho Middle School has an average of $23\frac{3}{8}$ students per teacher. Write this mixed number as a decimal.

______

28. Alan bought 20 yards of fencing. He used 5.9 yards to surround one flower garden and 10.3 yards to surround another garden. Write the amount remaining as a fraction in simplest form.

______

29. In a survey, 9 out of 15 students named math as their favorite class. Express this rate as a decimal.

______

30. CCSS **Use Math Tools** The frequency table shows the favorite college football teams of middle school students. What fraction of the students chose the Sooners? Write the fraction as a decimal.

______

______

Team	Tally	Frequency
Buckeyes	|||	3
Gators	~~||||~~ |	6
Sooners	~~||||~~	5
Tigers	||	2
Lions	||||	4

Standardized Test Practice

31. Chase shaded 0.25 of the design. Which fraction in simplest form represents the shaded part of the design?

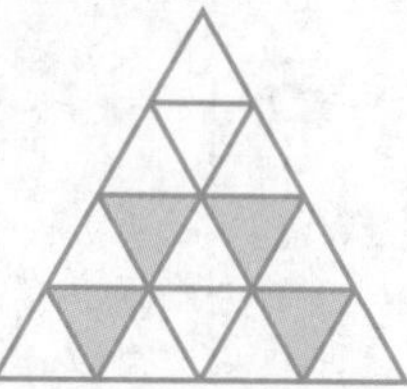

Ⓐ $\frac{1}{2}$

Ⓑ $\frac{25}{100}$

Ⓒ $\frac{4}{16}$

Ⓓ $\frac{1}{4}$

32. The formula $d = v + \frac{1}{20}v^2$ can be used to find the distance d required to stop a certain model car traveling at v miles per hour. Which of the following represents $\frac{1}{20}$?

Ⓕ 0.05

Ⓖ 0.21

Ⓗ 0.4

Ⓘ 1.2

33. **Short Response** Lori ran the distances shown in the table. Write the total distance, in miles, as a fraction in simplest form.

Day	Distance (mi)
Monday	0.35
Wednesday	0.2
Friday	0.25

34. Which decimal represents the shaded portion of the figure below?

Ⓐ 0.25

Ⓑ 0.333

Ⓒ 0.375

Ⓓ 0.4

CCSS Common Core Review

Simplify each fraction. 4.NF.1

35. $\frac{20}{100} = \frac{\square}{\square}$

36. $\frac{35}{100} = \frac{\square}{\square}$

37. $\frac{72}{100} = \frac{\square}{\square}$

38. Jasper made 230 flyers for a dance. He handed two flyers out to each student. How many students received flyers? 4.NBT.6

39. Look for a pattern and complete the table. 5.NBT.2

Multiplication Problem	Product
36 × 100	3,600
36 × 10	
36 × 1	
36 × 0.1	
36 × 0.01	

Inquiry Lab
Model Percents

HOW can you model a percent?

Content Standards
Preparation for 6.RP.3c

Mathematical Practices
1, 3, 4

Mosaics Jackie is using 1-inch tiles to make the mosaic shown at the right. She needs a total of 100 tiles. What percent of the tiles are green?

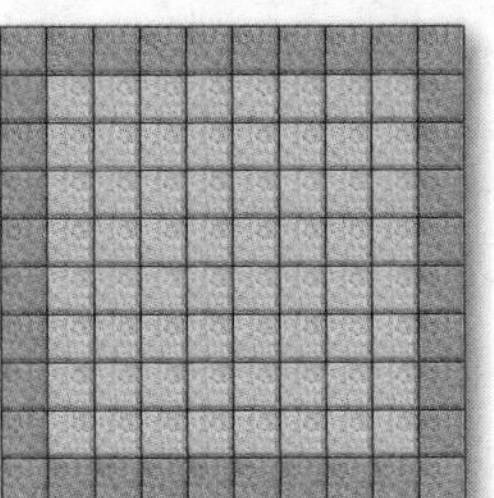

What do you know? ______________________

____

What do you need to find? ______________________

____

Investigation 1

A 10 × 10 grid can be used to represent *hundredths.* It can also represent percents. The word *percent* (%) means *out of one hundred.* For example, 50% means 50 out of one hundred.

Step 1 Use a 10 × 10 grid to model the percent of tiles in the mosaic that are green.

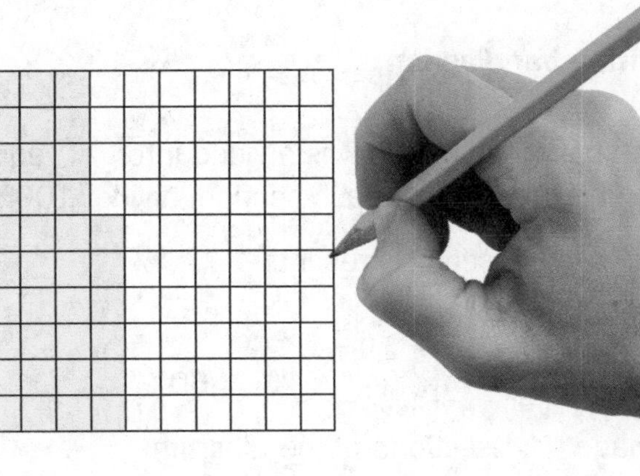

Step 2 In the mosaic, ☐ tiles out of 100 are green.

So, ☐% of the squares are green.

Investigation 2

Model 18% with a 10 × 10 grid.

Step 1 18% means ____ out of 100.

Step 2 Shade the squares filling one column at a time.

Shade ____ squares out of 100.

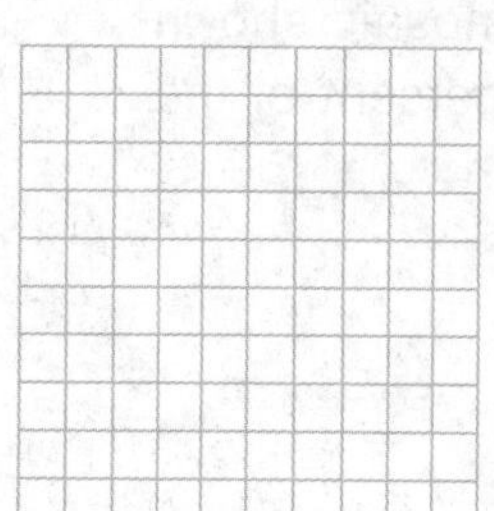

Investigation 3

Percents can also be modeled with bar diagrams. The entire bar represents 100%. The bar diagram below is divided into 10 equal sections, each representing 10%. The shaded region represents 40%.

0% 40% 100%

10%	10%	10%	10%	10%	10%	10%	10%	10%	10%

Model 60% with a bar diagram.

Step 1 The bar diagram below is divided into ____ equal sections.
To find the value of each section, divide. 100% ÷ 5 = 20%.

So, each section represents ____%.

Step 2 ____% + ____% + ____% = 60%

Shade ____ sections of the diagram.

0% 100%

Work with a partner. Identify each percent modeled.

1. ________

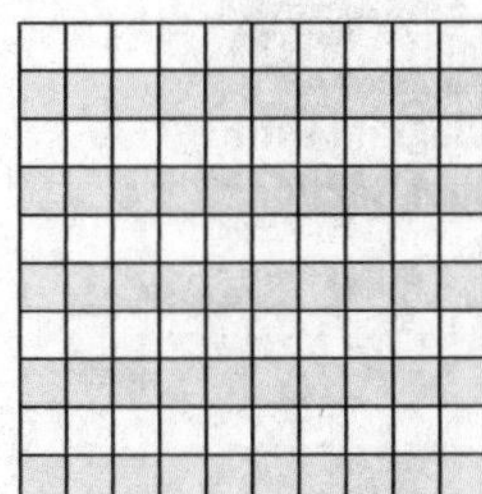

2. ________

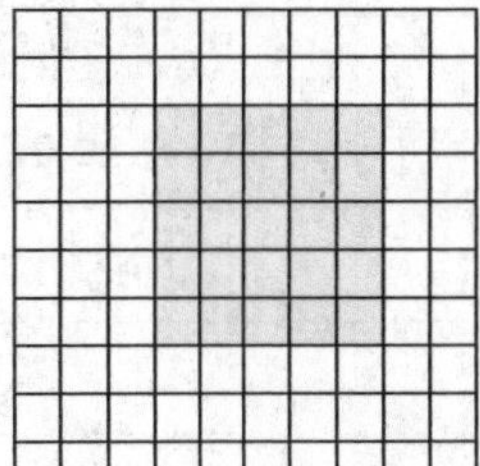

3. ________

0% 50% 100%

4. ________

0% 50% 100%

Work with a partner. Model each percent.

5. 37%

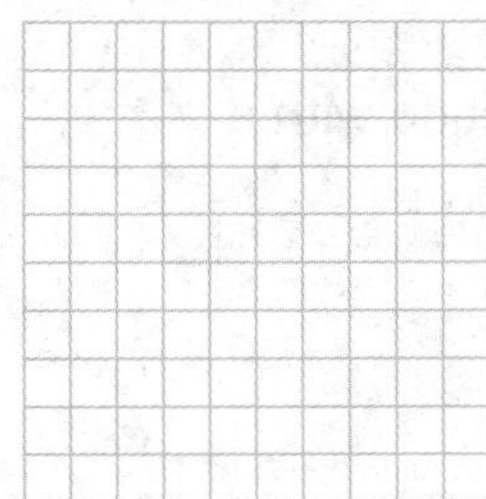

6. 8%

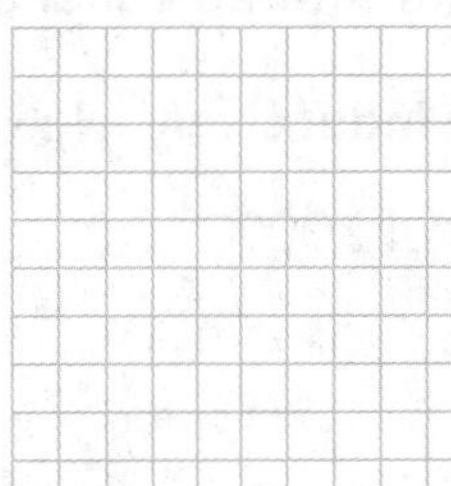

7. 45%

0% 50% 100%

8. 5%

0% 50% 100%

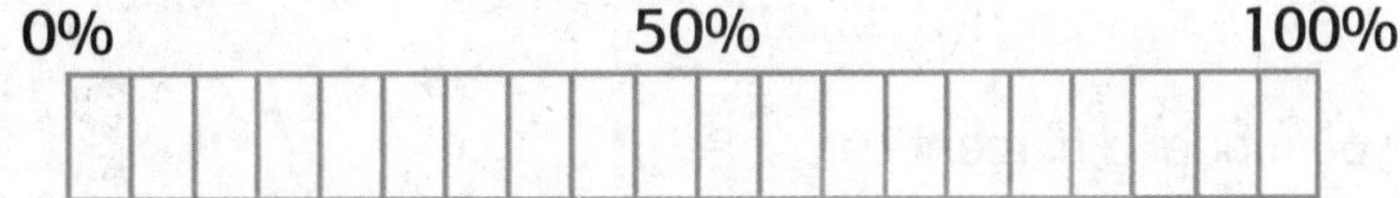

Analyze

Work with a partner to determine the number of shaded sections for each model. The first one is done for you.

	Percent	Number of Shaded Sections using each Model		
		10 × 10 Grid	Bar Diagram with 10 Equal Sections	Bar Diagram with 20 Equal Sections
	45	45	4.5	9
9.	15			
10.	30			
11.	55			
12.	70			
13.	85			
14.	65			

15. Refer to Exercise 10. Identify the fraction of each model that would be shaded. Then circle the method that corresponds to the fraction that is written in simplest form.

a. 10 × 10 grid ________

b. bar diagram with 10 equal sections ________

c. bar diagram with 20 equal sections ________

16. CCSS **Reason Inductively** How can you write a percent as a fraction with a denominator of 100? ________________________________

____

Reflect

17. CCSS **Model with Mathematics** Write a real-world problem that involves a percent. Then model the percent used in the problem. ________

____

18. Inquiry HOW can you model a percent? ________________

____

Lesson 2

Percents and Fractions

What You'll Learn

Scan the lesson. Predict two things you will learn about writing a fraction as a percent.

-
-

Essential Question

WHEN is it better to use a fraction, a decimal, or a percent?

Vocabulary

percent

Common Core State Standards

Content Standards
Preparation for 6.RP.3c

Mathematical Practices
1, 3, 4, 5

Real-World Link

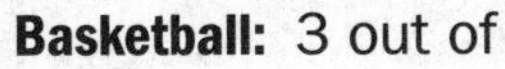

Sports Students were asked to choose their favorite sport to play.

1. For each sport, shade a 10 × 10 grid that represents the number of students that chose the sport.

Basketball: 3 out of 20

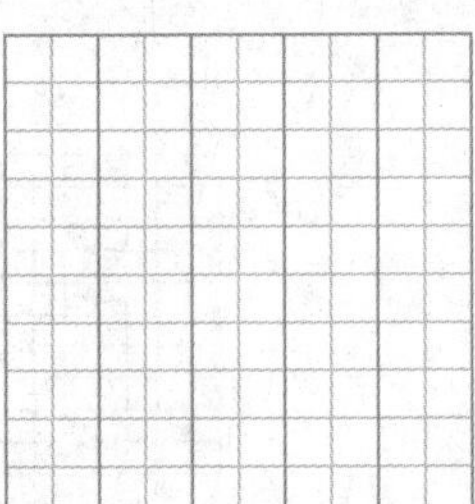

Football: 3 out of 25

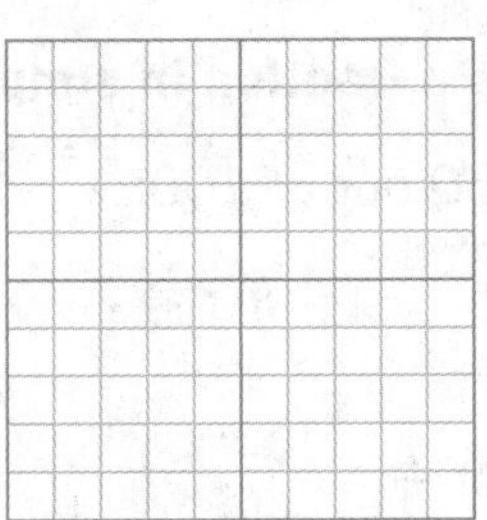

Gymnastics: 1 out of 20

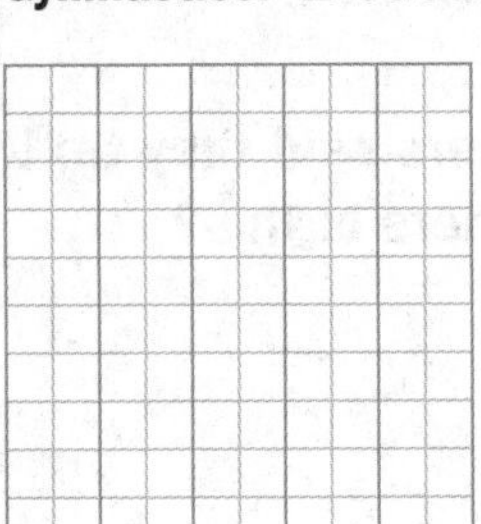

Swimming: 9 out of 100

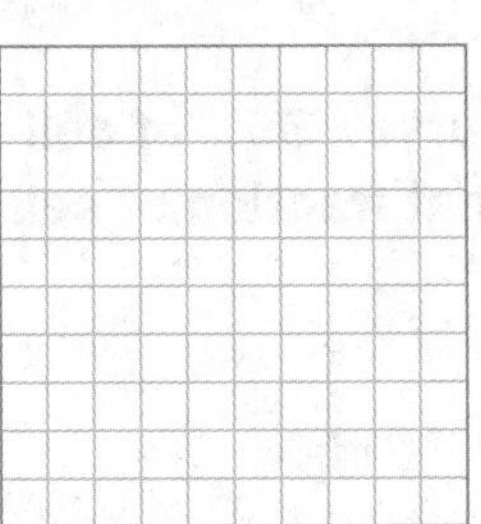

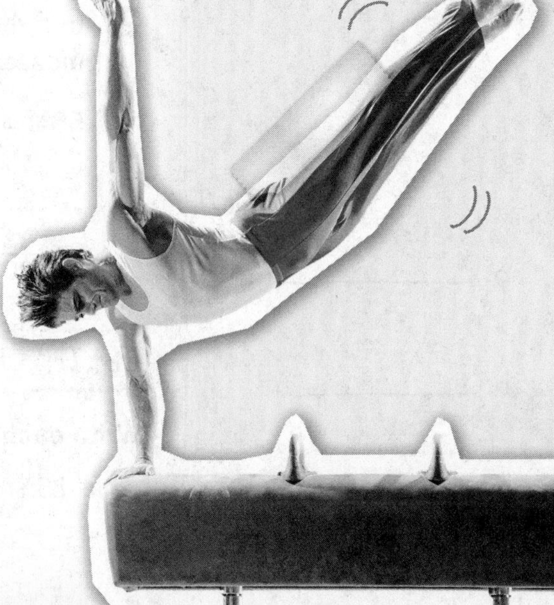

2. What fraction with a denominator of 100 to represent the number of students who chose each sport?

Basketball: $\frac{\square}{\square}$ Football: $\frac{\square}{\square}$

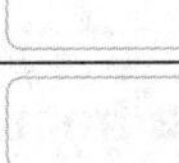

Gymnastics: $\frac{\square}{\square}$ Swimming: $\frac{\square}{\square}$

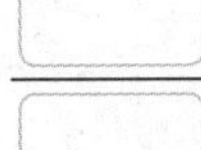

Key Concept Percents as Fractions

Work Zone

Words A **percent** is a ratio that compares a number to 100.

Example 45% ⇒ 45 out of 100 or $\frac{45}{100}$

Models

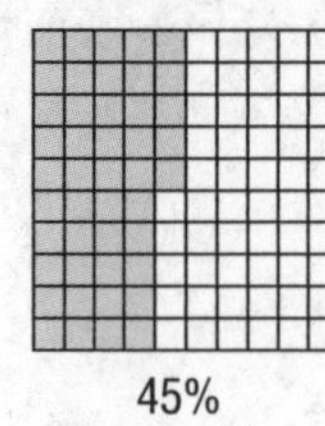

45%

To write a percent as a fraction, first write the percent as a rate per 100. Then simplify.

Examples

Watch Tutor

1. Write 50% as a fraction in simplest form.

50% means *50 out of 100.*

$50\% = \frac{50}{100}$ Definition of percent

$= \frac{\overset{1}{\cancel{50}}}{\underset{2}{\cancel{100}}}$ or $\frac{1}{2}$ Simplify. Divide the numerator and the denominator by the GCF, 50.

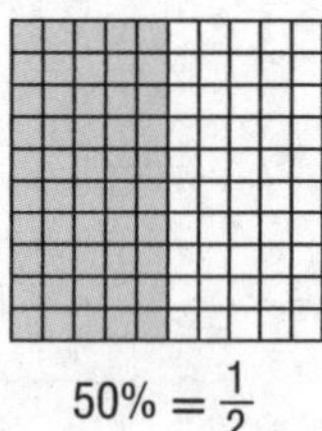

$50\% = \frac{1}{2}$

2. In a recent survey, 55% of cell phone owners said they text message. What fraction of cell phone owners is this?

$55\% = \frac{55}{100}$ Definition of percent

$= \frac{11}{20}$ Simplify.

So, $\frac{11}{20}$ of cell phone owners text message.

Check for Reasonableness

In Example 2, you can conclude that $\frac{11}{20}$ is a reasonable answer because 55% is a little more than 50%, and $\frac{11}{20}$ is a little more than $\frac{10}{20}$ or $\frac{1}{2}$.

a. ____________

b. ____________

c. ____________

Got It? Do these problems to find out.

Write each percent as a fraction in simplest form.

a. 75% **b.** 90% **c.** 38%

Example

3. The table shows the percent of each movie type rented during a month. What fraction of the rentals were action movies?

Types of Movies	
action	35%
children's	5%
comedy	45%
drama	5%
horror	5%
romance	5%

$35\% = \frac{35}{100}$ Definition of percent

$= \frac{\overset{7}{\cancel{35}}}{\underset{20}{\cancel{100}}}$ Divide the numerator and denominator by the GCF, 5.

Action movies were rented $\frac{7}{20}$ of the time.

Got It? **Do this problem to find out.**

d. Write the fraction of rentals that were horror movies.

d. ______

Fractions as Percents

To write a fraction as a percent, find an equivalent ratio with 100 as a denominator.

Example

4. Write the fraction $\frac{6}{8}$ as a percent.

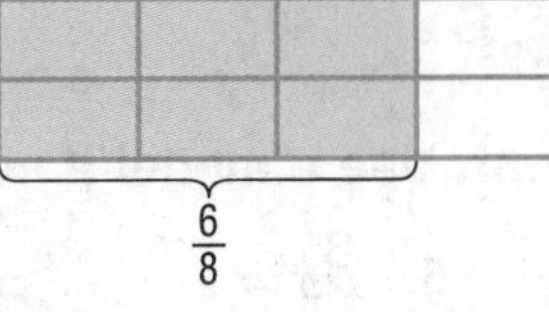

$\frac{6}{8} = \frac{3}{4}$ Simplify by dividing by the GCF, 2.

$\frac{3}{4} = \frac{\blacksquare}{100}$ Write equivalent ratios. One ratio is the fraction. The other ratio is the unknown value compared to 100.

$\frac{3}{4} = \frac{75}{100}$ (× 25) Since 4 × 25 = 100, multiply 3 by 25 to find the unknown value.

So, $\frac{75}{100}$ or 75% of the rectangle is shaded.

Got It? **Do this problem to find out.**

e. Write the fraction $\frac{9}{12}$ as a percent.

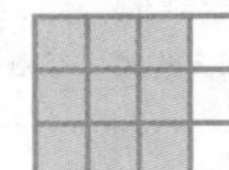

e. ______

Example

5. Mitch made 12 out of 40 shots during the championship game. What percent of his shots did Mitch make?

$\frac{12}{40} = \frac{3}{10}$ Simplify $\frac{12}{40}$ by dividing the numerator and denominator by the GCF, 4.

$\frac{3}{10} = \frac{\blacksquare}{100}$ (×10) Write equivalent ratios.

$\frac{3}{10} = \frac{30}{100}$ Since 10 × 10 = 100, multiply 3 by 10 to find the unknown value.

So, $\frac{12}{40} = \frac{30}{100}$ or 30%.

Got It? **Do this problem to find out.**

f. Alana spelled 19 out of 25 words correctly. What percent of the words did Alana spell correctly?

f. ________

Guided Practice

Write each percent as a fraction in simplest form. (Examples 1–3)

1. 15% = ________

2. 80% = ________

3. 33% = ________

Write each fraction as a percent. Use a model if needed. (Example 4)

4. $\frac{3}{10}$ = ________

5. $\frac{3}{20}$ = ________

6. $\frac{2}{5}$ = ________

7. Elsa ran 7 out of 10 days. What percent of the days did she run? (Example 5)

8. **Building on the Essential Question** Why is it helpful to write a fraction as a percent?

Rate Yourself!

How confident are you about percents and fractions? Check the box that applies.

For more help, go online to access a Personal Tutor.

FOLDABLES Time to update your Foldable!

Name ________________________ My Homework ________________

Independent Practice

Go online for Step-by-Step Solutions

Write each percent as a fraction in simplest form. (Examples 1–3)

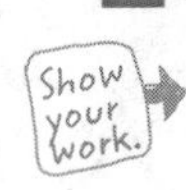

1. 2% = ______

2. 20% = ______

3. 85% = ______

4. 4% = ______

Write each fraction as a percent. Use a model if needed. (Example 4)

5. $\frac{2}{10}$ = ______

6. $\frac{3}{4}$ = ______

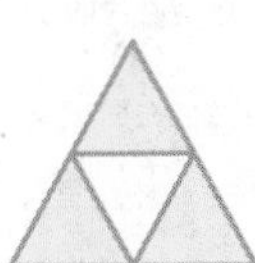

7. $\frac{7}{20}$ = ______

8. $\frac{11}{25}$ = ______

9. During his workout, Elan spent 28% of the time on the treadmill. What fraction of his workout was on the treadmill? (Examples 1–3)

______

10. A cat spends about 7 out of 10 hours sleeping. About what percent of a cat's day is spent sleeping? (Example 5)

______

11. A survey showed that 82% of youth most often use the Internet at home. What fraction of youth surveyed most often use the Internet somewhere else?

______

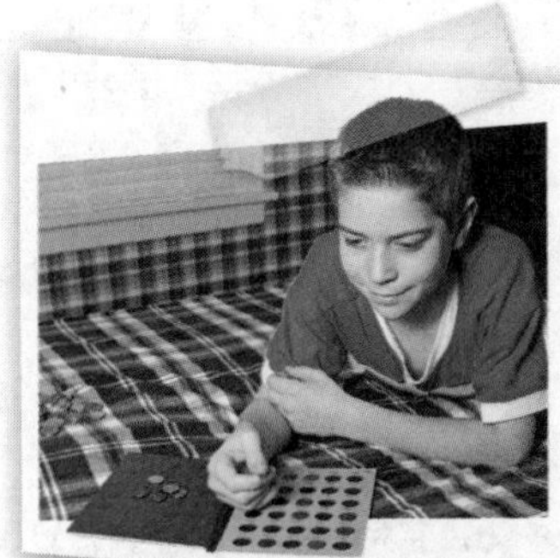

12. Cedro collects state quarters. He has 42 out of 50 available quarters. What is 42 out of 50 as a percent?

______

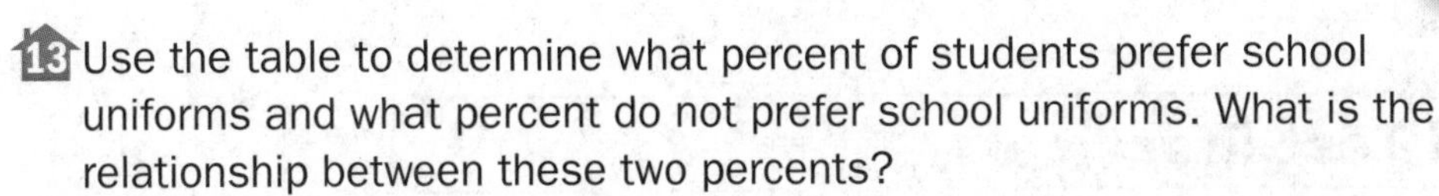

13. Use the table to determine what percent of students prefer school uniforms and what percent do not prefer school uniforms. What is the relationship between these two percents?

______

Prefer School Uniforms	
No	Yes
卌 卌 卌 |	||||

14. **Multiple Representations** The table shows the percent of Earth's atmosphere that is each element.

Element	Percent
Nitrogen	78
Oxygen	21
Other	1

a. **Bar Diagram** Model 21% using a bar diagram.

0% 100%

b. **Number** Write the percent of Earth's atmosphere that is nitrogen as a fraction in simplest form. ______

H.O.T. Problems Higher Order Thinking

15. **Reason Inductively** Write three fractions that can be written as percents between 50% and 75%. Justify your solution.

______

16. **Persevere with Problems** For each model below, write the shaded region as a percent and as a fraction.

a. 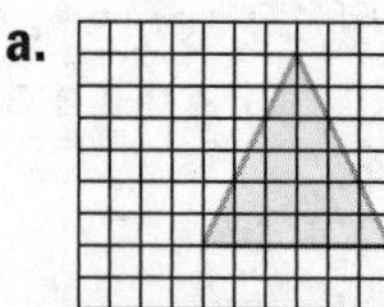 ______ b. 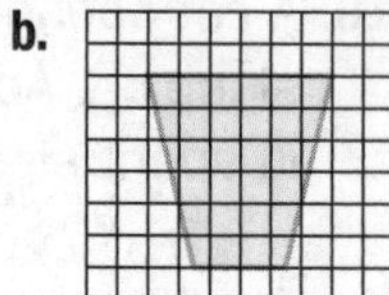 ______ c. 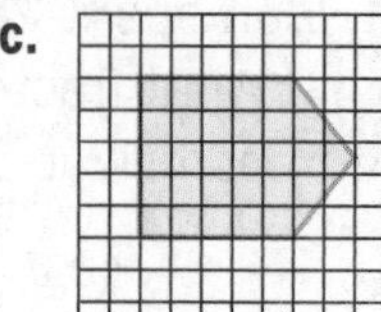 ______

17. **Which One Doesn't Belong?** Identify the number that does not belong with the other three. Explain your reasoning.

$\frac{9}{20}$	$\frac{45}{100}$	45%	$\frac{8}{45}$

______

18. **Persevere with Problems** Complete each blank to find an expression that is equal to 16%.

a. ______ for every 100 b. ______ for every 50

c. 1 for every ______ d. 0.5 for every ______

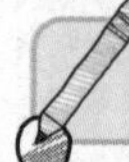

Standardized Test Practice

19. Of Bernie's insect collection, 15 out of 25 insects are butterflies. Which of the following is *not* another way of expressing 15 out of 25?

Ⓐ $\frac{3}{5}$ Ⓑ 0.6 Ⓒ 60% Ⓓ $\frac{5}{3}$

Extra Practice

Write each percent as a fraction in simplest form.

20. 14% = $\frac{7}{50}$

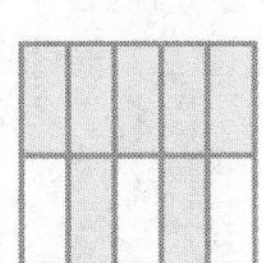

$14\% = \frac{14}{100}$

$\frac{14}{100} = \frac{7}{50}$ (÷ 2)

21. 47% = ______

22. 86% = ______

23. 88% = ______

Write each fraction as a percent. Use a model if needed.

24. $\frac{7}{10}$ = ______

25. $\frac{21}{25}$ = ______

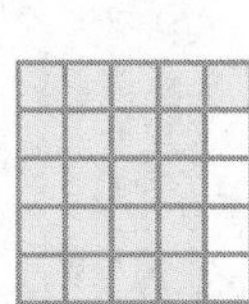

26. $\frac{3}{5}$ = ______

27. $\frac{18}{25}$ = ______

28. In a recent year, 22% of E-mail users said they spend less time using E-mail because of spam. What fraction of E-mail users is this?

______

29. About $\frac{19}{20}$ of celery is water. What percent is this?

______

30. CCSS **Use Math Tools** Mrs. Lane took a survey of the types of pants her students were wearing. She collected the data at the right. What percent of her students were wearing shorts?

Type of Pants	Number of Students
Jeans	14
Shorts	9
Capris	2

______

31. **STEM** The circle graph shows the fraction of each type of weather during September.

a. What percent of the days were sunny? ______

b. What percent of the days were rainy? ______

c. What percent of the days were sunny or rainy? ______

d. What percent of the days were cloudy or partly cloudy?

______

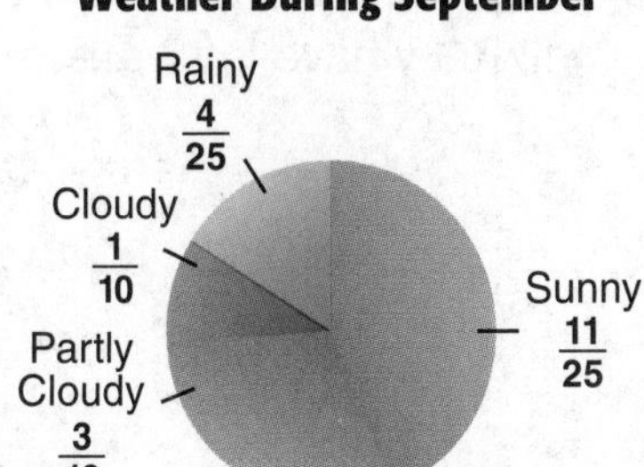

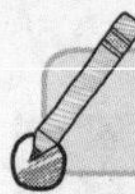

Standardized Test Practice

32. On Friday, 65% of the students at Plainview Middle School bought a hot lunch in the cafeteria. What fractional part of the school did *not* buy a hot lunch in the cafeteria?

Ⓐ $\frac{1}{65}$
Ⓑ $\frac{13}{20}$
Ⓒ $\frac{7}{20}$
Ⓓ $\frac{6}{5}$

33. The average brain is about 2% of a person's total body weight. Which fraction is equivalent to 2%?

Ⓕ $\frac{1}{2}$
Ⓖ $\frac{1}{5}$
Ⓗ $\frac{1}{50}$
Ⓘ $\frac{1}{500}$

34. Short Response The table shows the number of votes for the new school mascot. What percent of the students chose a mascot other than Tigers?

Mascot	Number of Students
Polar Bears	6
Tigers	14
Vikings	19
Eagles	11

CCSS Common Core Review

Multiply. 5.NBT.2

35. 0.685 × 100 = ______

36. 0.09 × 10 = ______

37. 3.255 × 100 = ______

38. Refer to the table. Which lap has the slowest time? 5.NBT.3b

Lap	Time (minutes)
1	1.59
2	1.85
3	1.64

39. Ruby has $10. She buys the items shown. How much will Ruby have left? 5.NBT.7

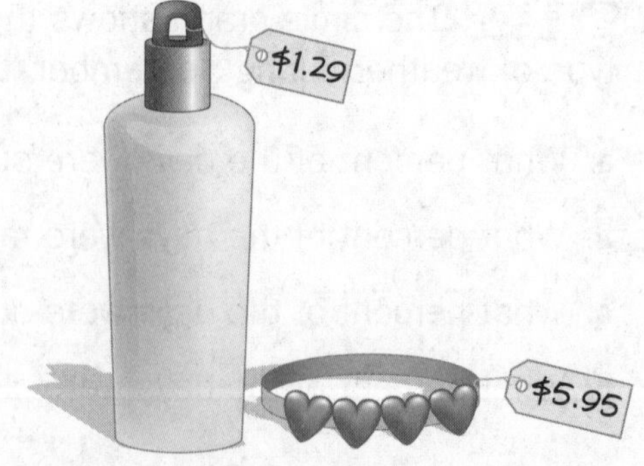

Lesson 3

Percents and Decimals

What You'll Learn

Scan the lesson. List two real-world scenarios in which you would write decimals as percents.

- ______
- ______

Essential Question

WHEN is it better to use a fraction, a decimal, or a percent?

Common Core State Standards

Content Standards
Preparation for 6.RP.3c

Mathematical Practices
1, 3, 4, 5, 6

Real-World Link

Watch

School A recent survey tells the favorite subjects of students at Martin Middle School.

Math: 28%	Science: 21%	Social Studies: 15%
Art: 16%	English: 13%	Other: 7%

1. Write a fraction with a denominator of 100 to represent the percent for each subject.

Math: $\frac{\square}{100}$ Science: $\frac{\square}{100}$

Art: $\frac{\square}{100}$ Social Studies: $\frac{\square}{100}$

English: $\frac{\square}{100}$ Other: $\frac{\square}{100}$

2. Write each fraction from Exercise 1 as a decimal.

Math: ______ Science: ______

Art: ______ Social Studies: ______

English: ______ Other: ______

3. **Make a Conjecture** Look back at Exercise 2. Compare the decimals to the percents. Explain how to write a percent as a decimal. ______

Key Concept

Write Percents as Decimals

Words To write a percent as a decimal, divide by 100 and remove the % sign. This is the same as moving the decimal point two places to the left.

Example 48% = 48%
= 0.48

Work Zone

Another way to write a fraction as a decimal is to write the percent as a fraction. Then write the fraction as a decimal.

Examples

Write each percent as a decimal.

1. 56%

Method 1 **Write the percent as a fraction.**

$56\% = \frac{56}{100}$ Rewrite the percent as a fraction with a denominator of 100.

$= 0.56$ Write *56 hundredths* as a decimal.

Method 2 **Move the decimal point.**

56% = 56% Move the decimal point two places to the left.

= 0.56 Remove the percent sign.

2. 8%

$8\% = \frac{8}{100}$ Rewrite the percent as a fraction with a denominator of 100.

$= 0.08$ Write *8 hundredths* as a decimal.

3. 2%

2% = 02% Move the decimal point two places to the left.

= 0.02 Remove the percent sign.

Got It? **Do these problems to find out.**

a. 32% **b.** 6% **c.** 93%

a. ______

b. ______

c. ______

Write Decimals as Percents

Key Concept

Words To write a decimal as a percent, multiply by 100 and add a % sign. This is the same as moving the decimal point two places to the right.

Example $0.36 = 0.36\%$
$= 36\%$

Another way to write a decimal as a percent is to write the decimal as a fraction with a denominator of 100. Then write the fraction as a percent.

STOP and Reflect

Why does it help to write a decimal as a fraction with a denominator of 100 when writing decimals as percents?

Examples

4. Write 0.38 as a percent.

Method 1 **Write the decimal as a percent.**

$0.38 = \frac{38}{100}$ Write *38 hundredths* as a fraction.

$= 38\%$ Write the fraction as a percent.

Method 2 **Move the decimal point.**

$0.38 = 0.38$ Move the decimal point two places to the right.

$= 38\%$ Add the percent sign.

5. Write 0.2 as a percent.

$0.2 = \frac{2}{10}$ Write *2 tenths* as a fraction.

$\frac{2}{10} = \frac{20}{100}$ (×10, ×10) Write the equivalent fraction with a denominator of 100.

$= 20\%$ Write the fraction as a percent.

Got It? Do these problems to find out.

Write each decimal as a percent.

d. 0.47 **e.** 0.73 **f.** 0.5

d. ______

e. ______

f. ______

Example

6. The United States produces more corn than any other country, producing 0.4 of the total corn crops. Write 0.4 as a percent.

$0.4 = 0.40$ Annex a zero.

$= 0.40\%$ Multiply by 100 and add a % sign.

$= 40\%$ Simplify.

Check $0.4 = \frac{40}{100}$ Write the decimal as a fraction with a denominator of 100.

$= 40\%$ ✓ Write the fraction as a percent.

Guided Practice

Write each percent as a decimal. (Examples 1–3)

1. 27% = ______

2. 15% = ______

3. 4% = ______

Write each decimal as a percent. (Examples 4 and 5)

4. 0.3 = ______

5. 0.82 = ______

6. 0.51 = ______

7. STEM About 0.7 of the human body is water. What percent is equivalent to 0.7? (Example 6) ______

8. **Building on the Essential Question** What is the relationship between percents and decimals?

Rate Yourself!

How well do you understand percents and decimals? Circle the image that applies.

Clear

Somewhat Clear

Not So Clear

For more help, go online to access a Personal Tutor.

FOLDABLES Time to update your Foldable!

Name ______________________ My Homework ______________________

Independent Practice

Go online for Step-by-Step Solutions

Write each percent as a decimal. (Examples 1–3)

1. 35% = ______ 2. 2% = ______ 3. 31% = ______ 4. 95% = ______

Show your work.

Write each decimal as a percent. (Examples 4 and 5)

5. 0.22 = ______ 6. 0.79 = ______ 7. 0.1 = ______ 8. 0.16 = ______

9. **Financial Literacy** A bank offers an interest rate of 4% on a savings account. Write 4% as a decimal. (Examples 1–3)

10. When making a peanut butter and jelly sandwich, 96% of people put the peanut butter on first. Write 96% as a decimal. (Examples 1–3)

11. In a recent year, 0.12 of Americans downloaded a podcast from the Internet. What percent is equivalent to 0.12? (Example 6)

12. In a recent year, the number of homes with digital cameras grew 0.44 from the previous year. Write 0.44 as a percent. (Example 6)

13. **Financial Literacy** The formula $I = prt$ gives the simple interest I earned on an account where an amount p is deposited at an interest rate r for a certain number of years t. Use the table to order the accounts from least to greatest interest earned after 5 years.

Accounts at First Savings Bank

Account	p ($)	r (%)
A	350	4
B	500	3.5
C	280	4.25

14. **Persevere with Problems** Daphne wants to buy a coat that costs $80. The store that sells the coat has multiple locations. The sales tax in each county is shown in the table. How much more would the coat cost in Delaware county than Fairfield county?

County	Tax Rate (%)
Delaware	7.25
Fairfield	6.5
Franklin	6.75

15. Dante took three tests on Friday. He got a 92% on his English test, an 88% on his math test and a 90% on his science test. Write each percent as a decimal in order from least to greatest.

H.O.T. Problems Higher Order Thinking

16. **Reason Inductively** Write a decimal between 0.5 and 0.75. Then write it as a fraction in simplest form and as a percent.

17. **Persevere with Problems** How would you write $43\frac{3}{4}\%$ as a decimal?

18. **Model with Mathematics** Write a problem about a real-world situation in which you would either write a percent as a decimal or write a decimal as a percent.

Standardized Test Practice

19. Each square below is divided into sections of equal size. Which square has 75% of its total area shaded?

Ⓐ 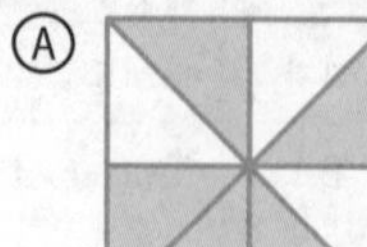

Ⓒ 

Ⓑ 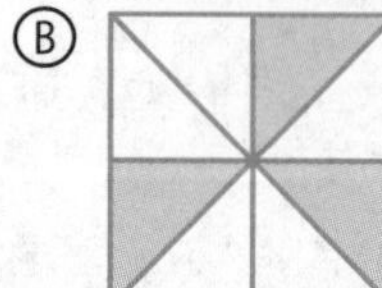

Ⓓ 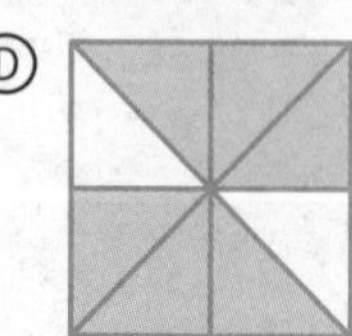

Name ______ My Homework ______

Extra Practice

Write each percent as a decimal.

20. 17% = 0.17

$17\% = \frac{17}{100}$

$= 0.17$

Homework Help

21. 3% = ______

22. 1% = ______

23. 11% = ______

Write each decimal as a percent.

24. 0.99 = ______

25. 0.62 = ______

26. 0.6 = ______

27. 0.87 = ______

28. In one day at a store, 7% of the sales were from shoes. Write 7% as a decimal.

______

29. In one hour on a certain street, 65% of the cars that passed were black. Write 65% as a decimal.

______

30. In a recent year, 0.57 of those registered to vote in the United States voted in an election. Write 0.57 as a percent.

______

31. In a recent study, 0.82 of Americans own a cell phone. What percent is equivalent to 0.82?

______

32. CCSS **Be Precise** In the United States, sales tax is added to items that you purchase. The rate of sales tax varies by state and sometimes by county or region. Use the table to order the counties from least to greatest sales tax.

______

______

County Sales Tax	
County	**Sales Tax**
A	6.75%
B	0.0625
C	$\frac{7}{100}$

Standardized Test Practice

33. When you buy a sweater, 6.75% sales tax is added to the price of the sweater. What is 6.75% written as a decimal?

Ⓐ 67,500 Ⓒ 0.675

Ⓑ 67.5 Ⓓ 0.0675

34. The table shows the free throw percentage during practice for three members of a basketball team. What is Zoe's percentage written as a decimal?

Player	Free Throws (%)
Abby	45
Sofia	68
Zoe	52

Ⓕ 52 Ⓗ 0.52

Ⓖ 5.2 Ⓘ 0.052

35. **Short Response** Tamika is buying the baseball hat shown below. What decimal represents 25%?

36. At baseball practice, Neil caught 23 out of 25 hits in the outfield. Which of the following is *not* another way of expressing 23 out of 25?

Ⓐ $\frac{23}{25}$ Ⓒ 0.92

Ⓑ 23% Ⓓ 92%

Common Core Review

Fill in each ◯ with <, >, or = to make a true statement. 5.NBT.3b

37. 2.50 ◯ 2.5

38. 0.006 ◯ 0.1

39. 0.015 ◯ 0.005

40. The table shows results for the 100 meter sprint. Who had the fastest time? 5.NBT.3b

Athlete	Time (s)
Bryson	12.14
Malik	11.84
Marcell	11.94
Wyatt	12.44

41. Aliah ate 0.75 sandwich. Her brother ate 1.5 sandwiches. Who ate more? 5.NBT.3b

Lesson 4

Percents Greater than 100% and Percents Less than 1%

What You'll Learn

Scan the lesson. List two headings you would use to make an outline of the lesson.

-

-

Essential Question

WHEN is it better to use a fraction, a decimal, or a percent?

Common Core State Standards

Content Standards
Preparation for 6.RP.3c

Mathematical Practices
1, 3, 4, 5

Real-World Link

Plants There are over 220,000 species of plants on Earth. Of those, 590 are carnivorous. Plants such as a Venus Fly Trap catch their prey as food.

1. Write the fraction of species of carnivorous plants in simplest form.

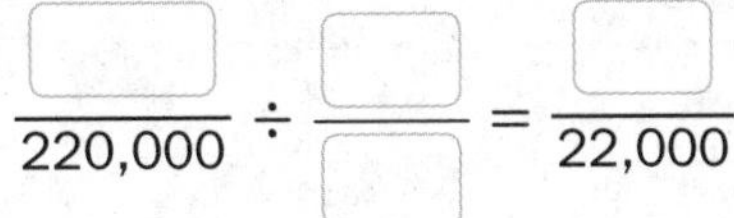

$\frac{\square}{220{,}000} \div \frac{\square}{\square} = \frac{\square}{22{,}000}$

2. Write your answer to Exercise 1 as a decimal rounded to the thousandth. Use division to find your answer.

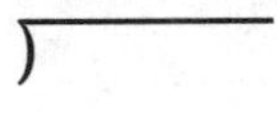

$\square \approx \square$

3. Write your answer to Exercise 2 as a fraction.

4. **Make a Conjecture** Since 0.3 = 30% and 0.03 = 3%, what percent is equal to 0.003? Explain.

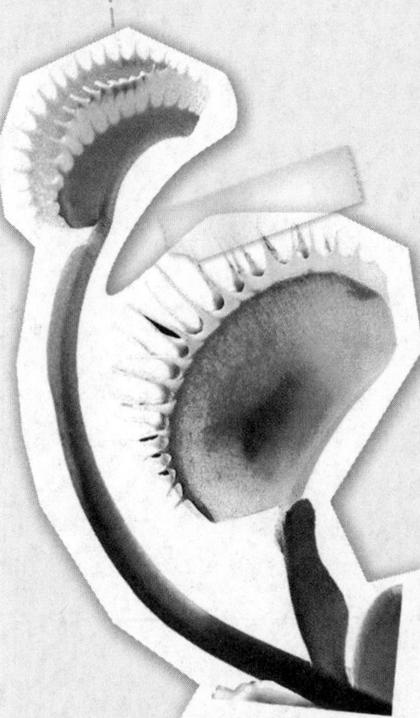

Work Zone

Percents as Decimals and Fractions

Percents greater than 100% or less than 1% can also be written as decimals or as fractions.

Examples

1. Write 0.2% as a decimal and as a fraction in simplest form.

$0.2\% = 00.2$ Divide by 100 and remove % symbol.

$= 0.002$ Decimal form

$= \frac{2}{1,000}$ or $\frac{1}{500}$ Fraction form

Percents

A percent less than 1% equals a number less than 0.01 or $\frac{1}{100}$. A percent greater than 100% equals a number greater than 1.

2. Write 170% as a mixed number in simplest form and as a decimal.

$170\% = \frac{170}{100}$ Definition of percent

$= 1\frac{70}{100}$ or $1\frac{7}{10}$ Mixed number form

$= 1.7$ Decimal form

Show your work.

a. ______

b. ______

c. ______

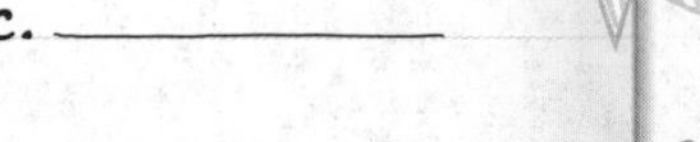

Got It? Do these problems to find out.

Write each percent as a decimal and as a mixed number or fraction in simplest form.

a. 0.25% **b.** 300% **c.** 530%

Real World

Example

3. Jimmy's savings increased by 250%. Write 250% as a mixed number in simplest form and as a decimal.

$250\% = \frac{250}{100}$ Definition of a percent

$= 2\frac{50}{100}$ or $2\frac{1}{2}$ Mixed number form

$= 2.5$ Decimal form

So, Jimmy more than doubled his savings.

Got It? Do this problem to find out.

d. The stock price for a corporation increased by 0.11%. Write 0.11% as a decimal and as a fraction in simplest form.

d. ______

Mixed Numbers and Decimals as Percents

To write a decimal as a percent, multiply by 100 and add a percent sign. To write a mixed number as a percent, first write the mixed number as an improper fraction.

Alternative Method

$1 = 100\%$

$\frac{1}{4} = 25\%$

So, $1\frac{1}{4} = 125\%$.

Example

4. Write $1\frac{1}{4}$ as a percent.

$1\frac{1}{4} = \frac{5}{4}$ Write $1\frac{1}{4}$ as an improper fraction.

$\frac{5}{4} = \frac{■}{100}$ Find an equivalent fraction.

$\frac{5}{4} = \frac{125}{100}$ (× 25) Since 4 × 25 = 100, multiply 5 by 25 to find an equivalent fraction.

So, $1\frac{1}{4}$ is $\frac{125}{100}$ or 125%.

Got It? Do these problems to find out.

Write each mixed number as a percent.

e. $2\frac{9}{10}$

f. $3\frac{2}{5}$

e. ______

f. ______

Examples

STOP and Reflect

Is the decimal 6.7 equal to 67%? Explain below.

5. Write 1.68 as a percent.

$1.68 = 1.68$ Multiply by 100.

$= 168\%$ Add % symbol.

6. Write 0.0075 as a percent.

$0.0075 = 0.0075$ Multiply by 100.

$= 0.75\%$ Add % symbol.

Got It? Do these problems to find out.

g. 2.5 **h.** 0.004 **i.** 0.0016

g. ______

h. ______

i. ______

Example

7. STEM The cheetah is the fastest land mammal in the world. The peregrine falcon is the fastest bird in the world. Its speed is 2.1 times as fast as the cheetah. Write this number as a percent.

$2.1 = 2.10$ Multiply by 100.

$= 210\%$ Add % symbol.

The peregrine falcon's speed is 210% of the cheetah's speed.

Got It? **Do this problem to find out.**

j. STEM The slowest land mammal is the sloth. Its speed is about 0.0016 that of a cheetah. Write this number as a percent.

j. ______

Guided Practice

Write each percent as a decimal and as a mixed number or fraction in simplest form. (Examples 1–3)

1. 325% = ______

2. 480% = ______

3. 0.6% = ______

Write each mixed number or decimal as a percent. (Examples 4–6)

4. $1\frac{4}{5}$ = ______

5. 0.0015 = ______

6. 2.75 = ______

7. A manufacturing company finds that 0.0019 of the light bulbs it makes are defective. Write this as a percent. (Example 7) ______

8. **Building on the Essential Question** How are percents greater than 100% used in real-world contexts?

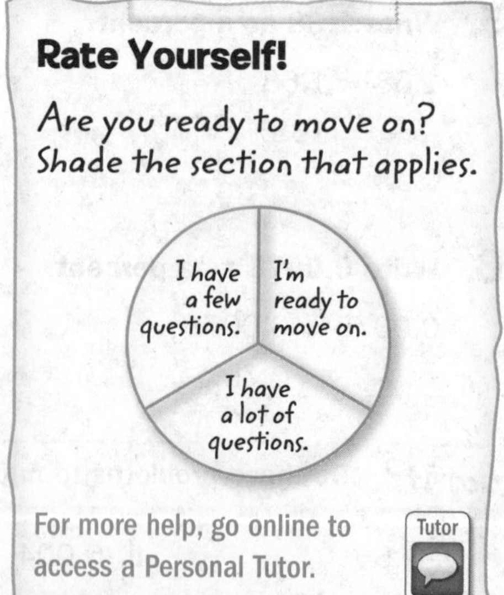

Name ______________________ My Homework ______________________

Independent Practice

Go online for Step-by-Step Solutions eHelp 

Write each percent as a decimal and as a mixed number or fraction in simplest form. (Examples 1–3)

1. 350% = ______

2. 600% = ______

3. 0.15% = ______

4. 0.55% = ______

Write each mixed number as a percent. (Example 4)

5. $2\frac{1}{2}$ = ______

6. $9\frac{3}{4}$ = ______

7. $4\frac{1}{5}$ = ______

8. $7\frac{3}{10}$ = ______

Write each decimal as a percent. (Examples 5 and 6)

9. 8.5 = ______

10. 2.64 = ______

11. 0.009 = ______

12. 0.0034 = ______

13. The size of a large milk shake is 1.4 times the size of a medium milk shake. Write 1.4 as a percent. (Example 7)

14. **STEM** Fresh water from lakes accounts for only 0.001 of the world's water supply. Write this decimal as a percent. (Example 7)

15. In a recent year, the United States Census Bureau reported that 0.3% of the population in the United States was Japanese. Write this percent as a decimal and as a fraction. Then interpret its meaning as a ratio of the United States population.

16. Adrienne answered all 21 multiple-choice questions correctly on her science test. If her teacher decided to let one of the questions count as a bonus, worth the same number of points as the other problems on the test, what was Adrienne's test score? Write your answer as a decimal and as a percent.

17. **Use Math Tools** Refer to the table at the right.

Elements in the Human Body	
Element	**Percent**
Magnesium	0.05
Potassium	0.35
Sodium	0.15
Sulfur	0.25

a. Write the percent of magnesium found in the human body as a decimal.

b. Which element makes up $\frac{1}{400}$ of the human body?

H.O.T. Problems Higher Order Thinking

18. **Find the Error** Raj is writing $\frac{3}{2,000}$ as a percent. Find his mistake and correct it.

19. **Persevere with Problems** The speed of a giraffe is 250% of the speed of a squirrel. If a squirrel's speed is 12 miles per hour, find the speed of a giraffe.

20. **Model with Mathematics** Write a real-world problem involving a percent greater than 100%. Then solve the problem.

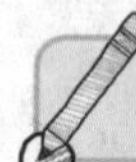

Standardized Test Practice

21. A certain stock increased its value by 467% over 10 years. Which number is *not* equivalent to 467%?

Ⓐ 4.67

Ⓑ 0.467

Ⓒ $4\frac{67}{100}$

Ⓓ $\frac{467}{100}$

Extra Practice

Write each percent as a decimal and as a mixed number or fraction in simplest form.

22. 475% = $4.75; 4\frac{3}{4}$

Homework Help

$475\% = \frac{475}{100}$

$= 4\frac{75}{100}$ or $4\frac{3}{4}$

$= 4.75$

23. 400% = ______

24. 0.05% = ______

25. 0.04% = ______

Write each decimal as a percent.

26. 1.07 = ______

27. 35 = ______

28. 0.003 = ______

29. 0.0077 = ______

30. A collectible action figure sold for 193% of its original price. Write this percent as a decimal and as a mixed number or fraction in simplest form.

______

______

31. A car's tire pressure decreased by 0.098 of its original pressure. Write 0.098 as a percent.

______

Write each percent as a decimal.

32. $\frac{3}{4}\%$ = ______

33. $\frac{3}{25}\%$ = ______

CCSS **Use Math Tools** **One complete figure represents 100%. Write a percent to represent the shaded portion of each figure below.**

34. ______

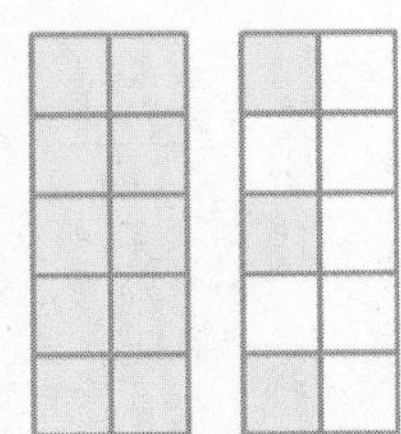

35. ______

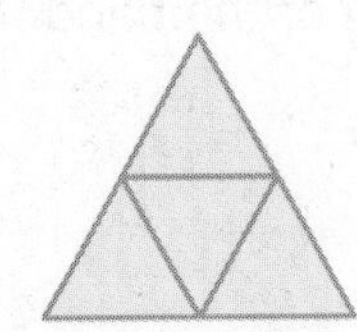

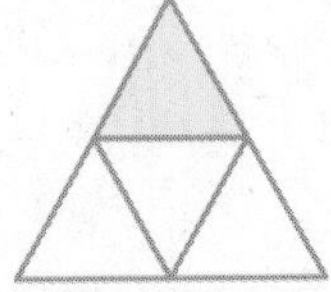

Standardized Test Practice

36. If one complete grid represents 100%, what percent of the figure below is shaded?

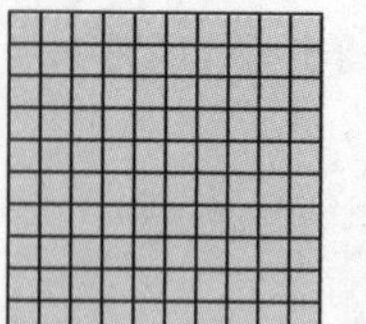 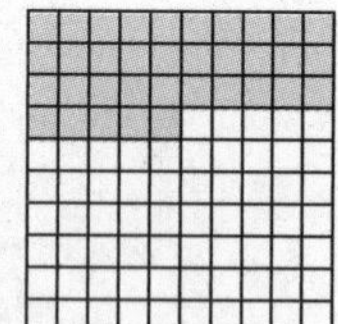

Ⓐ 25%
Ⓑ 100%
Ⓒ 125%
Ⓓ 135%

37. Short Response The medium-size sub sandwich is 1.33 times larger than the small-size sub sandwich. Write 1.33 as a percent.

____

38. About 0.036% of the water on Earth is found in lakes and rivers. What is 0.036% written as a fraction in simplest form?

Ⓕ $\frac{9}{25}$
Ⓖ $\frac{36}{100,000}$
Ⓗ $\frac{9}{25,000}$
Ⓘ $\frac{18}{50,000}$

39. What percent of one circle is modeled below?

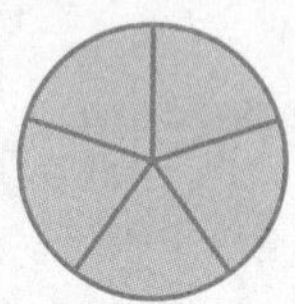 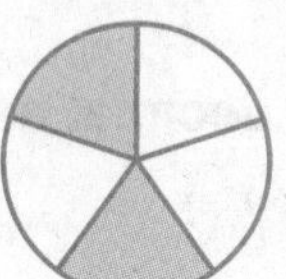

Ⓐ 40%
Ⓑ 100%
Ⓒ 120%
Ⓓ 140%

CCSS Common Core Review

Compare the fractions using <, >, or =. 4.NF.2

40. $\frac{3}{6}$ ◯ $\frac{1}{8}$

41. $\frac{10}{17}$ ◯ $\frac{11}{12}$

42. $\frac{7}{9}$ ◯ $\frac{5}{11}$

43. Evangeline walked $\frac{3}{10}$ of a mile on Monday, $\frac{5}{10}$ of a mile on Tuesday, and $\frac{25}{100}$ of a mile on Wednesday. Plot each distance on the number line. 4.NF.6

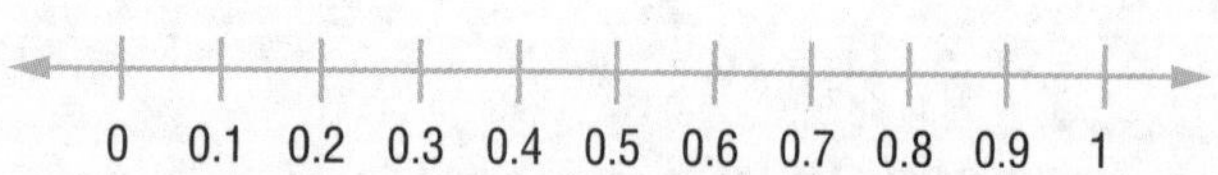

44. The flute players are $\frac{3}{10}$ of the band and the trumpet players are $\frac{1}{12}$ of the band. Is a greater fraction of the band flute players or trumpet players? 4.NF.2

____

____

Problem-Solving Investigation
Solve a Simpler Problem

Content Standards
6.RP.3, 6.RP.3c
Mathematical Practices
1, 3, 4

Case #1 First Place Pizza

The daily lunch report indicated that 80% of the 300 students at Midtown Middle School chose pizza for lunch.

How many students bought pizza for lunch?

Understand ***What are the facts?***

- The lunch report says 80% chose pizza.
- There are 300 students at the school.

Plan ***What is your strategy to solve this problem?***

Solve a simpler problem by finding 10% of the total students. Then use the result to find 80% of the total students.

Solve ***How can you apply the strategy?***

Complete the bar diagram. Fill in the value of each section.

300 students

0% 10% 20% 30% 40% 50% 60% 70% 80% 90% 100%

There are 300 ÷ 10, or 10 groups with ☐ students in each group.

Multiply. ☐ × 8 = ☐

So, ☐ students chose pizza for lunch.

Check ***Does the answer make sense?***

You know that 80% is close to 75%, which is $\frac{3}{4}$. Since $\frac{1}{4}$ of 300 is 75, $\frac{3}{4}$ of 300 is ☐. So, my answer is reasonable.

Analyze the Strategy

Reason Inductively Explain when you would use the *solve a simpler problem* strategy. ____________________

____

Case #2 Top Tip

Heidi's dad wants to leave an 18% tip for a $24.60 restaurant bill.

$24.60

0% 10% 20% 30% 40% 50% 60% 70% 80% 90% 100%

About how much money should he leave?

1 Understand

Read the problem. What are you being asked to find?

I need to estimate ______________________.

Underline key words and values. What information do you know?

Heidi's dad wants to leave an __________ *on a* __________ *bill.*

Is there any information that you do *not* need to know?

I do not need to know ______________________.

2 Plan

Choose a problem-solving strategy.

I will use the ______________________ *strategy.*

3 Solve

Use your problem-solving strategy to solve the problem.

Solve a simpler problem by finding 20% of $25.00. Use the result to estimate 18%. The whole is ______. Make a bar diagram that is divided into ______. Each part represents ______. The two shaded parts represent ______.

10% 10% 10% 10% 10% 10% 10% 10% 10% 10%

$25

Because the whole is $25.00, each part is ______.

So, 18% of $24.60 is about ______.

4 Check

Use information from the problem to check your answer.

$0.18 \times 24.60 =$ ______. So, $5 is a reasonable estimate.

Collaborate Work with a small group to solve the following cases. Show your work on a separate piece of paper.

Case #3 Books

Ebony estimates that she reads 100 books per year.

About how many books does she read per week?

Case #4 Candy

A candy factory can make 1,200 individually wrapped pieces of chocolate candy in one minute, 35% of which have caramel.

About how many pieces have caramel?

Case #5 Bracelets

Ruthie has 35 shape-bracelets, 60% of which are sea animals.

How many bracelets are sea animals?

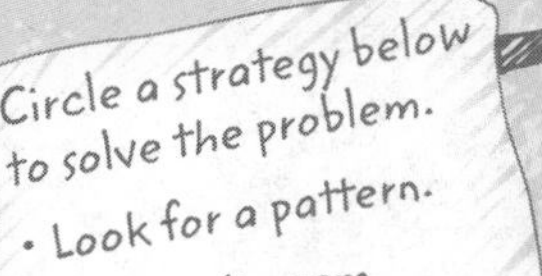

Case #6 Border

Part of a strip of border for a bulletin board is shown. All of the sections of the border are the same width.

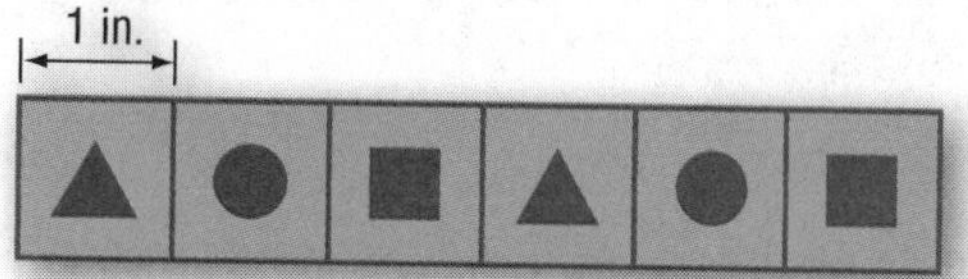

If the first shape on the strip is a triangle and the strip is 74 inches long, what is the last shape on the strip?

Mid-Chapter Check

Vocabulary Check

1. Define *percent*. Write $\frac{25}{100}$ as a percent then write $\frac{25}{100}$ as a decimal. (Lesson 2)

Skills Check and Problem Solving

Write each fraction as a decimal and each decimal as a fraction in simplest form. (Lesson 1)

2. $\frac{8}{20} =$ ______

3. $0.64 =$ ______

4. $\frac{3}{100} =$ ______

Write each percent as a decimal and each decimal as a percent. (Lessons 3 and 4)

5. $73\% =$ ______

6. $0.1 =$ ______

7. $254\% =$ ______

8. The number of chorus students increased by a factor of 1.2 from the previous year. Write 1.2 as a percent. (Lesson 4)

9. CCSS **Use Math Tools** The graph shows the pie sales during one week for Polly's Pies. (Lessons 2 and 3)

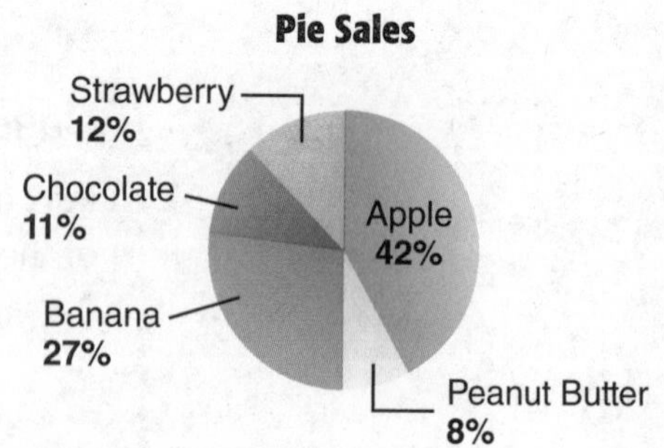

a. What fraction of the pies sold was apple?

b. Write the percent of strawberry pies sold as a decimal.

10. **Standardized Test Practice** Each circle is divided into sections of equal size. Which circle has 25% of its total area shaded? (Lesson 2)

Ⓐ 

Ⓑ 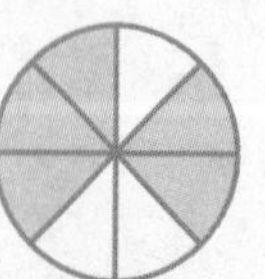

Ⓒ 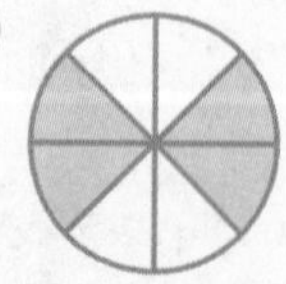

Ⓓ 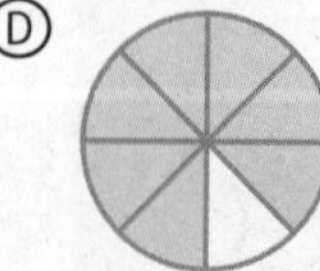

Compare and Order Fractions, Decimals, and Percents

What You'll Learn

Scan the lesson. Predict two things you will learn about comparing and ordering fractions.

-
-

Essential Question

WHEN is it better to use a fraction, a decimal, or a percent?

Vocabulary

least common denominator (LCD)

Common Core State Standards

Content Standards
Preparation for 6.RP.3c

Mathematical Practices
1, 2, 3, 4, 5, 6

Vocabulary Start-Up

The **least common denominator**, or LCD, is the least common multiple of the denominators of two or more fractions.

Complete the graphic organizer. Write the meaning of each word in the appropriate box. Provide examples.

Least	Common
Denominator	Multiple

least common denominator

Real-World Link

- Earnest is baking, but he wants to use only one measuring cup. He needs $\frac{1}{2}$ cup of sugar and $\frac{3}{4}$ cup of flour. What is the least common multiple of the denominators? ☐
- What size measuring cup should he use: $\frac{1}{2}$ cup, $\frac{1}{3}$ cup, or $\frac{1}{4}$ cup? Explain.

Work Zone

Compare and Order Fractions

To compare fractions, you can follow these steps.

1. Find the least common denominator (LCD) of the fractions. That is, find the least common multiple of the denominators.
2. Write an equivalent fraction for each fraction using the LCD.
3. Compare the numerators.

Example

Fill in each ◯ with <, >, or = to make a true statement.

1. $\frac{5}{8} \bigcirc \frac{7}{12}$

The LCM of the denominators, 8 and 12, is 24. So, the LCD is 24. Write an equivalent fraction with a denominator of 24 for each fraction.

$\frac{5}{8} = \frac{15}{24}$ (×3) $\frac{7}{12} = \frac{14}{24}$ (×2)

$\frac{15}{24} > \frac{14}{24}$, since 15 > 14. So, $\frac{5}{8} > \frac{7}{12}$.

Least Common Multiple

2 = 2
14 = 2 × 7
4 = 2 × 2
7 = 7
The LCM is 2 × 2 × 7 or 28.

Got It? Do these problems to find out.

a. $\frac{2}{3} \bigcirc \frac{4}{9}$ b. $\frac{5}{12} \bigcirc \frac{7}{8}$ c. $\frac{1}{6} \bigcirc \frac{5}{18}$

Example

2. **Order the fractions $\frac{1}{2}, \frac{9}{14}, \frac{3}{4}$, and $\frac{5}{7}$ from least to greatest.**

Rewrite each fraction using the LCD of 28.

$\frac{1}{2} = \frac{14}{28}$ (×14) $\frac{9}{14} = \frac{18}{28}$ (×2) $\frac{3}{4} = \frac{21}{28}$ (×7) $\frac{5}{7} = \frac{20}{28}$ (×4)

Since $\frac{14}{28} < \frac{18}{28} < \frac{20}{28} < \frac{21}{28}$, the order of the original fractions from least to greatest is $\frac{1}{2}, \frac{9}{14}, \frac{5}{7}, \frac{3}{4}$.

Got It? Do this problem to find out.

d. Order $\frac{1}{2}, \frac{5}{6}, \frac{2}{3}$, and $\frac{3}{5}$ from least to greatest.

d. ____________

Compare Fractions, Decimals, and Percents

It may be easier to compare fractions, decimals, and percents when they are all written as decimals.

$\frac{1}{5} = 0.2 = 20\%$	$\frac{2}{5} = 0.4 = 40\%$	$\frac{3}{5} = 0.6 = 60\%$	$\frac{4}{5} = 0.8 = 80\%$
$\frac{1}{8} = 0.125 = 12.5\%$	$\frac{3}{8} = 0.375 = 37.5\%$	$\frac{1}{3} = 0.\overline{3} = 33.\overline{3}\%$	$\frac{2}{3} = 0.\overline{6} = 66.\overline{6}\%$

Examples

Fill in each ◯ with <, >, or = to make a true statement.

3. $\frac{3}{4}$ ◯ **0.7**

$\frac{3}{4}$ ◯ 0.7 — Write the sentence.

0.75 ◯ 0.70 — Write $\frac{3}{4}$ as a decimal. Annex a zero to 0.7.

0.75 > 0.70 — Compare the hundredths place. 5 > 0

0.75

0.5 0.6 0.7 0.8 0.9 1

Since 0.75 is to the right of 0.7 on the number line, $\frac{3}{4} > 0.7$.

4. Lucita made 85% of her free throws. Henri made $\frac{7}{8}$ of his free throws. Who has the better average? Explain.

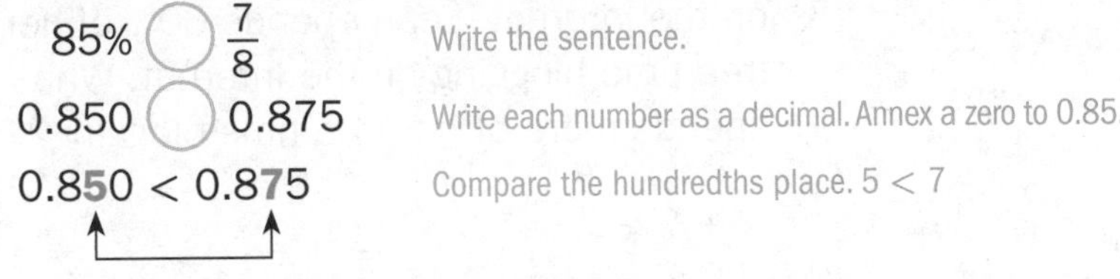

85% ◯ $\frac{7}{8}$ — Write the sentence.

0.850 ◯ 0.875 — Write each number as a decimal. Annex a zero to 0.85.

0.850 < 0.875 — Compare the hundredths place. 5 < 7

Since 0.850 < 0.875, Henri has the better average.

Check

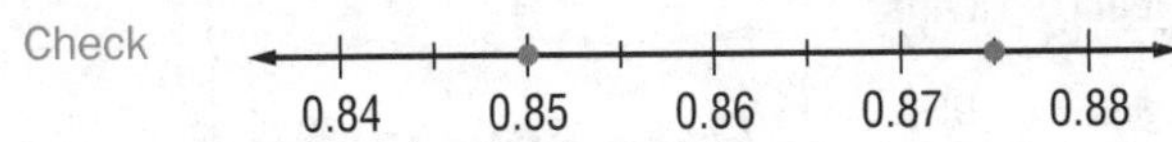

Since 0.85 is to the left of 0.875, the answer is correct. ✓

Got It? Do these problems to find out.

e. $\frac{2}{3}$ ◯ 0.6 **f.** 0.7 ◯ $\frac{8}{11}$ **g.** $\frac{1}{5}$ ◯ 0.2

h. 42% ◯ 0.44 **i.** 7% ◯ $\frac{7}{10}$ **j.** 6.5 ◯ 650%

Example

5. The table shows the school carnival attendance. Which grade has the greatest part of the class attending the carnival?

Grade	Attendance
6	$\frac{5}{8}$
7	0.5
8	58.3%

Order the numbers from least to greatest. Express each number as a decimal with the same number of places.

$\frac{5}{8} = 0.625$ $\quad 0.5 = 0.500 \quad$ $58.3\% = 0.583$

Graph the numbers on a number line.

0.500 0.525 0.550 0.575 0.600 0.625

From least to greatest, the numbers are 0.5, 58.3%, and $\frac{5}{8}$.

Since $\frac{5}{8}$ represents Grade 6, Grade 6 has the greatest part of the class attending the school carnival.

Got It? Do this problem to find out.

Show your work.

k. Hiroshi found that $\frac{3}{5}$ of his class prefers vanilla ice cream, 26% prefers chocolate, and 0.14 prefers strawberry. Which kind of ice cream do students prefer the least?

k. ________

Guided Practice

1. Order the fractions $\frac{4}{5}, \frac{1}{2}, \frac{9}{10}$, and $\frac{3}{4}$ from least to greatest. (Examples 1 and 2)

____

Show your work.

2. Cora spends $\frac{2}{3}$ of her free time blogging on the Internet. Leah spends 60% of her free time blogging on the Internet. Who spends more of her free time blogging? (Examples 3 and 4) ________

3. The table shows the wins for some middle school football teams. Which team has the greatest fraction of wins? (Example 5)

Team	Wins
Eagles	95%
Wolves	$\frac{9}{10}$
Mustangs	0.89

____

4. **Building on the Essential Question** How do you compare fractions, decimals, and percents?

____

____

____

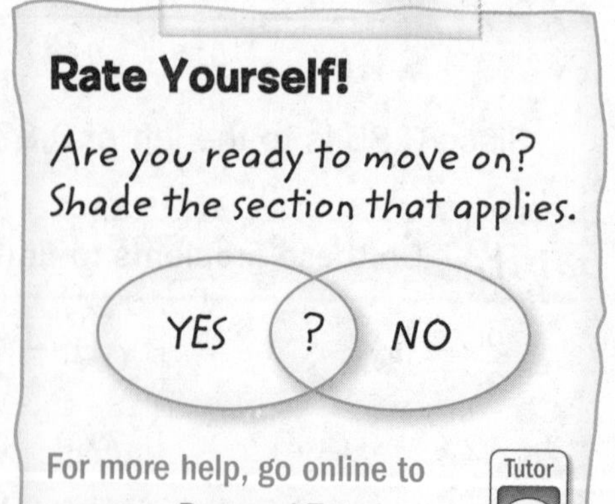

Name ______________________ My Homework ______________________

Independent Practice

Go online for Step-by-Step Solutions

Fill in each ◯ with <, >, or = to make a true statement. (Examples 1 and 3)

1. $\frac{1}{3}$ ◯ $\frac{3}{5}$

2. $\frac{7}{12}$ ◯ $\frac{1}{2}$

3. $\frac{1}{4}$ ◯ 0.4

4. 0.7 ◯ $\frac{7}{9}$

Order the fractions from least to greatest. (Example 2)

5. $\frac{1}{2}, \frac{2}{3}, \frac{1}{4}, \frac{5}{6}$

6. $\frac{2}{3}, \frac{2}{9}, \frac{5}{6}, \frac{11}{18}$

7. Darius spends 35% of his time doing math homework. Alex spends $\frac{2}{5}$ of his time doing math homework. Who spends more homework time on math? Explain. (Example 4)

8. Three snack bars contain $\frac{1}{5}$, 0.22, and 19% of their Calories from fat. Which snack bar contains the least amount of Calories from fat? (Example 5)

9. **CCSS Model with Mathematics** Use the graphic novel frame below for Exercises a–b.

a. Write each score as a decimal. ______________________

b. Compare the three scores. ______________________

10. **Be Precise** Complete the graphic organizer. Write the original numbers to complete the statement.

Number	Steps to Write the Number as a Decimal with Three Places		Decimal
$\frac{3}{8}$	Divide the ________ by the ________.	→	0.375
0.3	The number is a decimal. Annex ________ zeros.	→	0.300
38.7%	Move the ________ point ________ places to the left. Remove the ________ symbol.	→	0.387

So, ________ < ________ < ________.

11. Order the portion of responses listed in the table from least to greatest.

____

Number of Times Eating Fast Food per Week	0	1-2	3-4	5+
Portion of Responses	17%	$\frac{11}{20}$	0.2	8%

H.O.T. Problems Higher Order Thinking

12. **Reason Abstractly** Specify three fractions with different denominators that have an LCD of 24. Then arrange the fractions in order from least to greatest.

____

13. **Persevere with Problems** Order $\frac{3}{8}$, $\frac{3}{7}$, and $\frac{3}{9}$ from least to greatest without writing equivalent fractions with a common denominator. Explain your strategy. ________________________________

____

14. **Persevere with Problems** Are the fractions $\frac{3}{9}$, $\frac{3}{10}$, $\frac{3}{11}$, and $\frac{3}{12}$ arranged in order from least to greatest or from greatest to least? Explain.

____

____

Standardized Test Practice

15. Which of the following numbers has a value between 0 and 1?

Ⓐ $\frac{7}{8}$ Ⓑ 114% Ⓒ 1.14 Ⓓ $\frac{8}{7}$

Extra Practice

Fill in each ◯ with <, >, or = to make a true statement.

16. $\frac{7}{8}$ (>) $\frac{5}{6}$

Homework Help

$\frac{7}{8} = \frac{21}{24}; \frac{5}{6} = \frac{20}{24}$

$\frac{21}{24} > \frac{20}{24}$ so $\frac{7}{8} > \frac{5}{6}$

17. $\frac{14}{18}$ ◯ $\frac{7}{9}$

18. 0.75 ◯ $\frac{1}{2}$

19. $\frac{1}{3}$ ◯ 0.33

Order the fractions from least to greatest.

20. $\frac{1}{6}, \frac{2}{5}, \frac{3}{5}, \frac{3}{7}$

21. $\frac{5}{8}, \frac{3}{4}, \frac{1}{2}, \frac{9}{16}$

22. Shop Rite has jeans on sale for $\frac{3}{10}$ off. Save More has jeans on sale for 33% off. Which store has a better sale on jeans? Explain.

23. A city's population rose 3% one year, 0.08 the next year, and by $\frac{2}{50}$ the next year. Order these increases from least to greatest.

Order the numbers from least to greatest.

24. 0.4, $\frac{5}{8}$, 38%

25. $\frac{1}{2}$, 0.55, $\frac{5}{7}$

26. CCSS **Use Math Tools** The table shows the favorite subjects of students in a recent survey.

a. Did more students choose art or math? Explain.

b. Which subject did most students choose? Explain.

c. Order the subjects from least to greatest.

Favorite Subject

Subject	Portion of Students
Art	$\frac{4}{25}$
English	13%
Math	0.28
Other	7%
Science	$\frac{21}{100}$
Social Studies	0.15

Standardized Test Practice

27. Fairview Elementary started a recycling program. The display shows the portion of each item that is recycled at the school.

Which of the following lists the items recycled from least to greatest?

Ⓐ plastic, glass, aluminum, paper

Ⓑ glass, plastic, aluminum, paper

Ⓒ paper, glass, plastic, aluminum

Ⓓ aluminum, paper, glass, plastic

28. A plumber needs to drill a hole that is just slightly larger than $\frac{3}{16}$ inch in diameter. Which measure is the smallest but still larger than $\frac{3}{16}$ inch?

Ⓕ $\frac{3}{32}$ inch

Ⓖ $\frac{5}{16}$ inch

Ⓗ $\frac{13}{64}$ inch

Ⓘ $\frac{17}{32}$ inch

29. Short Response Mr. Tucker has four sockets in his tool chest that are labeled $\frac{3}{4}$ in., $\frac{3}{16}$ in., $\frac{11}{32}$ in., and $\frac{1}{8}$ in. Order the sizes of sockets from smallest to largest.

____

CCSS Common Core Review

Round each decimal to the nearest hundredth. 5.NBT.4

30. 0.623 ≈ ________

31. 4.288 ≈ ________

32. 5.105 ≈ ________

33. In a survey, $\frac{9}{25}$ of students ride the bus to school and $\frac{19}{50}$ walk to school. What fraction of students ride the bus or walk to school? 4.NF.3d

____

34. The student council bought 7 bags of apples for their fall party. How much did they pay for the apples? 5.NBT.7

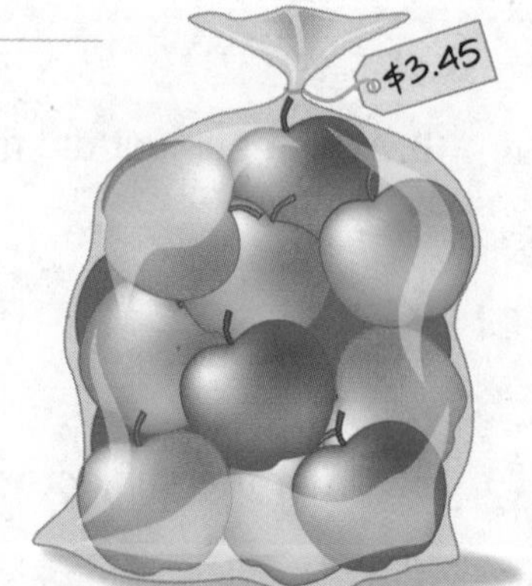

____

____

Lesson 6

Estimate with Percents

What You'll Learn

Scan the lesson. Predict two things you will learn about estimating with percents.

-
-

Essential Question

WHEN is it better to use a fraction, a decimal, or a percent?

Common Core State Standards

Content Standards
6.RP.3, 6.RP.3c

Mathematical Practices
1, 3, 4, 5

Real-World Link

Movies Josefina surveyed 298 students and found that 52% like scary movies. Estimate the number of students that like scary movies.

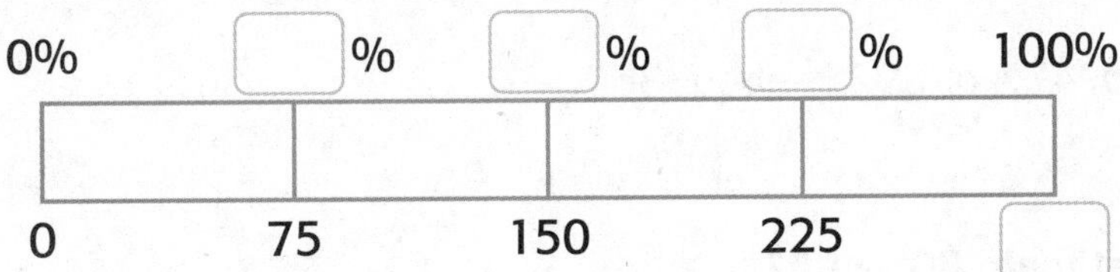

1. Write the common percents from 0% to 100% at the top of the bar diagram.
2. What common percent is 52% close to? ______
   Shade the bar diagram above to show your answer.
3. Round 298 to the nearest hundred. 298 ≈ ______
   Write your answer in the box below 100%.
4. Use the bar diagram to estimate 52% of 298. Explain.

5. Use the bar diagram below to estimate 73% of 400. ______

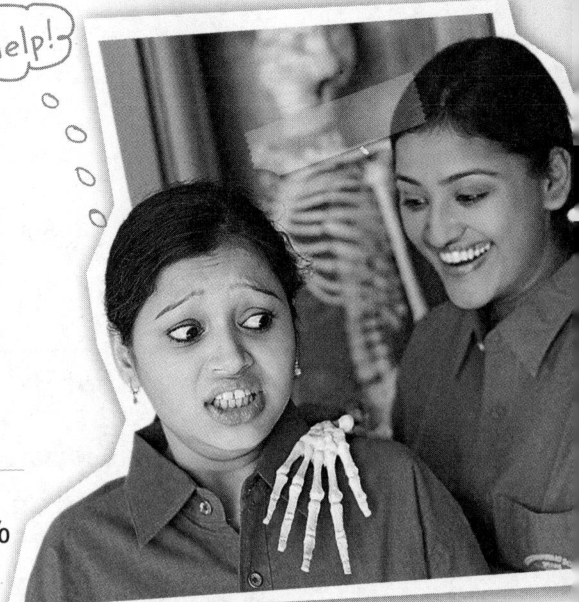

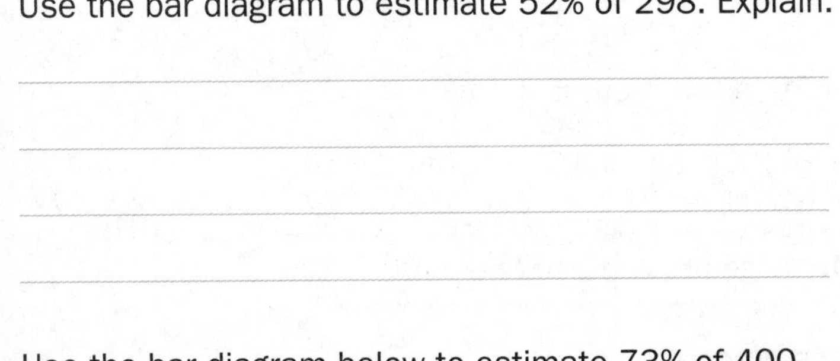

Work Zone

Estimate the Percent of a Number

Estimating with percents will provide a reasonable solution to many real-world problems. Choose compatible numbers when estimating the percent of a number.

Examples

1. Estimate 47% of 692.

47% is close to 50% or $\frac{1}{2}$. Round 692 to 700.

$\frac{1}{2}$ of 700 is 350. $\frac{1}{2}$ or *half* means to divide by 2.

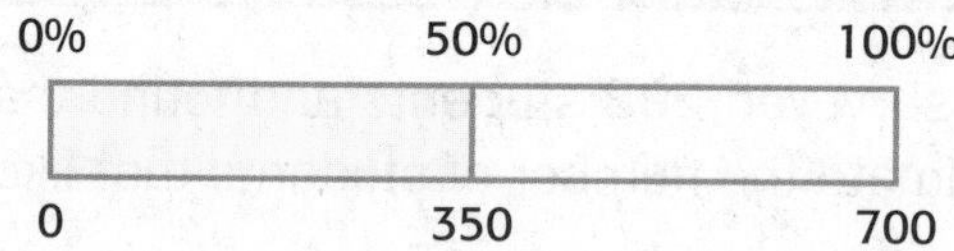

So, 47% of 692 is about 350.

2. Estimate 60% of 27.

60% is $\frac{3}{5}$.

Round 27 to 25 since it is divisible by 5.

$\frac{1}{5}$ of 25 is 5. $\frac{1}{5}$, or *one fifth*, means divide by 5.

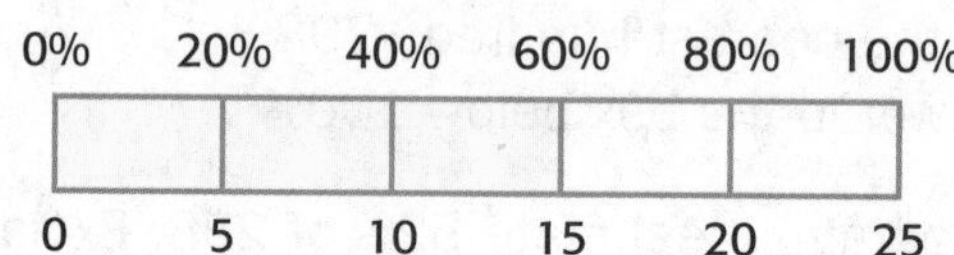

So, $\frac{3}{5}$ of 25 is 3 × 5 or 15.

So, 60% of 27 is about 15.

Show your work.

a. ____________

b. ____________

c. ____________

Got It? **Do these problems to find out.**

Estimate each percent.

a. 48% of 76 **b.** 18% of 42 **c.** 73% of 41

Example

3. STEM **Polar bears can eat as much as 10% of their body weight in less than one hour. If an adult male polar bear weighs 715 pounds, about how much food can he eat in one hour?**

To determine how much food a polar bear can eat in one hour, you need to estimate 10% of 715.

Method 1 **Find equivalent ratios.**

$10\% = \frac{1}{10}$ and $715 \approx 700$

$\frac{1}{10} = \frac{■}{700}$ Write the equivalent ratios.

$\frac{1}{10} = \frac{■}{700}$ ($\times 70$, $\times 70$) Since $10 \times 70 = 700$, multiply 1 by 70.

The unknown value is 70.

Method 2 **Use mental math.**

$10\% = \frac{1}{10}$ and $715 \approx 700$

$\frac{1}{10}$ of 700 is 70.

So, a polar bear can eat about 70 pounds of food in one hour.

Got It? **Do this problem to find out.**

d. Kayleigh decided to donate 30% of her savings. If she has $238 in her savings account, about how much will she donate?

d. ____________

Estimate Using the Rate per 100

You can also estimate with percents using a rate per 100.

STOP and Reflect

When would you use mental math to estimate the percent of a number? Explain below.

Examples

4. **Estimate 17% of 198.**

17% = 17 out of 100 Write the percent as a rate per 100.

$198 \approx 200$ Round to the nearest hundred.

Since 200 is 100 + 100, add 17 + 17 to estimate 17% of 198.

34 is about 17% of 198.

5. An airline records the snack orders of passengers. Last year 9% of all passengers ordered ginger ale to drink. There are 408 passengers on the flight to Houston, Texas. About how many passengers does the airline expect to order ginger ale on this flight?

Estimate 9% of 408.

$9\% = 9$ out of 100 — Write the percent as a rate per 100.

$408 \approx 400$ — Round to the nearest hundred.

Since 400 is 100×4, multiply 9×4 to estimate 9% of 408.

36 is about 9% of 408. So, about 36 passengers will order ginger ale.

Show your work.

e. ________

f. ________

g. ________

Got It? **Do these problems to find out.**

Estimate using a rate per 100.

e. 27% of 307

f. 76% of 192

g. Last year 24% of the zoo visitors were under the age of 3. Last week, the zoo had 996 visitors. About how many of the zoo visitors were under the age of 3?

Guided Practice

Estimate each percent. (Examples 1 and 2)

1. 19% of \$53 ≈ ________

2. 21% of 96 ≈ ________

3. 59% of 16 ≈ ________

4. A purse that originally cost \$29.99 is on sale for 50% off. About how much is the sale price of the purse? (Example 3)

____

5. Mr. Marcucci received a bonus of \$496 from his employer. He has to pay 33% of his bonus to taxes. How much will Mr. Marcucci pay in taxes? (Examples 4 and 5)

____

6. **Building on the Essential Question** When is an estimate more useful than an exact answer?

____

____

Rate Yourself!

How confident are you about estimating with percents? Shade the ring on the target.

For more help, go online to access a Personal Tutor.

Name ______________________ My Homework ______________________

Independent Practice

Go online for Step-by-Step Solutions 

Estimate each percent. (Examples 1 and 2)

1. 47% of \$118 ≈ ________

2. 19% of 72 ≈ ________

3. 42% of 16 ≈ ________

4. 67% of 296 ≈ ________

Estimate using a rate per 100. (Example 4)

5. 24% of 289 ≈ ________

6. 67% of 208 ≈ ________

7. **STEM** Penguins spend almost 75% of their lives in the sea. An Emperor Penguin in the wild has a life span of about 18 years. About how many years does this penguin spend in the sea? (Example 3)

____

8. In Nathan's baseball card collection, 58% of the cards are players from the National League. He has 702 baseball cards. About how many baseball cards are players from the National League? Use a rate per 100 to estimate. (Example 5)

____

9. **CCSS Model with Mathematics** Refer to the graphic novel frame below for Exercises a–b.

a. Suppose Angel is shooting baskets and makes 40% of the 15 shots. Does he win a prize? Explain your reasoning.

____

b. About what percent of the baskets need to be made in order to win a prize? ________

10. About 42% of Alaska's population lives in the city of Anchorage. If Alaska has a total population of 648,818, about how many people live in Anchorage?

 11. During the basketball season, Tyrone made 37 baskets out of 71 attempts. About what percent of his shots did he miss?

CCSS **Use Math Tools** **Estimate the percent that is shaded in each figure.**

12. 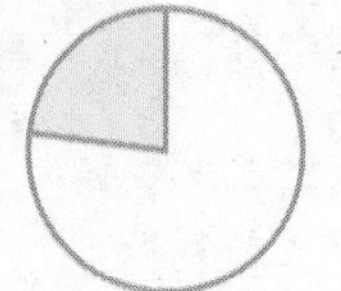

13. 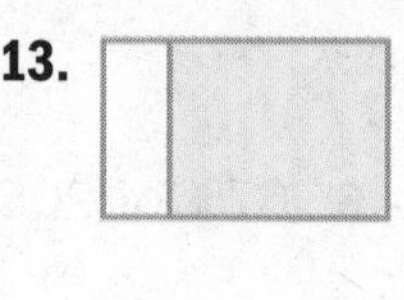

14. 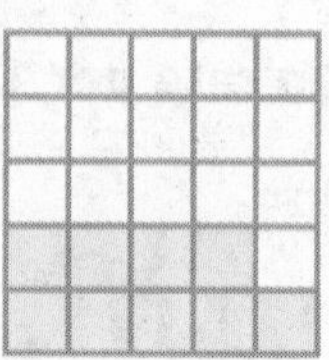

H.O.T. Problems Higher Order Thinking

15. CCSS **Reason Inductively** Rachel wants to buy a shirt regularly priced at \$32. It is on sale for 40% off. Rachel estimates that she will save $\frac{2}{5}$ of \$30 or \$12. Will the actual amount be more or less than \$12? Explain.

16. CCSS **Persevere with Problems** Order 10% of 20, 20% of 20, and $\frac{1}{5}$% of 20 from least to greatest.

17. CCSS **Construct an Argument** A classmate is trying to estimate 42% of \$122. Explain how your classmate should solve the problem.

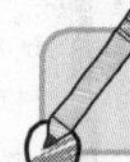

Standardized Test Practice

18. Evan wants to buy a digital camera that sells for \$199.99. If he uses his discount card, he will save 18%. About how much will he save using the discount card?

Ⓐ \$20

Ⓑ \$40

Ⓒ \$50

Ⓓ \$160

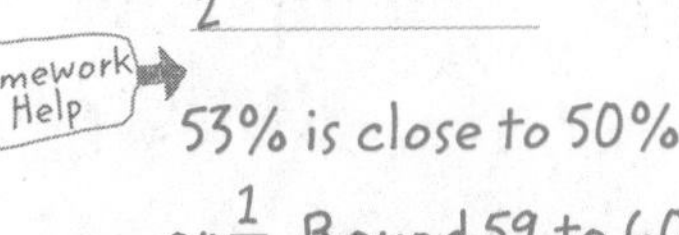

Extra Practice

Estimate each percent.

19. 53% of 59 ≈ $\frac{1}{2}$ of 60 is 30.

Homework Help: 53% is close to 50% or $\frac{1}{2}$. Round 59 to 60.

20. 35% of 147 ≈

21. 26% of 125 ≈

22. 79% of 82 ≈

Estimate using a rate per 100.

23. 19% of 288 ≈

24. 74% of 315 ≈

25. 61% of 407 ≈

26. 89% of 195 ≈

27. Trevon spent 8 hours and 15 minutes at an amusement park yesterday. He spent 75% of the time at the park on rides. About how much time did he spend on rides?

28. A group of friends went on a hiking trip. They planned to hike a total of 38 miles. They want to complete 25% of the hike by the end of the first day. About how many miles should they hike the first day?

29. Briana has just finished her sixth grade scrapbook. In her scrapbook, 47% of the pages include her twin sister, Bethany. The scrapbook has 896 photos. About how many photos include Bethany? Use a rate per 100 to estimate.

30. The community garden has 596 vegetables. In the garden, 64% of the vegetables are green vegetables. About how many vegetables in the garden are green? Use a rate per 100 to estimate.

Use Math Tools **Estimate the percent that is shaded in each figure.**

31.

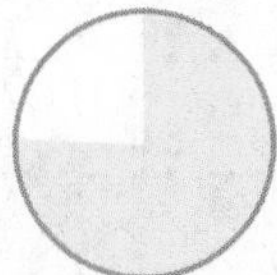

32.

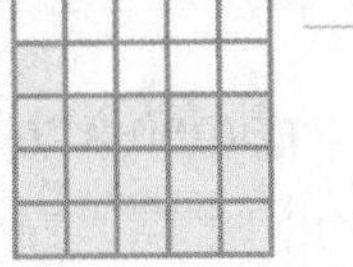

Standardized Test Practice

33. Refer to the graph. If 4,134 people were surveyed, which of the following can be used to estimate the number of 18- to 24-year-olds that own a portable MP3 player?

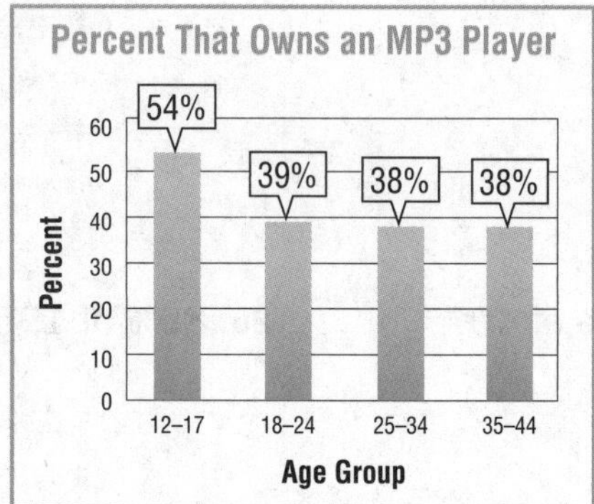

Ⓐ $\frac{1}{2}$ of 4,000 = 2,000

Ⓑ $\frac{2}{5}$ of 4,000 = 1,600

Ⓒ $\frac{1}{3}$ of 4,000 = 1,300

Ⓓ $\frac{1}{5}$ of 4,000 = 800

34. Short Response In a survey of teens, 21% said their friends like to read and talk about books. About how many teens out of 1,095 would say their friends read and talk about books?

35. After a group of 24 parts were tested, 5 were found to be defective. About what percent of the parts tested were defective?

Number Tested	Number Defective
24	5

Ⓕ 5%

Ⓖ 20%

Ⓗ 25%

Ⓘ 33%

Common Core Review

Write each fraction as a decimal. 4.NF.6

36. $\frac{22}{100}$ = ______

37. $\frac{7}{100}$ = ______

38. $\frac{67}{100}$ = ______

39. $\frac{15}{100}$ = ______

40. $\frac{12}{100}$ = ______

41. $\frac{6}{100}$ = ______

42. At a clothing store, T-shirts are on sale for $9.97 each. What is the cost of 3 T-shirts? 5.NBT.7 ______

43. The Sylvester family planted a garden with the dimensions shown. What is the area of the garden? 5.NBT.7

7.5 ft

3 ft

______

Inquiry Lab

Percent of a Number

HOW can you model the percent of a number?

Content Standards
6.RP.3, 6.RP.3c

Mathematical Practices
1, 3, 4

Movies There were 180 people in a movie theater. Twenty percent of them received the student discount and 10% received the senior citizen discount. The rest did not receive a discount. How many people did not receive a discount?

What do you know? ______________________________

____

What do you need to find? ______________________________

____

Investigation

Model the situation using two bar diagrams.

Step 1 Use a bar diagram to represent 100%. Then use another bar diagram of equal length to represent 180 people.

student discount (first two sections); senior discount (third section)

percent	**10%**	**10%**	**10%**	**10%**	**10%**	**10%**	**10%**	**10%**	**10%**	**10%**	100%
people	**18**	**18**	**18**	**18**	**18**	**18**	**18**	**18**	**18**	**18**	180

Step 2 Divide each bar into 10 equal parts. Think: 180 ÷ 10 = ☐

So, each part of 180 represents ☐ people.

Step 3 Determine how many people did not receive a discount. Shade 2 sections of each bar diagram to represent the student discount. Shade 1 section of each bar diagram to represent the senior discount.

There are ☐ unshaded sections in each bar diagram.

☐ × ☐ = ☐

So, ☐ people at the movie did not receive a discount.

Collaborate

Model with Mathematics **Work with a partner. Find the percent of each number using two bar diagrams.**

1. 50% of 80 children = ________

		100%
		children

2. 25% of $32 = ________

3. 80% of 40 points = ________

4. 30% of 70 teachers = ________

Reflect

5. **Reason Inductively** Explain how to use two bar diagrams to find 45% of $60.

6. **Inquiry** HOW can you model the percent of a number?

Lesson 7

Percent of a Number

What You'll Learn

Scan the lesson. List two real-world scenarios in which you would use the percent of a number.

-
-

Essential Question

WHEN is it better to use a fraction, a decimal, or a percent?

Common Core State Standards

Content Standards
6.RP.3, 6.RP.3c

Mathematical Practices
1, 3, 4, 5

Real-World Link

Snacks In a survey, 200 students chose their favorite snacks. Use the table to find the number of students who chose each snack.

Snack	Percent	Fraction	Equivalent Fraction	Number of Responses
Fruit	23%	$\frac{23}{100}$	$\frac{46}{200}$	46 out of 200
Cheese	15%	$\frac{\square}{100}$	$\frac{\square}{200}$	☐ out of 200
Veggies	17%	$\frac{\square}{100}$	$\frac{\square}{200}$	☐ out of 200
Cookies	15%	$\frac{\square}{100}$	$\frac{\square}{200}$	☐ out of 200
Chips	18%	$\frac{\square}{100}$	$\frac{\square}{200}$	☐ out of 200
No Snack	12%	$\frac{\square}{100}$	$\frac{\square}{200}$	☐ out of 200

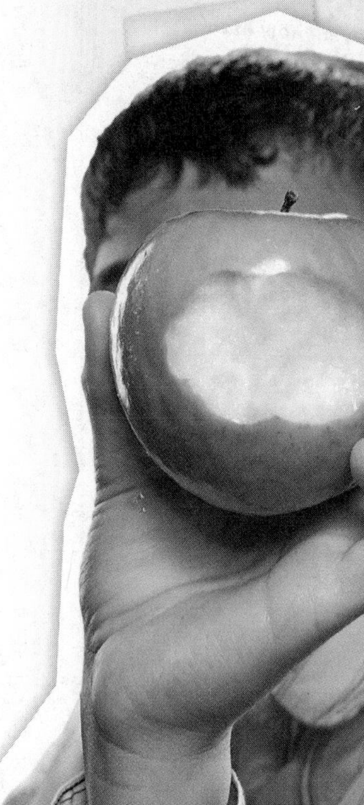

Check Add the number of responses in the last column.

46 + ☐ + ☐ + ☐ + ☐ + ☐ = 200 ✓

1. How does finding the percent as a rate per 100 help you find the number of responses out of 200?

Work Zone

Find the Percent of a Number

You can use fractions and decimals to find the percent of a number. To find the percent of a number, write the percent as a fraction with a denominator of 100. Then multiply the fraction by the number.

Example

1. **Refer to the circle graph. Suppose there are 300 students at York Middle School. Find the number of students that have cheese as a snack.**

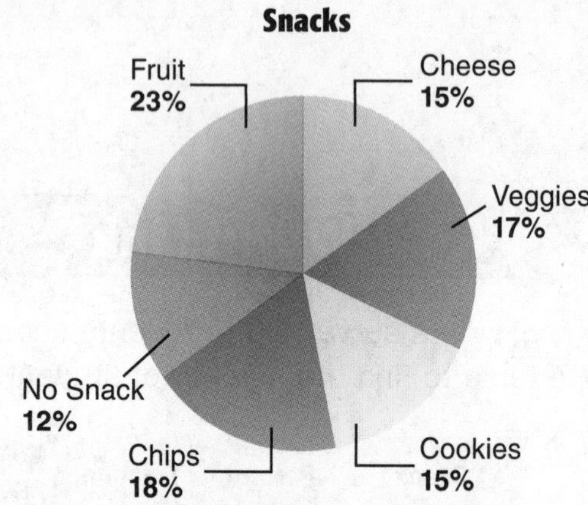

Method 1 **Write the percent as a fraction.**

$15\% = \frac{15}{100}$ or $\frac{3}{20}$ Write the percent as a rate per 100.

$\frac{3}{20}$ of $300 = \frac{3}{20} \times 300$ Multiply.

$= 45$

Method 2 **Write the percent as a decimal.**

$15\% = 0.15$

0.15 of $300 = 0.15 \times 300$

$= 45$

So, 45 students have cheese as a snack.

Check Use a bar diagram.

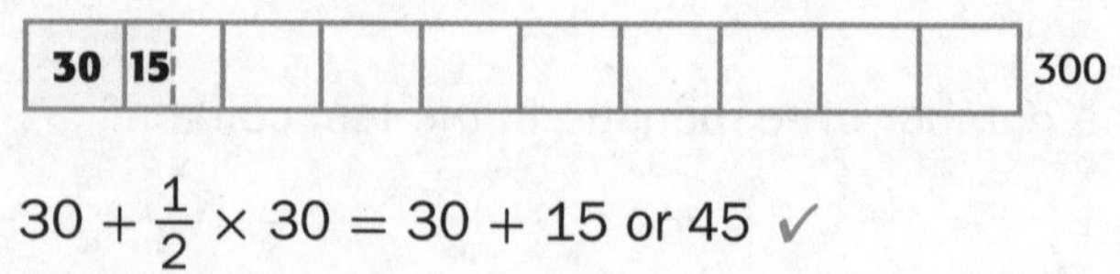

$30 + \frac{1}{2} \times 30 = 30 + 15$ or 45 ✓

STOP and Reflect

If 50 students were surveyed, how many students chose no snack?

Show your work.

Got It? **Do this problem to find out.**

a. Find the number of students at York Middle School that have chips as a snack.

a. ______

Percents Greater Than 100% and Less Than 1%

You may choose whether to write a percent as a fraction or as a decimal based on the problem.

Examples

2. Find 145% of 320.

$145\% = \frac{145}{100}$ or $\frac{29}{20}$ — Write 145% as a rate per 100. Then simplify.

$145\% \text{ of } 320 = \frac{29}{20} \times 320$ — Write the multiplication problem.

$= \frac{29}{\cancel{20}_1} \times \frac{\cancel{320}^{16}}{1}$ — Divide the numerator and denominator by 20.

$= 29 \times 16$ — Simplify.

$= 464$ — Multiply.

So, 145% of 320 is 464.

3. Find 220% of 65.

$220\% = \frac{220}{100}$ or $\frac{\square}{\square}$ — Write 220% as a fraction in simplest form.

$220\% \text{ of } 65 = \frac{\square}{\square} \times \square$ — Write the multiplication problem.

= ______ — Divide by the GCF.

= ____ × ____ — Simplify.

= ____ — Multiply.

So, 220% of 65 is ______.

4. Find 0.25% of 58.

$0.25\% = 0.0025$ — Write 0.25% as a decimal.

$0.25\% \text{ of } 58 = 0.0025 \times 58$ — Write the multiplication problem.

$= 0.145$ — Multiply.

So, 0.25% of 58 is 0.145.

b. ______

c. ______

d. ______

Got It? **Do these problems to find out.**

Find the percent of each number.

b. 128% of 550 **c.** 0.3% of 200 **d.** 0.85% of 600

Example

5. In a recent state Special Olympics meet, Franklin County sent a team of 70 players. Twenty percent of the team competed in soccer. How many athletes competed in soccer?

$20\% = 0.20$	Write 20% as a decimal.
$20\% \text{ of } 70 = 0.2 \times 70$	Write the multiplication problem.
$= 14$	Multiply.

So, 14 team members were soccer players.

Got It? Do this problem to find out.

Show your work.

e. In the same meet, 15% of the team from Delaware County competed in tennis. If there were 20 members on the team, how many competed in tennis?

e. ______

Guided Practice

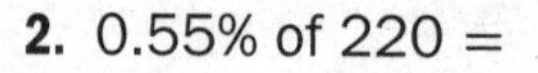

Find the percent of each number. (Examples 1–4)

1. 32% of 60 = ______
2. 0.55% of 220 = ______
3. 275% of 4 = ______

4. Troy wants to buy a jersey of his favorite MLS team. The jersey is 30% off the original price. If the original price of the jersey is $35, what is the amount Troy will save? (Example 5) ______

5. **Building on the Essential Question** How do you find a percent of a number?

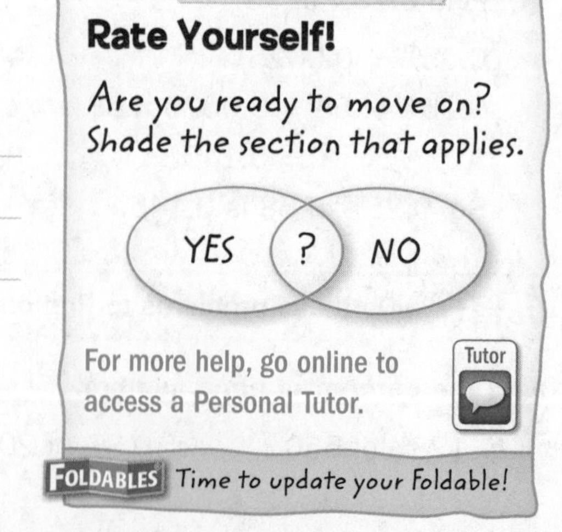

Rate Yourself!

Are you ready to move on? Shade the section that applies.

YES ? NO

For more help, go online to access a Personal Tutor.

FOLDABLES *Time to update your Foldable!*

Name ____________________ My Homework ____________________

Independent Practice

Go online for Step-by-Step Solutions 

The cafeteria at Midtown Middle School surveyed 575 students about their favorite food. Find the number of students that responded for each of the following. (Example 1)

1. chicken: 8% = ______

2. salad: 20% = ______

3. burgers: 16% = ______

4. fruit: 24% = ______

Find the percent of each number. (Examples 2–4)

5. 0.9% of 1,000 = ______

6. 0.46% of 80 = ______

7. 350% of 96 = ______

8. 222% of 55 = ______

9. The original price of a pair of shoes is $42. The sale price is 20% off the original price. What is the amount off the original price?

(Example 5) ______

10. Torri had $20 to buy a birthday present for her dad. She decided to buy a DVD for $18. The sales tax is 7%. Does she have enough money? Explain your reasoning.

______

11. Twenty-four students in Jamal's class are wearing tennis shoes. There are thirty students in his class. Jamal says that 70% of his class is wearing tennis shoes. Is Jamal correct? Explain your reasoning.

______

12. CCSS **Use Math Tools** Marisol keeps track of her weekly quiz grades as shown in the table.

a. Complete the table.

b. In which class did Marisol have the higher score?

Test	Number Correct	Score	Total
Math	68		85
Science		90%	70

______

c. Suppose Marisol scored a 96% on an English test. There were 50 questions on the test. How many did Marisol answer correctly?

______

13. **CCSS Use Math Tools** Use the graphic organizer to compare and contrast percents and fractions. Use the phrases *less than, equal to,* and *greater than* to complete each statement. Write an example in the space provided.

Percent	Shared Concept	Fraction
A whole is represented by a percent that is ______ 100%. Example: ______	whole	A whole is represented by a fraction with a numerator that is ______ the denominator. Example: ______
Part of a whole is represent by a percent that is ______ 100%. Example: ______	part of a whole	Part of a whole is represented by a fraction with a numerator that is ______ the denominator. Example: ______
An amount that is greater than one is represented by a percent that is ______ 100%. Example: ______	more than one	An amount that is greater than one is represented by a fraction with a numerator that is ______ the denominator. Example: ______

H.O.T. Problems Higher Order Thinking

14. **CCSS Model with Mathematics** Write and solve a real-world problem in which the percent of a number results in a number greater than the number itself.

15. **CCSS Justify Conclusions** Is 16% of 40 the same as 40% of 16? Explain your reasoning.

16. **CCSS Persevere with Problems** Find 15% of 15% of 15% of 500. How does this compare to finding 45% of 500?

Standardized Test Practice

17. At birth, a newborn's head is about 25% of its total length. If a baby is 18 inches long, about how many inches is the baby's head?

Ⓐ 3.5 inches
Ⓒ 4.5 inches
Ⓑ 4 inches
Ⓓ 5 inches

Name ______ My Homework ______

Extra Practice

Find the percent of each number.

18. 6% of 95 = 5.7

Homework Help: $6\% = 0.06$

$0.06 \times 95 = 5.7$

19. 15% of 110 = ______

20. 75% of 260 = ______

21. 28% of 575 = ______

22. 0.6% of 36 = ______

23. 108% of 148 = ______

24. 102% of 750 = ______

25. 0.03% of 1,500 = ______

26. Brenna completes 65% of her first serves. If she attempted 80 first serves last match, how many did she complete?

______

27. Jack is mixing a cleaning solution that is 12% bleach. After mixing the solution, Jack has 150 ounces of cleaning solution. How many ounces of bleach did Jack use?

______

28. What is 38% of 250?

______

29. 76% of 524 is what number?

______

30. What is 26% of 360?

______

31. 55% of 387 is what number?

______

32. CCSS **Use Math Tools** Mr. Blackwell tracks sales of ski equipment each week for a month. Complete the table to determine which week had the highest percent of ski equipment sales.

______

Week Number	Ski Equipment Sales ($)	Percent of Total Sales	Total Sales ($)
1		50	400
2	175		250
3		65	300
4	110		275

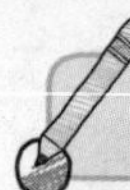

Standardized Test Practice

33. Which of the following does *not* represent how to find the amount taken off the price of the stereo?

Ⓐ 0.3×42 Ⓒ $42 \div 0.3$

Ⓑ $\frac{3}{10} \times 42$ Ⓓ 3×4.2

34. Carlos has read 45% of his book. If his book has 480 pages, how many pages has he read?

Ⓕ 45 pages Ⓗ 264 pages

Ⓖ 216 pages Ⓘ 435 pages

35. Short Response There are 450 cars in a car lot. About 28% of them are hybrid cars. About how many cars in a lot are hybrid cars? Explain your reasoning.

____

____

Common Core Review

Multiply. 5.NBT.7

36. $1.63 \times 20 =$ ______

37. $7.5 \times 12 =$ ______

38. $0.6 \times 15 =$ ______

39. $0.15 \times 50 =$ ______

40. $12 \times 1.2 =$ ______

41. $6 \times 0.8 =$ ______

42. Ella has 4 trading cards. Connor has 8 trading cards. How many times more cards does Connor have than Ella? 4.NBT.6 ______________________________

43. The art club had the members vote on three places to take a field trip. The results are in the table. If all of the members voted, what part of the club voted for the Carnegie Museum of Art? 5.NBT.7

Trip	Part of Club
Carnegie Museum of Art	
Fallingwater	0.20
Westemoreland Museum of American Art	0.48

Lesson 8

Solve Percent Problems

What You'll Learn

Scan the lesson. Predict two things you will learn about solving percent problems.

- ______________________________
- ______________________________

Essential Question

WHEN is it better to use a fraction, a decimal, or a percent?

Vocabulary

proportion
percent proportion

Common Core State Standards

Content Standards
6.RP.3, 6.RP.3c

Mathematical Practices
1, 2, 3, 4, 7

Vocabulary Start-Up

A **proportion** is an equation that shows that two ratios are equivalent. In a **percent proportion**, one ratio compares a part to the whole. The other ratio is the equivalent percent written as a fraction with a denominator of 100.

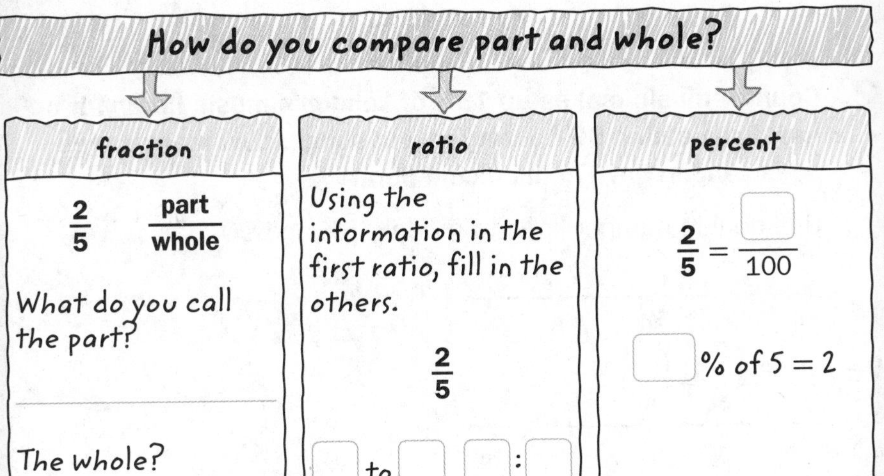

How do you compare part and whole?

fraction	ratio	percent
$\frac{2}{5}$ $\frac{\text{part}}{\text{whole}}$ What do you call the part? ______ The whole? ______	Using the information in the first ratio, fill in the others. $\frac{2}{5}$ ☐ to ☐ ☐ : ☐	$\frac{2}{5} = \frac{☐}{100}$ ☐ % of 5 = 2

Real-World Link

Basketball Kara is on her school basketball team. She has completed 9 out of 12 free throw shots successfully. Write the ratio as a percent and as a fraction in simplest form.

____

Work Zone

Use Number Lines to Find the Whole

If you know the part and the percent, you can find the whole, or the total. You have used bar diagrams to solve percent problems. Double number lines are another way to illustrate percents.

Examples

1. 10 is 25% of what number?

Use double number lines to model 25% and 10.

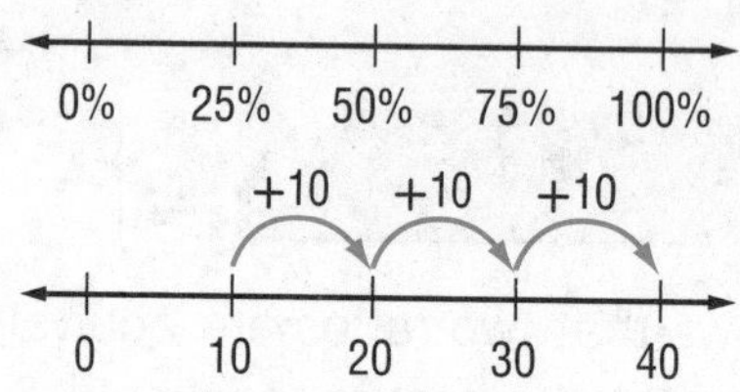

To model 25%, divide the number line into four parts.

Write 10 at the 25% mark. Add 10 at each mark to find the whole.

The number 40 is at the 100% mark.

So, 10 is 25% of 40.

2. Country music makes up 75% of Landon's music library. If he has downloaded 90 country music songs, how many songs does Landon have in his music library?

Use double number lines to model 75% and 90.

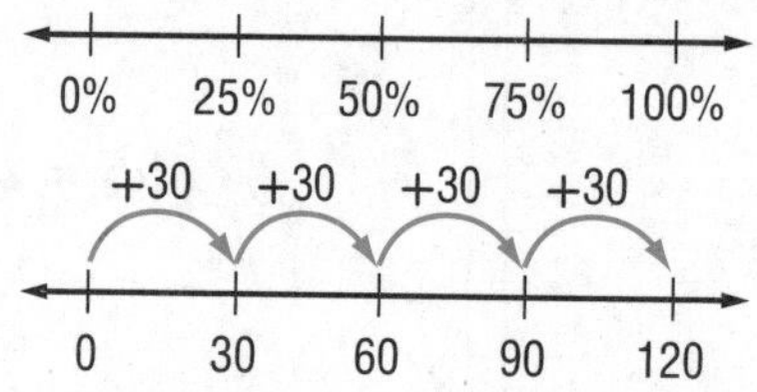

To model 75%, divide the number line into four parts.

$90 \div 3 = 30$. Add 30 at each mark to find the whole.

The number 120 is at the 100% mark.

So, Landon has 120 songs in his music library.

Check Look back at the number lines. The number 90 should line up with 75%. ✓

Show your work.

a. ___________

b. ___________

c. ___________

Got It? **Do these problems to find out.**

a. 30 is 50% of what number?

b. 60 is 20% of what number?

c. Peyton spent 60% of her money to buy a new television. If the television cost \$300, how much money did she have?

Use the Percent Proportion

The diagram uses a percent proportion to show that 75% of 32 is 24.

$$\begin{array}{l} \text{part} \longrightarrow \\ \text{whole} \longrightarrow \end{array} \frac{24}{32} = \frac{75}{100} \Big\} \text{ percent}$$

and Reflect

Write a percent proportion below to show that 50 is 25% of 200.

Examples

3. 15 is 30% of what number?

Words 15 is 30% of what number?

Proportion part → / whole → $\frac{15}{■} = \frac{30}{100}$ } percent

$\frac{15}{■} = \frac{30}{100}$ Write the proportion.

$\frac{15}{50} = \frac{30}{100}$ (÷2, ÷2) Since 15 is one half of 30, divide 100 by 2.

So, 15 is 30% of 50.

4. 225 is 75% of what number?

$\frac{\square}{■} = \frac{\square}{100}$ Write the proportion.

$\frac{\square}{\square} = \frac{\square}{100}$ Since 75 × ☐ = 225, multiply 100 by ☐.

So, 225 is 75% of ______.

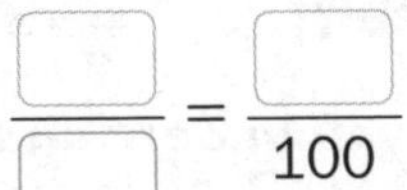

Do these problems to find out.

d. 75 is 15% of what number?

e. 9 is 36% of what number?

f. 7 is 70% of what number?

g. 7 is 35% of what number?

d. ______

e. ______

f. ______

g. ______

Example

5. Before 1982, pennies were 95% zinc and 5% copper. If 100 pennies minted in 1980 have an approximate mass of 15 grams of copper, what is the total mass of 100 pennies?

The percent is 5 and the part is 15. You need to find the whole.

$\frac{15}{\blacksquare} = \frac{5}{100}$ Write the proportion.

$\frac{15}{300} = \frac{5}{100}$ (×3) Since 5 × 3 = 15, multiply 100 by 3.

The total mass of 100 pennies is 300 grams.

Guided Practice

Use double number lines to find the whole. (Example 1)

1. 40 is 20% of what number? ______

Show your work.

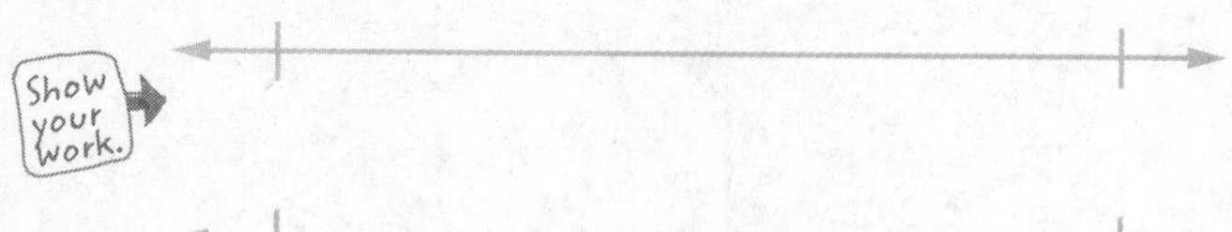

2. 90 is 25% of what number? ______

Write a percent proportion and solve each problem. (Examples 3 and 4)

3. 120 is 30% of what number?

4. 60 is 15% of what number?

5. In the first year of ownership, a new car can lose 20% of its value. If a car lost $4,200 of value in the first year, how much did the car originally cost? (Examples 2 and 5)

6. **Building on the Essential Question** How can you use proportions to solve percent problems?

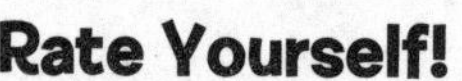

Rate Yourself!

How well do you understand percent problems? Circle the image that applies.

Clear

Somewhat Clear

Not So Clear

For more help, go online to access a Personal Tutor.

FOLDABLES *Time to update your Foldable!*

Name ______ My Homework ______

Independent Practice

Use double number lines to find the missing number. (Example 1)

1. 63 is 90% of what number? ______

2. 80 is 25% of what number? ______

Write a percent proportion and solve each problem. (Examples 3 and 4)

3. 22 is 44% of what number?

4. 450 is 75% of what number?

5. A store is having a sale where winter clothes are 60% of the original price. A sweater is on sale for $30. What was the original price of the sweater? (Examples 2 and 5)

6. Kai calculates that he spends 15% of a school day in science class. If he spends 75 minutes in science class, how many minutes does Kai spend in school? (Examples 2 and 5)

For Exercises 7–9, use the table.

7. If you have 3 cups of pineapple juice, how many total cups of punch can you make? ______

8. How many cups of sorbet are used in 8 cups of punch?

9. Elise does not like sorbet, so she omits that ingredient and adds 5 percent of each of the other ingredients. How many cups of punch will she have if she uses 6 cups of orange juice? ______

Punch Recipe	
Ginger Ale	40%
Orange Juice	25%
Pineapple Juice	20%
Sorbet	15%

10. **Identify Structure** Complete the following graphic organizers. Identify the missing information.

a.

$\frac{3}{4}$	part	3
	whole	

b.

47%	part	
	whole	100%

c.

12% of 225	part	12%
	whole	

d.

120 out of 400	part	
	whole	

e. How does identifying the part and the whole help you to write the percent proportion? ______________________________

____

H.O.T. Problems Higher Order Thinking

11. **Reason Abstractly** Write a percent proportion where the percent and the whole are known. Solve the problem to find the part. ______________________________

____

12. **Persevere with Problems** Using what you know about percents, explain why a commercial that says "80% of dentists use this toothpaste" might be misleading. ______________________________

____

____

____

13. **Reason Inductively** The purity of gold is listed in karats. Refer to the table. If a necklace is 75% gold, what karat is it? Explain your reasoning. ______________________________

____

Karats	Pure Gold (%)
24	100
12	50

14. **Construct an Argument** Omar scored an 82% on his first test of the quarter. Will a score of 38 out of 50 on the second test help or hurt his grade? Explain your reasoning. ______________________________

____

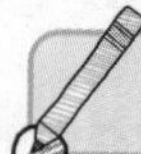

Standardized Test Practice

15. At a zoo, an Asian elephant eats roughly 5% of its body weight each day. If an Asian elephant eats 300 pounds of food a day, how much does it weigh?

Ⓐ 1,500 lb Ⓒ 18,000 lb

Ⓑ 6,000 lb Ⓓ 120,000 lb

Extra Practice

Use double number lines to find the missing number.

16. 140 is 70% of what number? 200

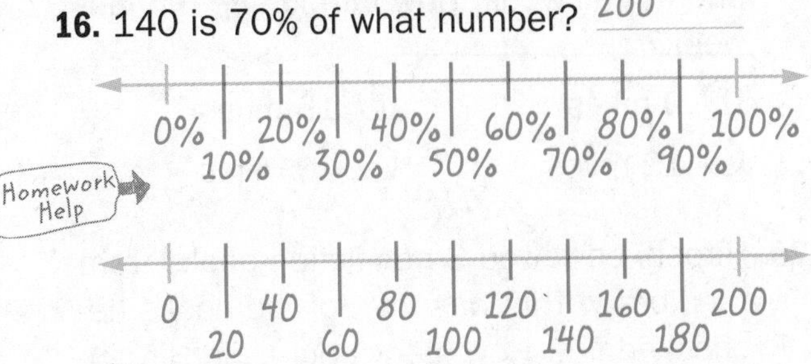

17. 240 is 60% of what number? ______

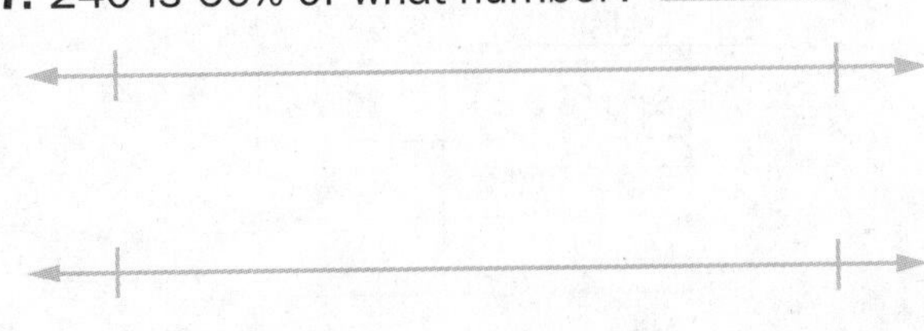

Write a percent proportion and solve each problem.

18. 95 is 95% of what number?

19. 270 is 90% of what number?

20. CCSS **Justify Conclusions** Action movies make up 85% of Devon's movie collection. If he has 20 movies, how many action movies are in Devon's collection? Explain your reasoning to a classmate.

21. The drama club held a car wash on Saturday and Sunday. They washed a total of 60 cars. If they washed 40% of the cars on Sunday, how many cars did they wash on Sunday?

22. A tiger can eat food that weighs up to 15% of its body weight. If a tiger can eat 75 pounds of food, how much does a tiger weigh?

23. According to the school survey, 12% of the students at Rockwood Junior High School speak Spanish. There are 36 students at the school who speak Spanish. How many students were surveyed?

24. Miley's Music has a sale on music CDs. All music CDs are discounted 15%. Mariana's receipt indicates that she saved $3 on her CD purchase. What is the full price of her music CD before the discount?

25. The interior paint color, Melon Madness, is 30% yellow. Raul used 72 ounces of yellow paint to mix the last batch. How many ounces of Melon Madness did he make in the last batch?

Standardized Test Practice

26. Refer to survey results shown below.

Favorite Subject	
English	23%
Science	30%
Social Studies	15%
Math	■
Music	12%

If 150 students were surveyed, how many students chose math as their favorite subject?

Ⓐ 7.5　　Ⓒ 30

Ⓑ 20　　Ⓓ 120

27. Isabel ran 9 miles with her friend Kaylee last week. This was 60% of the distance she ran all week. How far did she run last week?

Ⓕ 9 miles　　Ⓗ 15 miles

Ⓖ 12 miles　　Ⓘ 20 miles

28. Hiroshi borrowed 3 non-fiction books from the library. This was 25% of the books he checked out. How many books did Hiroshi check out?

Ⓐ 3 books　　Ⓒ 9 books

Ⓑ 6 books　　Ⓓ 12 books

29. Short Response The student council had a canned food drive. They collected a total of 63 cans on Wednesday. This is 21% of the total. How many canned goods were collected in all?

Day	Percent Collected
Wednesday	21
Thursday	46
Friday	33

____

CCSS Common Core Review

Find the equivalent fraction. 4.NF.1

30. $\frac{84}{120} = \frac{\square}{10}$

31. $\frac{60}{98} = \frac{30}{\square}$

32. $\frac{40}{64} = \frac{5}{\square}$

33. $\frac{32}{41} = \frac{96}{\square}$

34. $\frac{13}{15} = \frac{\square}{60}$

35. $\frac{24}{32} = \frac{12}{\square}$

36. A store has a sale for $\frac{3}{10}$ off gloves. Write $\frac{3}{10}$ as a decimal. 4.NF.5

____

37. Xavier runs 0.75 mile each day. How far has he run at the end of 6 days? 5.NBT.7

____

21ST CENTURY CAREER in Movies

Special Effects Animator

Are you fascinated by how realistic the special effects in movies are today? If you have creative talent and are good with computers, a career in special effects animation might be a great fit for you. Special effects animators use their artistic ability and expertise in computer-generated imagery (CGI) to simulate real-life objects like water and fire. They are also able to create fantastic images like flying superheroes, exploding asteroids, and monsters taking over cities.

Explore college and careers at ccr.mcgraw-hill.com

Is This the Career for You?

Are you interested in a career as a special effects animator? Take some of the following courses in high school.

- Digital Animation
- Calculus
- Geometry
- Physics
- Art/Sculpture

Turn the page to find out how math relates to a career in Movies.

The Effects are Amazing!

Special effects animators must specify when objects fade or change color. Table 1 shows when an object starts fading out. Table 2 shows the percent of an object's total lifetime that it has the initial color, cross-fading of colors, and the final color. Use the tables to solve each problem.

1. Express the part of total lifetime for each object in Table 1 as a fraction in simplest form. ________

2. At what percent of the light beam's total lifetime does it begin to fade out? ________

3. In Table 2, express the percents for the cross-fading of both objects as decimals.

____

4. Which best describes the part of the robot's lifetime in which it has the initial color: $\frac{3}{100}$, $\frac{3}{10}$, or $1\frac{3}{10}$? ________

5. What fraction of the tornado's lifetime does it have the initial color? ________

6. What fraction of the robot's lifetime does it have the final color? ________

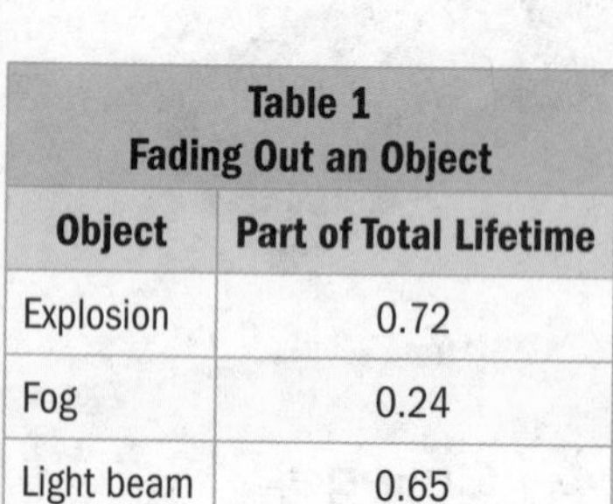

Table 1 Fading Out an Object	
Object	**Part of Total Lifetime**
Explosion	0.72
Fog	0.24
Light beam	0.65

Table 2 Changing Color of an Object			
Object	**Percent of Total Lifetime**		
	Initial Color	**Cross-Fading**	**Final Color**
Robot	30%	15%	55%
Tornado	12%	77%	11%

Career Project

It's time to update your career profile! Choose one of your favorite movies. Use the Internet to research how the movie's special effects were created. Write a brief description of the processes used by the special effects animators.

____

List several jobs that are created by the movie industry.

-
-
-
-
-

Chapter Review Check

Vocabulary Check Vocab

Unscramble each of the clue words. After unscrambling all of the terms, use the numbered letters to find the phrase.

INROALAT NUEMBR

☐☐☐☐☐☐☐☐ ☐☐☐☐☐☐

10 6 9 12 1

TEEPNCR

☐☐☐☐☐☐☐

4

LESTA MOCNOM EOMITNORNAD

☐☐☐☐☐ ☐☐☐☐☐☐ ☐☐☐☐☐☐☐☐☐☐☐

8 2

PIONORTOPR

☐☐☐☐☐☐☐☐☐☐

5 7

CTREENP ROIPPTORNO

☐☐☐☐☐☐☐ ☐☐☐☐☐☐☐☐☐☐

11 3 13

F☐☐☐☐☐☐☐☐ ☐☐☐ F☐☐

1 2 3 4 5 6 7 8 9 10 11 12 13

Complete each sentence using one of the unscrambled words above.

1. A ________________ is a ratio that compares a number to 100.

2. A ________________ is an equation that shows that two ratios are equivalent.

3. In a ________________, one ratio compares a part to a whole.

4. A number that can be written as a fraction is a ________________.

5. The ________________ is the least common multiple of the denominators of two or more fractions.

Key Concept Check

Use Your FOLDABLES®

Use your Foldable to help review the chapter.

Tape here

Fractions, Decimals, and Percents

Examples

Examples

Examples

Got it?

The problems below may or may not contain an error. If the problem is correct, write a "✓" by the answer. If the problem is not correct, write an "X" over the answer and correct the problem.

1. $\frac{4}{5} = 0.4$ (crossed out)

The first one is done for you.

$\frac{4}{5} \rightarrow 5\overline{)4.0}$ = 0.8; -40; 0

2. $0.55 = \frac{11}{20}$

3. $120\% = \frac{3}{25}$

Problem Solving

1. CCSS **Use Math Tools** The table shows how Eva budgets her weekly allowance. Write each decimal as a fraction in simplest form. (Lesson 1)

Category	Portion
Savings	0.4
Charity	0.15
Shopping	0.45

2. Jonah's savings increased by 0.15% in one month. Write 0.15% as a decimal and fraction in simplest form. (Lesson 4)

3. An executive of a marshmallow company said that the marshmallows are 80% air. What fraction of a marshmallow is air? (Lesson 2)

4. In a parking lot, 30% of the cars have a GPS. Write this as a decimal. (Lesson 3)

5. Miguel spent $\frac{1}{8}$ of his savings on a new video game system. Mila spent 12% of her savings on a DVD, and Manny spent 0.1 of his savings on a skateboard. Who spent the most of their savings? (Lesson 5)

6. CCSS **Be Precise** For each state in the table, find the number of square miles of each state that are covered by forests. (Lesson 7)

State	Percent of Land Covered by Forests	Area of State (square miles)
California	14	158,609
Kentucky	49	40,409
Ohio	27	44,825

7. In one year, Orlando received 62.51 inches of rain. In September, the city received 25% of that rainfall. About how much rain did Orlando receive in September? (Lesson 6)

8. The original price of a movie is $18. The sale price is 20% off the original price. What is the amount off the original price? (Lesson 8)

Reflect

Answering the Essential Question

Use what you learned about fractions, decimals, and percents to complete the graphic organizer.

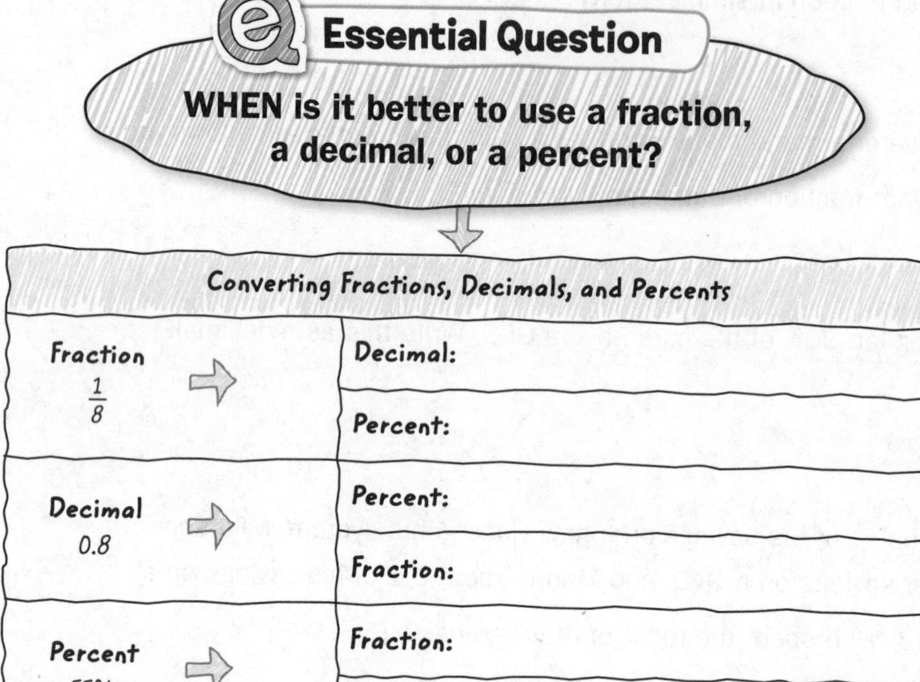

Converting Fractions, Decimals, and Percents	
Fraction $\frac{1}{8}$ →	Decimal:
	Percent:
Decimal 0.8 →	Percent:
	Fraction:
Percent 55% →	Fraction:
	Decimal:

Answer the Essential Question. WHEN is it better to use a fraction, a decimal, or a percent?

UNIT PROJECT

People Everywhere Federal, state, and local governments use population data that are gathered from the U.S. Census to help them plan what services communities need. In this project you will:

- **Collaborate** with your classmates to collect data and to compare populations within our state, county, or city.
- **Share** the results of your research in a creative way.
- **Reflect** on how you use mathematics to describe change and model real-world situations.

By the end of this project, you will have a better understanding of the population of people that live around you!

Collaborate

Go Online **Work with your group to research and complete each activity. You will use your results in the Share section on the following page.**

1. Use the U.S. Census Web site to find the total population for your state. Then compare the following:
   - the ratio of males to females
   - the ratio of people in your age group to those in any other age group

2. Research the population density of your state. Compare the results with all of the surrounding states. Plot your results on a number line. How does your state compare to others?

3. Research the growth rate of the population of your state over the past ten U.S. Census reports. Display your results in a table and a line graph. Describe any patterns in the line graph.

4. Explain in a journal entry or blog how you could use population data to predict the population of your state in 2020.

5. One way federal funds are distributed to states and counties is based on their population. Research the population of your county, as well as the surrounding counties. If $1,000,000 is to be divided among the counties, about what fraction of the money do you think your county would receive? Why?

Share

With your group, decide on a way to present what you have learned from each of the activities about the population of your state according to the U.S. Census. Some suggestions are listed below, but you can also think of other creative ways to present your information. Remember to show how you used math to complete each of the activities in this project!

- Create a presentation using the data you collected. Your presentation should include a spreadsheet, graph, and one other visual display.
- Write a persuasive letter to a local government official. In the letter, explain what you have learned in this project. Then lobby to have a new service provided for your community. This could be a new park, school, hospital, or whatever you think your community needs.

Check out the note on the right to connect this project with other subjects.

connect with Science

Environmental Literacy Select two states, other than your own, and research the types of landforms found in these states. Some questions to consider are:

- What different types of landforms are found in these states?
- How do the landforms found in these states compare with those found in your state?

Reflect

6. **Answer the Essential Question** How can you use mathematics to describe change and model real-world situations?

   a. How did you use what you learned about ratios and rates to describe change and model the real-world situations in this project?

   b. How did you use what you learned about fractions, decimals, and percents to describe change and model the real-world situations in this project?

UNIT 2

CCSS The Number System

Essential Question

HOW can mathematical ideas be represented?

Chapter 3

Compute with Multi-Digit Numbers

The standard algorithm used to multiply and divide whole numbers can be applied to operations with decimals. In this chapter, you will multiply and divide multi-digit decimals.

Chapter 4

Multiply and Divide Fractions

Models and equations can be used to represent real-world situations involving operations with fractions. In this chapter, you will multiply and divide fractions by whole numbers and by fractions.

Chapter 5

Integers and the Coordinate Plane

Integers, terminating decimals, and repeating decimals are rational numbers. In this chapter, you will compare and order rational numbers and graph points in four quadrants of the coordinate plane.

Unit Project Preview

Get Out the Map! Maps have been around for thousands of years. Over time maps have become more detailed and accurate. Today, using a map is as simple as turning on a GPS device and typing in an address.

Let's make a map of your neighborhood. Choose three or four locations in your neighborhood that you visit, such as a friend's house or the library. On the coordinate plane below, plot points to represent these locations in relation to your home.

At the end of Chapter 5, you'll complete a project that will teach you about using a map to travel somewhere new. So, bring your sense of adventure and get ready to explore someplace new!

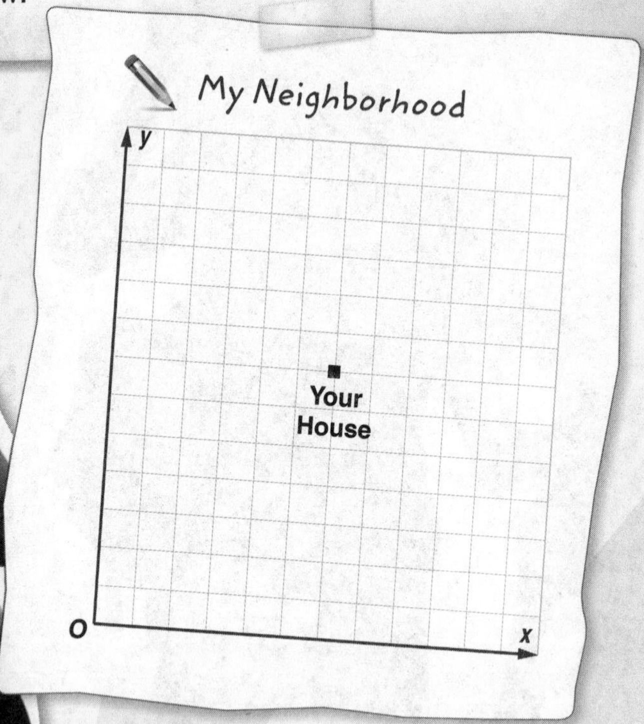

My Neighborhood

y

Your House

O

x

Chapter 3
Compute with Multi-Digit Numbers

Essential Question

HOW can estimating be helpful?

Common Core State Standards

Content Standards
6.NS.2, 6.NS.3

Mathematical Practices
1, 2, 3, 4, 5, 6

Math in the Real World

Skyscrapers A certain skyscraper in Chicago has 1,200,000 square feet of space. On average, there are 29,268 square feet of space on each floor. Estimate to find the number of floors in the building.

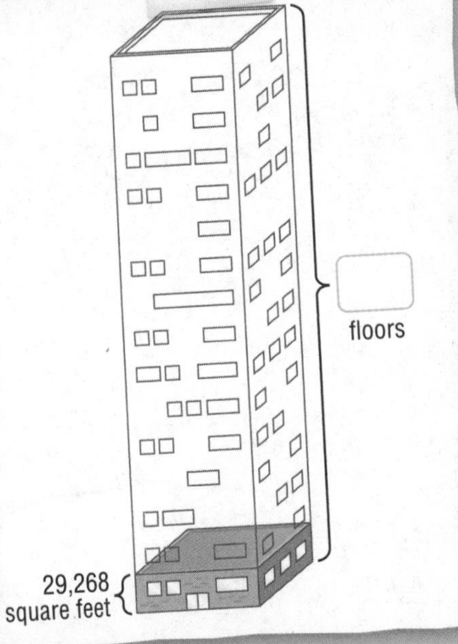

FOLDABLES® Study Organizer

1 Cut out the Foldable on page FL7 of this book.

2 Place your Foldable on page 250.

3 Use the Foldable throughout this chapter to help you learn about computing with multi-digit numbers.

What Tools Do You Need?

Vocabulary

compatible numbers

Review Vocabulary

Graphic Organizer One way to remember vocabulary terms is to connect them to an everyday meaning or an example. Use this information to complete the graphic organizer.

Everyday meaning:

Math meaning:

Product

Example:

Non-example:

When Will You Use This?

Watch Play it online!

Your Turn! **You will solve this problem in the chapter.**

Are You Ready?

Try the Quick Check below.
Or, take the Online Readiness Quiz.

Quick Review

Common Core Review 4.NBT.5, 5.NBT.6

Example 1

Find 13 × 15.

```
   13
 × 15
   65    Multiply the ones.
+ 130    Multiply the tens.
  195    Add.
```

Example 2

Find 323 ÷ 17.

```
     19
17)323    Divide the tens.
  −17
   153    Divide the ones.
 − 153
     0
```

Quick Check

Multiply **Find each product.**

1. 15 × 20 = ______

2. 19 × 51 = ______

3. 49 × 22 = ______

Divide **Find each quotient.**

4. 112 ÷ 8 = ______

5. 539 ÷ 11 = ______

6. 779 ÷ 19 = ______

7. A musician sold 64 million albums in 16 months. She sold the same amount in each month. How many albums did she sell in each month?

______

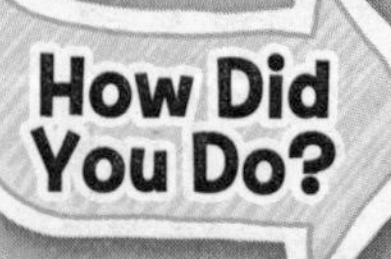

How Did You Do?

Which problems did you answer correctly in the Quick Check? Shade those exercise numbers below.

① ② ③ ④ ⑤ ⑥ ⑦

Lesson 1

Add and Subtract Decimals

What You'll Learn

Scan the lesson. Predict two things you will learn about adding and subtracting decimals.

- ______________________________
- ______________________________

Essential Question

HOW can estimating be helpful?

Common Core State Standards

Content Standards
6.NS.3

Mathematical Practices
1, 2, 3, 4, 5, 6

Real-World Link

Swimming One event in competitive swimming is the women's 100-meter butterfly. The table shows the times of different swimmers at a recent Olympics.

Women's 100-Meter Butterfly

Swimmer's	Time (s)
Lisbeth Trickett	56.73
Christine Magnuson	?
Gabriella Silva	58.10

You can use place value charts to compare the results.

1. It took Christine Magnuson 0.37 second longer to finish than Lisbeth Trickett. What was Magnuson's time, in seconds?

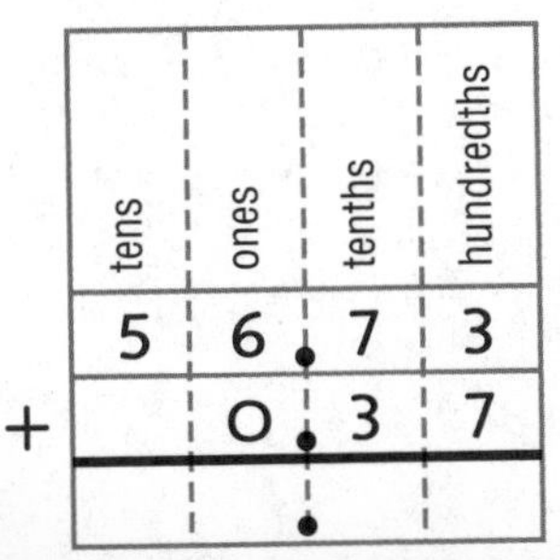

	tens	ones	tenths	hundredths
	5	6.	7	3
+		0.	3	7
		.		

2. At a high school meet, a swimmer swam the women's 100-meter butterfly in 72.34 seconds. How many seconds faster did Gabriella Silva swim her race?

	tens	ones	tenths	hundredths
	7	2.	3	4
−	5	8.	1	0
		.		

Work Zone

Add Decimals

To add decimals, line up the decimal points. Then, add digits in the same place-value position.

Examples

Tutor

1. Find the sum of 23.1 and 5.8.

Estimate $23.1 + 5.8 \approx 23 + 6$ or 29

$$\begin{array}{r} 23.1 \\ +\ 5.8 \\ \hline 28.9 \end{array}$$

Line up the decimal points.

Add as with whole numbers.

Check for Reasonableness $28.9 \approx 29$ ✓

The sum of 23.1 and 5.8 is 28.9.

Accuracy
To find an accurate sum or difference, line up the digits based on their place values. Use the decimal point to help align the correct digits before completing the operation.

2. Find the sum of 29.6 and 14.7.

Estimate $29.6 + 14.7 \approx$ ☐ + ☐ or ☐

Line up the decimal points.

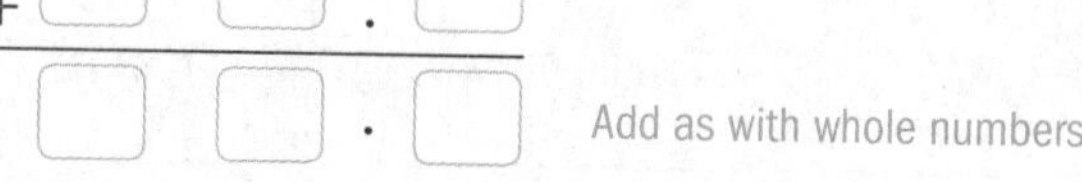

Add as with whole numbers.

Check for Reasonableness ☐ ≈ ☐

The sum of 29.6 and 14.7 is ☐.

Got It? Do these problems to find out.

Find each sum.

a. $54.7 + 21.4$ **b.** $14.2 + 23.5$ **c.** $17.3 + 33.5$

a. ______

b. ______

c. ______

Subtract Decimals

To subtract decimals, line up the decimal points. Then, subtract digits in the same place-value position. You may need to *annex*, or place zeros at the end of a decimal, in order to subtract.

Examples

3. Find the difference of 5.774 and 2.371.

Estimate $5.774 - 2.371 \approx 6 - 2$ or 4

$$\begin{array}{r} 5.774 \\ -\ 2.371 \\ \hline 3.403 \end{array}$$

Line up the decimal points.

Subtract as with whole numbers.

Check for Reasonableness $3.403 \approx 4$ ✓

So, $5.774 - 2.371 = 3.403$.

4. Find 6 − 4.78.

Estimate $6 - 4.78 \approx 6 - 5$ or 1

$$\begin{array}{r} 6.00 \\ -\ 4.78 \\ \hline 1.22 \end{array}$$

Annex zeros so that both numbers have the same number of decimal places.

Check for Reasonableness $1.22 \approx 1$ ✓

So, $6 - 4.78 = 1.22$.

5. Find 23 − 4.216.

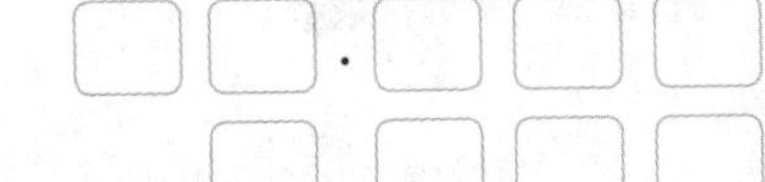

Annex the zeros so that both numbers have the same number of decimal places.

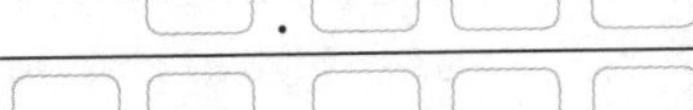

Subtract as with whole numbers.

So, 23 − 4.216 = ______.

d. ______

e. ______

f. ______

Got It? Do these problems to find out.

Find each difference.

d. 9.543 − 3.671 **e.** \$50.62 − \$39.81 **f.** 14 − 9.09

Example

6. Reagan is creating a video. The first video clip was 22.36 minutes long. The second video clip was 17.03 minutes long. What is the total length of the video?

Estimate $22.36 + 17.03 \approx 22 + 17$ or 39

$$\begin{array}{r} 22.36 \\ +17.03 \\ \hline 39.39 \end{array}$$

Line up the decimal points.

Add as with whole numbers.

Check for reasonableness $39.39 \approx 39$ ✓

So, the video is 39.39 minutes long.

Got It? **Do this problem to find out.**

g. ________

g. Jonathan is traveling for work. This morning his GPS indicated that the total distance to his destination is 589.4 miles. Before lunch he drove 208.62 miles. How much farther does Jonathan need to travel?

Guided Practice

Find each sum or difference. (Examples 1–5)

Show your work.

1. $14.7 + 87.9 =$ ________

2. $66.5 - 24.1 =$ ________

3. $52.1 - 31.47 =$ ________

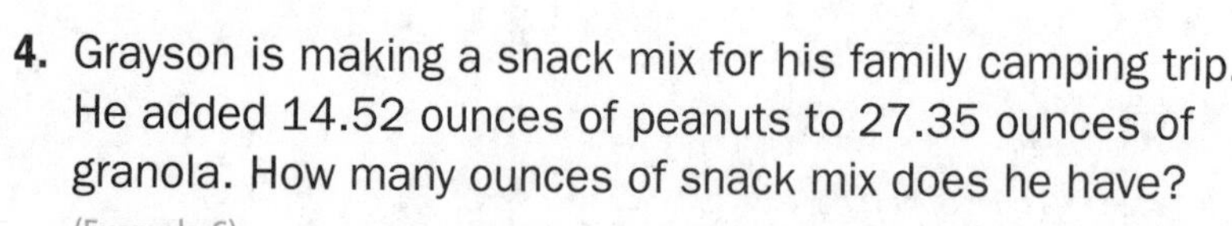

4. Grayson is making a snack mix for his family camping trip. He added 14.52 ounces of peanuts to 27.35 ounces of granola. How many ounces of snack mix does he have? (Example 6)

5. **Building on the Essential Question** How is estimation helpful when adding and subtracting decimals?

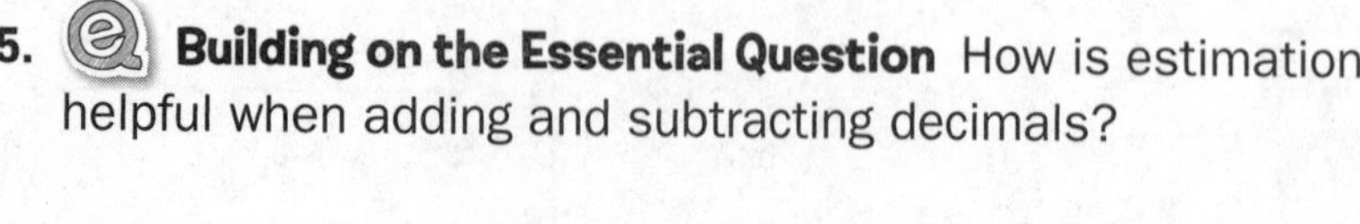

Rate Yourself!

☐ I understand how to add and subtract decimals.

Great! You're ready to move on!

☐ I still have questions about adding and subtracting decimals.

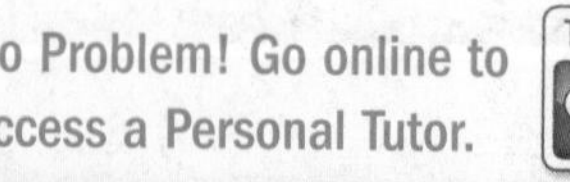

No Problem! Go online to access a Personal Tutor.

Name ______ My Homework ______

Independent Practice

Go online for Step-by-Step Solutions

Find each sum. (Examples 1 and 2)

1. 7.2 + 9.5 = ______

2. 1.34 + 2 = ______

3. 54.5 + 48.51 = ______

Show your work.

Find each difference. (Examples 3–5)

4. 5.6 − 3.5 = ______

5. 97 − 16.98 = ______

6. 58.67 − 28.72 = ______

7. The table shows the top three finishers in barrel racing. What is the time difference between Nicolas and Sancho? (Example 6)

______

Barrel Racing Results	
Rider	**Time (s)**
Nicolas	14.67
Becki	15.98
Sancho	16.40

8. In two months, Mica spent a total of $305.43 on groceries. She spent $213.20 in the first month. How much did she spend in the second month? (Example 6)

______

9. **Financial Literacy** A hat costs $10.95 and a T-shirt costs $14.20. How much change will you receive if you pay for both items with a $50 bill?

______

10. CCSS **Use Math Tools** The 4-by-100 meter relay is a track and field event involving four runners on each team.

a. What is the combined time of Carter and Frater?

______

b. How much faster did Powell run his leg of the race than Bolt?

______

c. What is the combined time of all the runners?

______

4 by 100 Meter Times	
Runner	**Time (s)**
Carter	13.4
Frater	12
Bolt	12
Powell	11.7

H.O.T. Problems Higher Order Thinking

11. **Find the Error** Luis is finding 8.9 − 3.72. Find his mistake and correct it.

12. **Reason Abstractly** Write two different pairs of decimals whose sums are 14.1. One pair should involve regrouping.

13. **Reason Inductively** Explain how you know that the sum of 12.6, 3.1, and 5.4 is greater than 20.

14. **Persevere with Problems** Josie found that 3.28 + 3.28 + 3.28 = 9.84. What is the missing factor in the related multiplication problem 3.28 × ☐ = 9.84? Explain.

15. **Reason Abstractly** Without subtracting 8.5 − 4.64, determine what digit will be in the hundredths place. Explain.

Standardized Test Practice

16. Keira is buying items for her kitchen. The store sells a large mixing bowl for $12.95, a spatula for $8.37, and measuring cups for $9.99. What is the total cost of these items?

Ⓐ $21.32

Ⓑ $29.12

Ⓒ $31.31

Ⓓ $41.31

Name ______________________ My Homework ______________________

Extra Practice

Find each sum.

17. 4.9 + 3.0 = 7.9

Homework Help

$$\begin{array}{r} 4.9 \\ +3.0 \\ \hline 7.9 \end{array}$$

18. 0.796 + 13 = ______

19. 15.63 + 24.36 = ______

Find each difference.

20. 19.86 − 4.94 = ______

21. 82 − 67.18 = ______

22. 14.39 − 12.16 = ______

23. **Financial Literacy** The current balance of Tami's checking account is $237.80. Find the new balance after Tami writes a check for $29.95.

______

24. The annual rainfall for Kayston Falls was 50.38 inches in 2012. In 2013, the annual rainfall was 55.76 inches. What is the difference in rainfall between the two years?

______

25. STEM The melting point of sodium is 97.8 degrees Celsius. The melting point of potassium is 63.65 degrees Celsius. How much higher is the melting point of sodium?

______

______

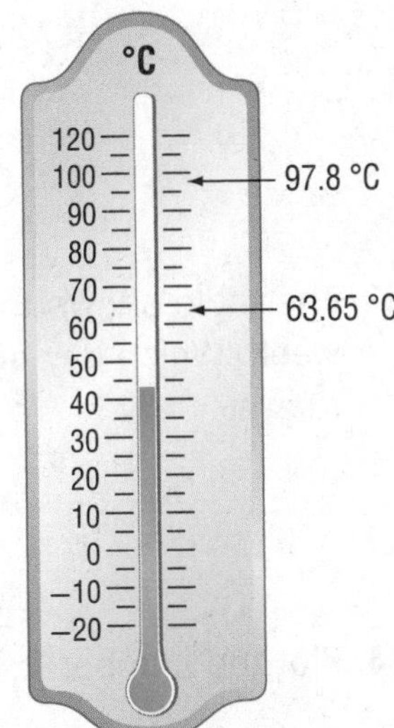

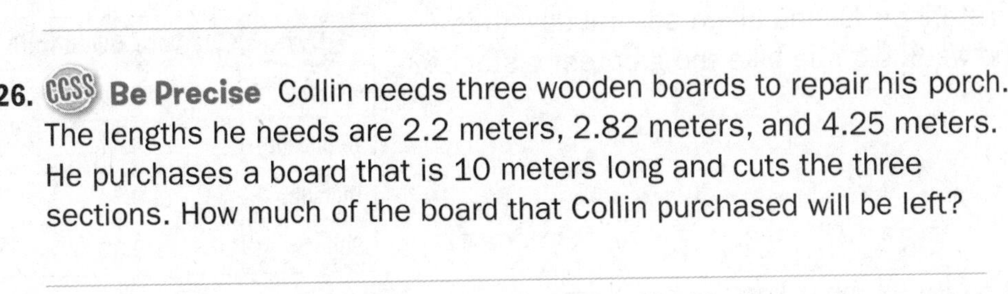

26. CCSS **Be Precise** Collin needs three wooden boards to repair his porch. The lengths he needs are 2.2 meters, 2.82 meters, and 4.25 meters. He purchases a board that is 10 meters long and cuts the three sections. How much of the board that Collin purchased will be left?

______

______

Standardized Test Practice

27. The table shows the top three finishers for a swimming event. What is the time difference between Kendrick and Andrew?

Boy's 50 Yard Freestyle	
Swimmer	**Time (s)**
Andrew	22.63
Kendrick	22.20
Ty	22.58

Ⓐ 0.38 s
Ⓑ 0.43 s
Ⓒ 44.83 s
Ⓓ 45.08 s

28. Amaya has a store credit of $50.86. She plans to purchase a video game for $24.97 and a golf club accessory for $6.99. How much store credit will she have left?

Ⓕ $43.87
Ⓖ $31.96
Ⓗ $25.89
Ⓘ $18.90

29. Short Response Dominic is traveling through the state of Oregon. He drove 66.4 miles from Salem to Eugene and then continued to Medford. The total distance of his trip was 233.25 miles. How far is the distance from Eugene to Medford?

____

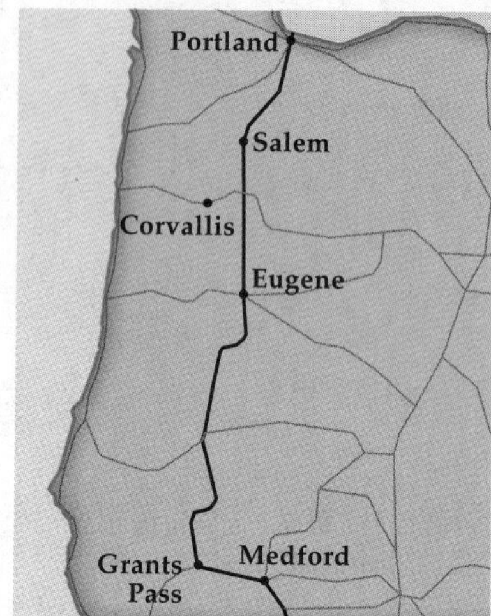

Common Core Review

Round each decimal to the nearest whole number. 5.NBT.4

30. 4.75 ≈ ______

31. 34.1 ≈ ______

32. 22.48 ≈ ______

33. The table shows the distances Juliana biked several days this week. Which day of the week did she bike the greatest distance? 5.NBT.3b

____

____

Day	Distance Biked (mi)
Monday	9.34
Wednesday	9.47
Thursday	9.74
Sunday	9.32

34. Plot the number 2.78 on the number line below. 4.NF.7

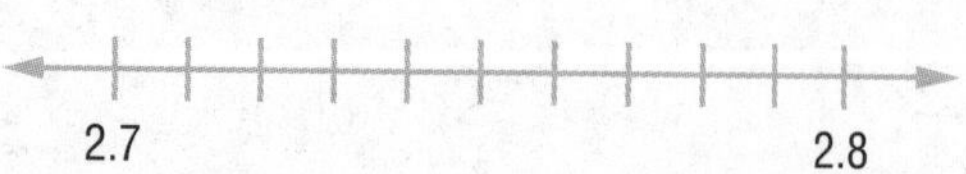

Lesson 2

Estimate Products

What You'll Learn

Scan the lesson. List two real-world scenarios in which you would estimate products.

- ______________________________
- ______________________________

Essential Question

HOW can estimating be helpful?

Common Core State Standards

Content Standards
Preparation for 6.NS.3

Mathematical Practices
1, 2, 3, 4, 5

Real-World Link

Skateboarding The record for the greatest distance traveled on skateboard in 24 hours was set in a recent year by James Peters. He traveled about 7.6 miles per hour.

1. Plot 7.6 on the number line.

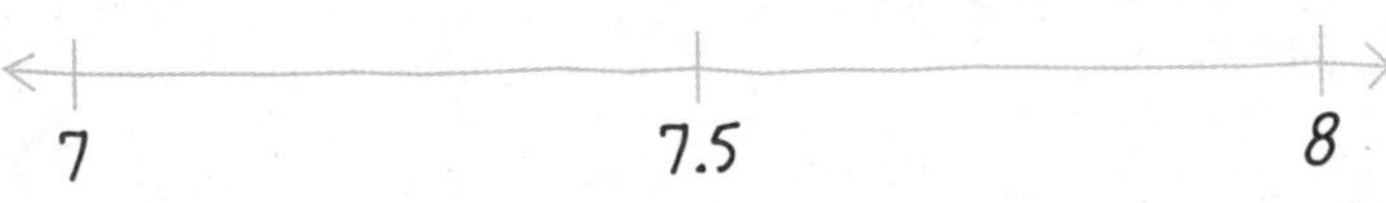

2. What whole number is 7.6 closest to? ☐

3. Estimate how many miles James Peters traveled in 24 hours.

   ☐ × ☐ = ______

4. Is your estimate higher or lower than the actual distance he traveled? Explain.

   ____

5. A new record was set later by Ted McDonald. He traveled about 10.1 miles per hour. About how much farther did Ted McDonald travel?

   ☐ × 24 = ______

   ☐ − ☐ = ______

   So, Ted McDonald traveled about ______ miles farther.

Work Zone

Estimate Products Using Rounding

To estimate products of decimals, round each number. First underline the digit to be rounded. Then look at the digit to the right of the place being rounded.

- If the digit is 4 or less, the underlined digit remains the same.
- If the digit is 5 or greater, add 1 to the underlined digit.
- After rounding, all place values to the right of the underlined digit have a value of zero.

After the numbers are rounded, multiply.

Rounding Decimals

When rounding decimals, such as 99.96 to the tenths, the 9 must round up. So, 99.96 rounded to the nearest tenth is 100.0.

Examples

1. Estimate 8.7 × 2.8.

Round to the nearest whole number to make it easier to compute mentally.

$$\begin{array}{r} 8.7 \\ \times\ 2.8 \\ \hline \end{array} \rightarrow \begin{array}{r} 9 \\ \times\ 3 \\ \hline 27 \end{array}$$

Round 8.7 to 9.
Round 2.8 to 3.

The product is about 27.

2. Estimate 42.6 × 37.2.

Round to the greatest place value to make it easier to compute mentally.

42.6 ≈ ____

37.2 ≈ ____

Multiply.

The product is about ____.

Got It? Do these problems to find out.

Estimate each product.

a. 9.6 × 1.8   b. 4.2 × 3.1   c. 68.4 × 21.3

a. ____

b. ____

c. ____

Examples

3. A greyhound can travel 39.3 miles per hour. At this speed, about how far could a greyhound travel in 6.5 hours?

$$\begin{array}{r} 39.3 \\ \times\ 6.5 \\ \hline \end{array} \rightarrow \begin{array}{r} 40 \\ \times\ 7 \\ \hline 280 \end{array}$$

Round 39.3 to 40.
Round 6.5 to 7.

The greyhound could travel about 280 miles in 6.5 hours.

4. Suppose one U.S. dollar is equal to 5.8 Egyptian pounds. About how many Egyptian pounds would you receive for $48.50?

Round to the greatest place value to make it easier to compute mentally.

5.8 ≈ ☐

48.50 ≈ ☐

Multiply.

×

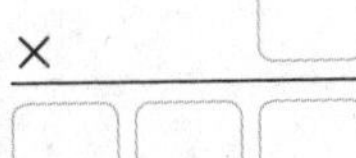

So, $48.50 is equal to about ☐ Egyptian pounds.

STOP and Reflect

You could estimate 18.76 × 5 by multiplying 19 × 5 or by multiplying 20 × 5. Which answer will be more exact? Explain below.

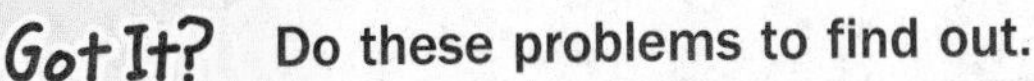

Got It? Do these problems to find out.

Show your work.

d. **STEM** Earth is rotating around the Sun about 18.6 miles per second. About how many miles does it travel in 4.8 seconds?

e. The average walking speed of a person is 4.8 kilometers per hour. Estimate the number of kilometers could you walk in 3 hours?

f. **STEM** A King Palm can grow about 2.1 feet a year. Estimate the height of the King Palm, in yards, after 15 years.

d. ______

e. ______

f. ______

Example

5. Patrice has $20 to buy 5 binders. Binders cost $4.29 each. Does she have enough money? Explain your reasoning.

Estimate.

$5 \times \$4 = \$20$ Estimate 4.29 as 4.

$5 \times \$5 = \$25$ Estimate 4.29 as 5.

The actual cost is between $20 and $25. So, Patrice does not have enough money to buy the binders.

Guided Practice

Estimate each product. (Examples 1 and 2)

1. $5.8 \times 4 \approx$ ______

2. $13.92 \times 2.7 \approx$ ______

3. $94.89 \times 3.11 \approx$ ______

4. **Financial Literacy** A grocery store sells American cheese for $3.89 per pound. About how much would 1.89 pounds of the cheese cost? (Examples 3 and 4)

______

5. Greg has 52 megabytes left on his MP3 player. He wants to download 7 songs that each use 7.9 megabytes of memory. He estimates that he will need 56 megabytes of memory. Is his estimate is reasonable? Explain your reasoning. (Example 5)

______

______

6. **Building on the Essential Question** How do you determine which place value to use when estimating products?

______

______

______

______

Rate Yourself!

How confident are you about estimating products? Check the box that applies.

For more help, go online to access a Personal Tutor.

Independent Practice

Go online for Step-by-Step Solutions

Estimate each product. (Examples 1 and 2)

1. $9.7 \times 3.3 \approx$ ______

2. $3.4 \times 5.6 \approx$ ______

3. $17.5 \times 8.4 \approx$ ______

4. $44.8 \times 5.1 \approx$ ______

5. $28.21 \times 8.02 \approx$ ______

6. $71.92 \times 2.01 \approx$ ______

7. On average, the U.S. produces 36.5 million tons of fruit each year. About how much fruit does it produce in 2.25 years? (Examples 3 and 4)

8. Lisha is making headbands using ribbon. She would like to make 12 headbands. Each one requires 15.5 inches of ribbon. She estimates that she will need to buy 160 inches of ribbon. Is her estimate reasonable? Explain your reasoning. (Example 5)

9. **Financial Literacy** Hannah's hourly wage at the ice cream shop is $5.85. The table shows the number of hours she worked. She estimates her earnings to be $120. Without calculating her actual earnings, determine if her estimate is more or less than her actual earnings. Explain your reasoning.

Day	Hours Worked
Monday	3.5
Tuesday	4.25
Wednesday	3.75
Thursday	2.5
Friday	4.75

10. **STEM** A car releases 19.6 pounds of carbon dioxide for every 1 gallon of gasoline burned. Estimate the number of pounds of carbon dioxide released if 14.5 gallons are burned.

11. **Model with Mathematics** Refer to the graphic novel frame below for Exercises a–b.

a. How much more does Raj need until he has enough to buy the video game system? ______

b. Raj estimates that if he works for 20 hours, he will have enough to buy the video game system. Is he correct? Explain. ______

H.O.T. Problems Higher Order Thinking

12. **Reason Abstractly** Name three decimals with a product that is about 40.

______

13. **Persevere with Problems** A scooter can travel between 22 and 28 miles on each gallon of gasoline. If one gallon of gasoline costs between $3.75 and $3.95 per gallon, about how much will it cost to travel 75 miles? ______

14. **Justify Conclusions** Suppose your friend multiplied 1.2 and 2.6 and got 31.2 as the product. Is your friend's answer reasonable? Justify your response.

______

______

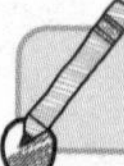

Standardized Test Practice

15. Green peppers are on sale for $2.89 per pound. Mrs. Moseley bought 1.75 pounds of peppers. About how much did she pay for the peppers?

Ⓐ less than $4
Ⓑ between $5 and $6
Ⓒ between $6 and $7
Ⓓ more than $7

Name ______________________ My Homework ______________________

Extra Practice

Estimate each product.

16. 26.3 → 26
× 9.7 → × 10
26 × 10 = 260

Homework Help

17. 33.6
× 82.1

18. 99.1
× 11.2

19. **STEM** A single year on Saturn is equal to 29.4 years on Earth. About how many Earth-years are equal to 3.2 years on Saturn?

20. Miguel received a $50 gift card to a bookstore. He would like to buy 3 books that cost $15.75 each including tax. He estimates that he cannot buy all three books because each book costs about $20, and all three books would cost $60. Is his estimate reasonable? Explain your reasoning.

Use estimation to determine whether each answer is reasonable. If the answer is reasonable, write yes. If not, write no and provide a reasonable estimate.

21. 22.8 × 4.7 = 107.16 ______

22. 2.1 × 4.9 × 7.2 = 105.84 ______

23. 7.8 × 1.1 × 4.2 = 50 ______

24. 43.8 × 2.8 × 3.1 = 371.8 ______

25. CCSS **Use Math Tools** The table shows some nutritional facts about orange juice. Estimate each value for 1 quart of orange juice. (*Hint:* 4 cups is equal to 1 quart.)

Orange Juice (1 cup)	
Calories	112
Vitamin C	96.9 mg
Carbohydrates	26.8 g
Calcium	22.4 mg

Standardized Test Practice

26. Mario and Andrew's hourly charge for mowing lawns is shown.

Mario	Andrew
$8.25/h	$5.85/h

Suppose Mario and Andrew each worked 20 hours. About how much more money did Mario earn?

Ⓐ $30 Ⓒ $60
Ⓑ $40 Ⓓ $70

27. Short Response Javier bought 4 pencil toppers at the school store for $3.69 each. He estimated how much he needed to pay and gave the cashier $16. Is Javier's estimation reasonable? Explain your reasoning.

____

28. The Student Book Club is ordering 12 copies of a book. The books cost $8.99 each. About how much will the order cost?

Ⓕ $90 Ⓗ $120
Ⓖ $108 Ⓘ $135

29. Medina's school lunch menu is shown.

Friday			
Pizza	$1.75	Fruit Punch	$0.75
Fish and Fries	$2.25	Milk	$0.80
Salad	$1.15	Pudding	$0.85

Which of the following is a reasonable estimate for the cost of two slices of pizza, a salad, and fruit punch?

Ⓐ $4 Ⓒ $8
Ⓑ $6 Ⓓ $10

Common Core Review

Multiply. 5.NBT.5

30. 65×18 ______

31. 15×23 ______

32. 198×75 ______

33. Marissa spent $15.63 at the bookstore. She paid with a $20 bill.

How much change will she receive? 5.NBT.7 ______________

____

34. Cristian is placing photos onto scrapbook paper for his photo album. The scrapbook paper is 12 inches long and 12 inches wide. What is the area of the paper?

(*Hint:* area = length × width) 4.MD.3 ______________

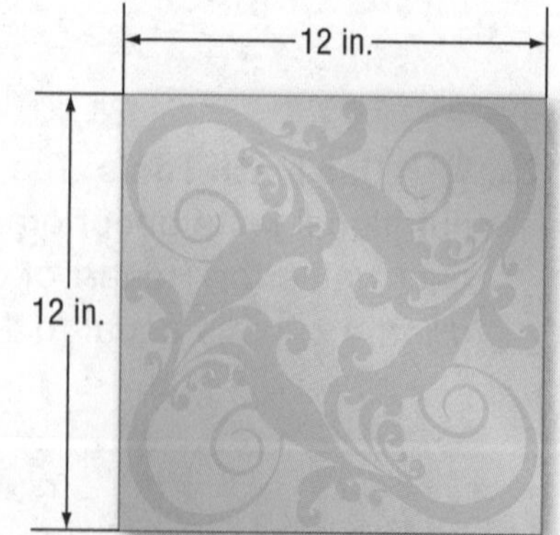

Lesson 3

Multiply Decimals by Whole Numbers

What You'll Learn

Scan the lesson. List two headings you would use to make an outline of the lesson.

- ______________________________

- ______________________________

Essential Question

HOW can estimating be helpful?

Common Core State Standards

Content Standards
6.NS.3

Mathematical Practices
1, 3, 4, 5, 6

Real-World Link

Watch

Plants Bamboo is one of the fastest growing plants. It can grow about 4.9 feet in height per day. It is a favorite food of panda bears. You can use repeated addition to find the total height a bamboo plant can grow over a number of days. Complete the table below. The first one is done for you.

	Number of Days	Repeated Addition	Multiplication
	2	4.9 + 4.9 = 9.8	2 × 4.9 = 9.8
1.	3	+ + =	× 4.9 =
2.	4	+ + + =	× 4.9 =
3.	5	+ + + + =	× 4.9 =

4. Use the pattern in the table to predict 6 × 4.9.

   Check by using repeated addition. ______________

5. **Make a Conjecture** Look back at Exercises 1–4. Compare the number of decimal places in each factor to the number of decimal places in the product. How do you determine the placement of the decimal point in a product?

Work Zone

Multiply Decimals

Using repeated addition can help you place the decimal point. The whole number represents the number of times the decimal is used as an addend. So, place the decimal point in the product the same number of places from the right as the decimal factor.

Examples

1. Find 4 × 0.83.

Estimate $4 \times 1 = 4$

0.83	0.83	0.83	0.83

4 groups of 0.83

0.83 ← two decimal places
× 4
3.32 Place the decimal point two places from the right.

Check for Reasonableness $3.32 \approx 4$ ✓

2. Find 3 × 14.2.

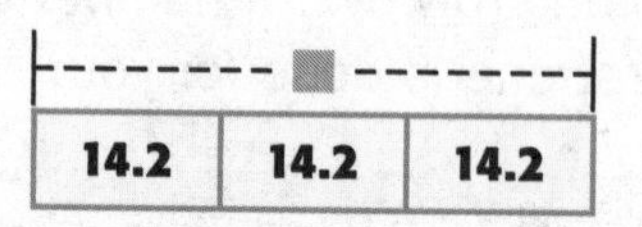

Estimate $3 \times 14 = 42$

14.2	14.2	14.2

3 groups of 14.2

14.2 ← one decimal place
× 3
42.6 Place the decimal point one place from the right.

Check for Reasonableness $42.6 \approx 42$ ✓

3. Find 4 × 0.95.

Estimate 4 × ☐ = ☐

0.95	0.95	0.95	0.95

☐ groups of ☐

Use the bar diagram to find the product.

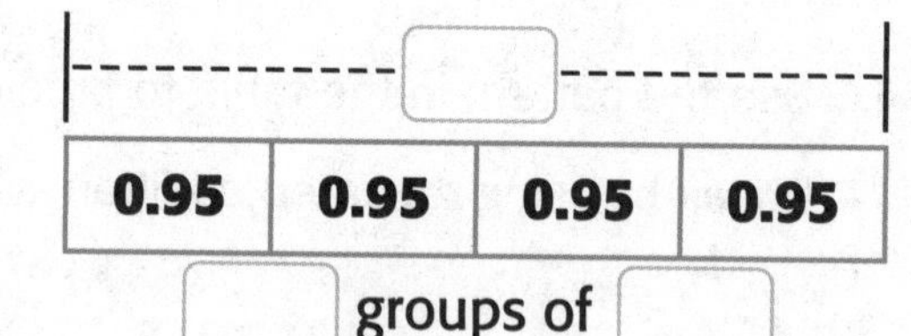

☐ ☐
0 . 9 5
× 4
☐ ☐ ☐

0.95 has ☐ decimal places.

Place the decimal point ☐ places from the right.

Check for Reasonableness ☐ ≈ ☐ ✓

a. ______

b. ______

c. ______

Got It? **Do these problems to find out.**

a. 5×0.25 **b.** 8×4.47 **c.** 9×2.63

Annex Zeros in the Product

If there are not enough decimal places in the product, you need to annex zeros to the left. To annex a zero means to place a zero at the beginning or end of a decimal.

Examples

4. Find 2 × 0.018.

$$\begin{array}{r} \overset{1}{0.018} \\ \times \quad 2 \\ \hline 0.036 \end{array}$$

0.018 ← three decimal places

0.036 ← Annex a zero on the left of 36 to make three decimal places.

Check by Adding

$$\begin{array}{r} 0.018 \\ +0.018 \\ \hline \square.\square\square\square \end{array}$$ ✓

5. Find 4 × 0.012.

$$\begin{array}{r} 0.012 \\ \times \quad 4 \\ \hline \square.\square\square\square \end{array}$$

0.012 has ☐ decimal places.

Annex a ☐ to make ☐ decimal places.

Check by Adding

$$\begin{array}{r} 0.012 \\ 0.012 \\ 0.012 \\ +0.012 \\ \hline \square.\square\square\square \end{array}$$ ✓

Do these problems to find out.

d. 3 × 0.02 **e.** 0.12 × 8 **f.** 11 × 0.045

Show your work.

d. ________

e. ________

f. ________

Example

6. A batch of trail mix calls for 1.2 pounds of dry cereal. Nigela is making 5 batches of trail mix. She already has 2.2 pounds of cereal. How many more pounds of dry cereal does she need?

Step 1 Multiply.

$$\begin{array}{r} 1.2 \\ \times\ 5 \\ \hline 6.0 \end{array}$$

1.2 ← one decimal place

6.0 ← one decimal place

Step 2 Subtract.

$$\begin{array}{r} 6.0 \\ -\ 2.2 \\ \hline 3.8 \end{array}$$

So, Nigela will need 3.8 more pounds of dry cereal.

Guided Practice

Multiply. (Examples 1–5)

Show your work.

1. $2.7 \times 6 =$ ______
2. $0.52 \times 3 =$ ______
3. $5 \times 0.09 =$ ______
4. $4 \times 0.027 =$ ______
5. $0.071 \times 8 =$ ______
6. $0.065 \times 18 =$ ______

7. A bee hummingbird has a mass of 1.8 grams. How many grams are 6 hummingbirds and a 4-gram nest? (Example 6)

______

8. Justin buys 12 pencils for $0.56 each. He pays with a $10 bill. How much change will he receive? (Example 6)

______

9. **Building on the Essential Question** How can estimating products help you to place the decimal correctly? ______

______

______

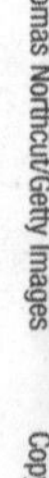

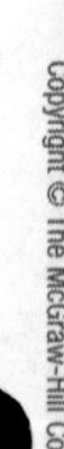

Name ______________________ My Homework ______________________

Independent Practice

Go online for Step-by-Step Solutions

Multiply. (Examples 1–5)

1. 1.2 × 7 = ______

2. 0.7 × 9 = ______

3. 2 × 1.3 = ______

4. 0.8 × 9 = ______

5. 3 × 0.02 = ______

6. 0.0036 × 19 = ______

7 The table shows the number of gallons of gasoline the Beckleys purchased on their road trip. What was the total cost for gas for the trip? (Example 6)

Fuel	
Number of Gallons	**Cost per Gallon ($)**
12	4.89
17	4.72
15	5.09

8. Sharon buys 14 folders for $0.75 each. How much change will she receive if she pays with $15? (Example 6) ______

9 **STEM** The hottest temperature recorded in the world, in degrees Fahrenheit, can be found by multiplying 13.46 by 10. Find this temperature. Justify your procedure. ______

10. CCSS **Justify Conclusions** Asher recently bought the poster shown at the right. What is its area? Explain your reasoning to a classmate. (*Hint:* Use area = length × width.) ______

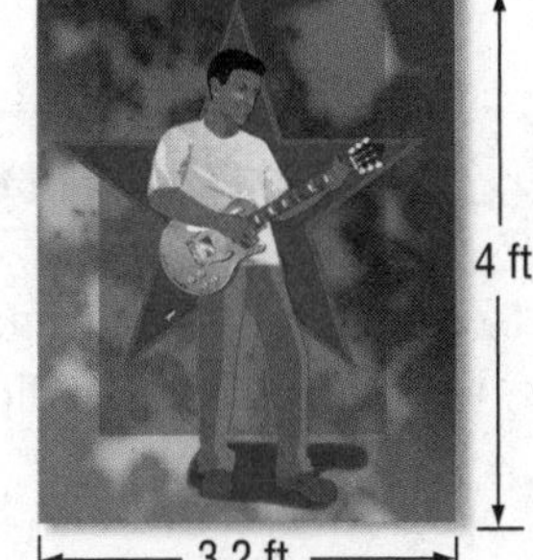

11. **Use Math Tools** The thickness of each type of coin is shown in the table. How much thicker is a stack of a dollar's worth of nickels than a dollar's worth of quarters? Explain your answer.

Coin	Thickness (mm)
penny	1.55
nickel	1.95
dime	1.35
quarter	1.75

H.O.T. Problems Higher Order Thinking

12. **Model with Mathematics** Write a real-world problem involving multiplication by a decimal factor. Then solve the problem.

13. **Persevere with Problems** Discuss two different ways to find the value of the expression 5.4 × 1.17 × 100 that do not require you to first multiply 5.4 × 1.17.

14. **Reason Inductively** Use the product of 123 × 47 to find the product of 123 × 0.47. Explain the difference in the two products.

15. **Construct an Argument** Your friend thinks that 1.5 × 8 = 1.20 because you do not count the zero when placing the decimal point. Is your friend correct? Justify your reasoning.

Standardized Test Practice

16. Anita bought 3 bags of sugar. Each bag weighed 36.8 ounces. How many ounces of sugar did she buy?

Ⓐ 11.04 Ⓒ 110.4

Ⓑ 73.6 Ⓓ 120.8

Extra Practice

Multiply.

17. 1.7 × 5 = 8.5

3
1.7
× 5
8.5

18. 0.9 × 4 = ______

19. 2.4 × 8 = ______

20. 3 × 0.5 = ______

21. 7 × 0.012 = ______

22. 0.0198 × 75 = ______

23. The mass of a certain monarch butterfly is 0.56 gram. What is the mass of 4 monarch butterflies?

24. The height of Mount Everest, in meters, can be found by multiplying 8.85 by 1,000. Find the height of Mount Everest. Explain your answer.

25. A sheet of printer paper is 8.5 inches by 11 inches. What is the area of the paper? (*Hint:* area = length × width)

26. Sofia bought 12 pens for $0.59 each. She paid with a $10 bill. How much change will she receive?

27. CCSS **Be Precise** One kilometer is about 0.62 mile. It is 12 kilometers from Noah's house to the ice skating rink. About how many miles is it from Noah's house to the ice skating rink?

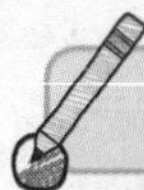

Standardized Test Practice

28. Short Response The school store is selling the following items.

Item	Price
Pennant	\$2.49
Bumper Sticker	\$1.79
Magnet	\$0.89

If Miguel buys two pennants, two bumper stickers, and four magnets, how much will he spend for all the items?

29. Short Response Find the area of the rug shown.

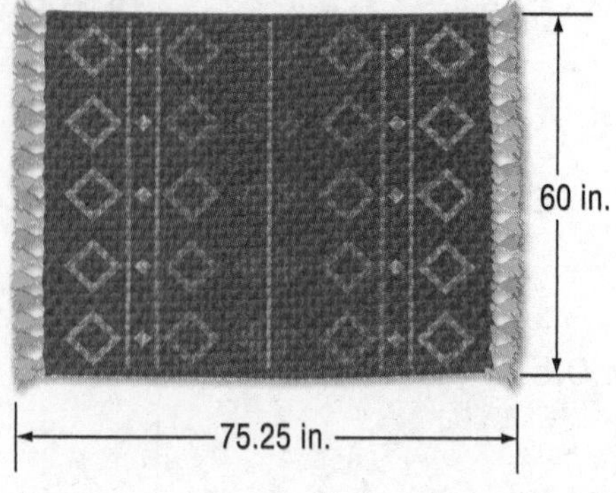

30. The table shows the admission prices to an amusement park.

Admission Prices	One-Day Pass	Two-Day Pass
Adult	\$39.59	\$43.99
Child (ages 3–9)	\$30.59	\$33.99

What is the total price of one-day passes for two adults and three children?

Ⓐ \$140.36 Ⓑ \$170.95 Ⓒ \$179.95 Ⓓ \$189.95

Common Core Review

Round each decimal to the nearest whole number. 5.NBT.4

31. 5.7 ≈ ______

32. 0.05 ≈ ______

33. 13.49 ≈ ______

34. Use number patterns and powers of ten to complete the table. 5.NBT.2

Factor		Factor		Product
2.9	×	10	=	
3.44	×		=	344
	×	100	=	870
10.25	×		=	102.5
156.23	×	10	=	

35. Several students from Southbend Middle School are visiting the Smithsonian American Art Museum. Mrs. Mabika divided the students into 5 equal groups. There are 3 boys and 4 girls in each group. Fill in the missing numbers to find the total number of students. 5.OA.2

☐ × (☐ + 4) = ☐ students

Lesson 4

Multiply Decimals by Decimals

What You'll Learn

Scan the lesson. List two real-world scenarios in which you would multiply a decimal by a decimal.

- ______________________________
- ______________________________

Essential Question

HOW can estimating be helpful?

Common Core State Standards

Content Standards
6.NS.3

Mathematical Practices
1, 2, 3, 4, 5, 6

Real-World Link

Planets The table shows the weight of a 1-pound object on each planet.

Planet	Weight (Pounds)
Mercury	0.3
Venus	0.9
Earth	1
Mars	0.3
Jupiter	2.3
Saturn	1
Uranus	0.8
Neptune	1.1

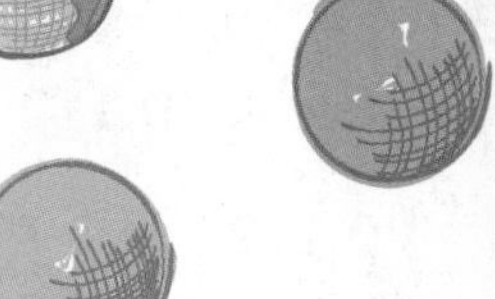
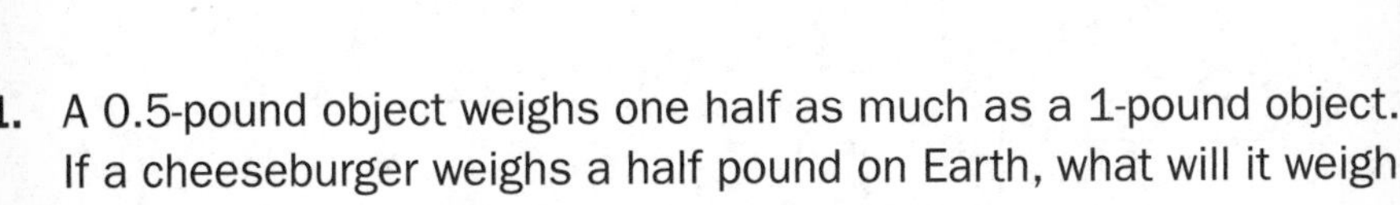

1. A 0.5-pound object weighs one half as much as a 1-pound object. If a cheeseburger weighs a half pound on Earth, what will it weigh on Jupiter? ______________

2. What would a barbell that weighs 5 pounds on Earth weigh on Jupiter? ______________

3. Use the results from Exercises 1 and 2 to find 0.05×2.3. Explain your answer. ______________

Work Zone

Multiply Decimals

When multiplying a decimal by a decimal, multiply as with whole numbers. To place the decimal point, find the sum of the number of decimal places in each factor. The product has the same number of decimal places.

STOP and Reflect

How is the product of 4.2 × 6.7 similar to the product of 42 × 67? How are they different? Explain below.

Examples

1. Find 3.6 × 0.05.

Estimate 3.6 × 0.05 ⟶ 4 × 0 or 0

3.6	← one decimal place
× 0.05	← two decimal places
0.180	← three decimal places

The product is 0.180 or 0.18.

Once you place the decimal point, you can drop the zero at the right.

2. Find 0.112 × 7.2.

Estimate 0.112 × 7.2 ≈ ☐ × ☐ or ☐

0.112 has ☐ decimal places.

7.2 has ☐ decimal place.

So the product has ☐ + ☐, or ☐ decimal places.

		0.	1	1	2
×				7.	2
			☐	☐	☐
+		☐	☐	☐	
	☐	☐	☐	☐	☐

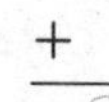
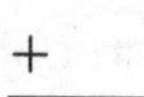

The product is ______.

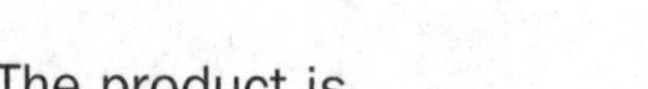

Check for reasonableness ______ ≈ ______ ✓

Got It? Do these problems to find out.

a. 5.7 × 2.8 **b.** 4.12 × 0.05 **c.** 0.014 × 3.7

a. ______

b. ______

c. ______

Annex a Zero

If there are not enough decimal places in the product, you need to annex zeros to the left.

Examples

3. Find 1.4 × 0.067.

$$\begin{array}{r} 0.067 \\ \times \ 1.4 \\ \hline 268 \\ + \ 67 \\ \hline 0.0938 \end{array}$$

0.067 ← three decimal places

× 1.4 ← one decimal place

0.0938 ← Annex a zero to make four decimal places.

4. Find 0.45 × 0.053.

The product will have ☐ decimal places. Annex zeros, if needed.

0. 4 5

× 0. 0 5 3

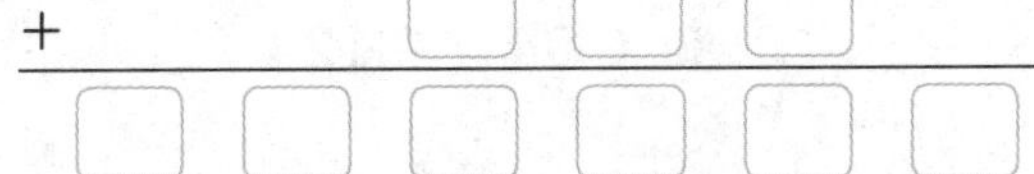

+

Check Multiply related whole numbers.

4 5

× 5 3

Move the decimal to the left 5 places. What is the number? ______

Is the answer the same? ☐

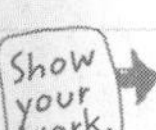

Got It? Do these problems to find out.

d. 0.04 × 0.32 **e.** 0.26 × 0.205 **f.** 1.33 × 0.06

d. ______

e. ______

f. ______

Example

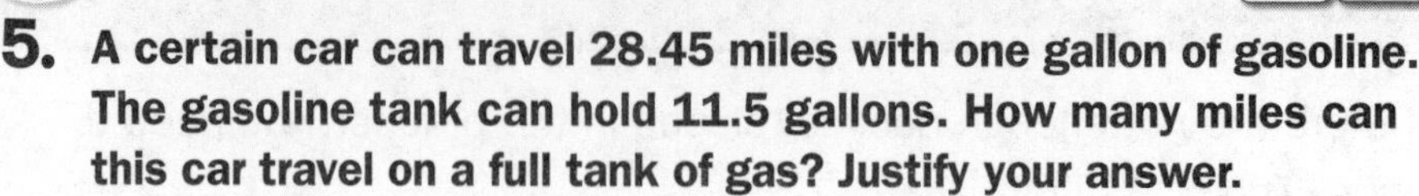

5. A certain car can travel 28.45 miles with one gallon of gasoline. The gasoline tank can hold 11.5 gallons. How many miles can this car travel on a full tank of gas? Justify your answer.

Estimate $28.45 \times 11.5 \rightarrow 30 \times 12$ or 360

```
   28.45     ← two decimal places
 × 11.5      ← one decimal place
  14225
  2845
+ 2845
 327.175     ← The product has three decimal places.
```

The car could travel 327.175 miles. Since $327.175 \approx 360$, the answer is reasonable.

Guided Practice

Multiply. (Examples 1–4)

Show your work.

1. $0.6 \times 0.5 =$ ______
2. $27.43 \times 1.089 =$ ______
3. $0.98 \times 7.3 =$ ______
4. $2.7 \times 1.35 =$ ______
5. $0.03 \times 0.09 =$ ______
6. $0.04 \times 2.12 =$ ______

7. A mile is equal to approximately 1.609 kilometers. How many kilometers is 2.5 miles? Justify your answer. (Example 5) ______

8. **Building on the Essential Question** Why is estimating not as helpful when multiplying very small numbers such as 0.007 and 0.053? ______

Rate Yourself!

Are you ready to move on? Shade the section that applies.

I have a few questions. | I'm ready to move on. | I have a lot of questions.

For more help, go online to access a Personal Tutor.

Independent Practice

Go online for Step-by-Step Solutions

Multiply. (Examples 1–4)

1. $0.7 \times 0.4 =$ ________

2. $0.4 \times 3.7 =$ ________

3. $0.52 \times 2.1 =$ ________

4. $6.2 \times 0.03 =$ ________

5. $14.7 \times 11.361 =$ ________

6. $0.28 \times 0.08 =$ ________

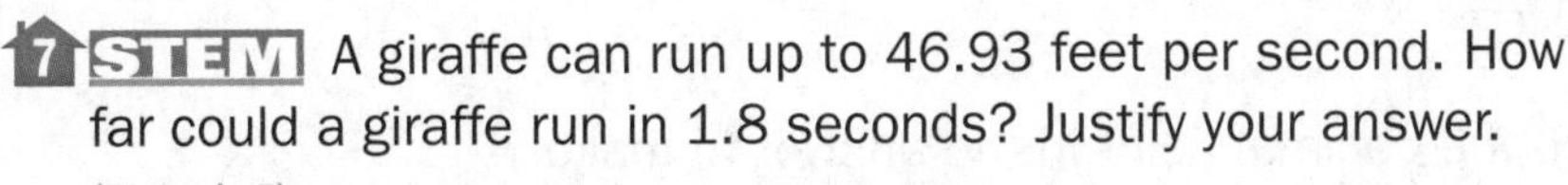

7. **STEM** A giraffe can run up to 46.93 feet per second. How far could a giraffe run in 1.8 seconds? Justify your answer. (Example 5)

8. A nutrition label indicates that one serving of apple crisp oatmeal has 2.5 grams of fat. How many grams of fat are there in 3.75 servings? Justify your answer. (Example 5)

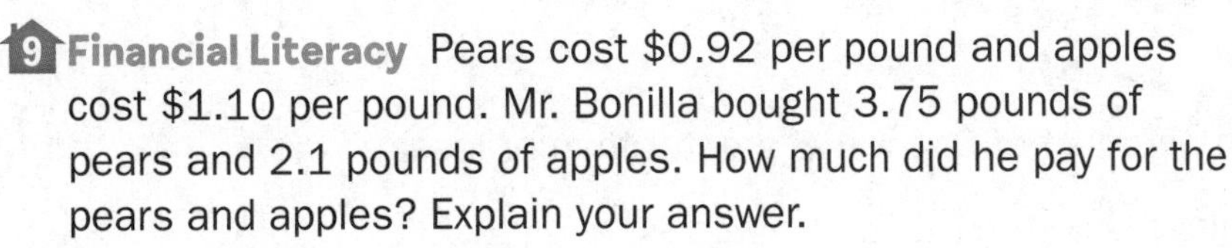

9. **Financial Literacy** Pears cost \$0.92 per pound and apples cost \$1.10 per pound. Mr. Bonilla bought 3.75 pounds of pears and 2.1 pounds of apples. How much did he pay for the pears and apples? Explain your answer.

Multiply.

10. $25.04 \times 3.005 =$ ________

11. $1.03 \times 1.005 =$ ________

12. $5.12 \times 4.001 =$ ________

13. **Use Math Tools** Complete the graphic organizer to show the relationship between decimal factors and their products.

×	2	0.2	0.02	0.002
3	6	0.6	0.06	
0.3		0.06	0.006	0.0006
0.03	0.06	0.006		
0.003		0.0006		

How do you determine the number of zeros to annex in the product of 0.002 and 0.003? ______

______

H.O.T. Problems Higher Order Thinking

14. **Reason Abstractly** Write a multiplication problem in which the product is between 0.05 and 0.75. ______

15. **Justify Conclusions** Place the decimal point in the answer to make it correct. Explain your reasoning. 3.9853 × 8.032856 = 32013341...

______

______

16. **Construct an Argument** Determine whether the following statement is *always*, *sometimes*, or *never* true. Give examples to justify your answer.

The product of two decimals less than 1 is less than either of the factors.

______

______

17. **Reason Inductively** Is the product of 0.4 × 1.8 greater than or less than 0.4? Explain your reasoning. ______

______

18. **Persevere with Problems** Evaluate the expression 0.3(3 − 0.5). ______

Standardized Test Practice

19. What is the area of the rectangle?

Ⓐ 14.04 cm^2

Ⓑ 10.248 cm^2

Ⓒ 8.992 cm^2

Ⓓ 7.868 cm^2

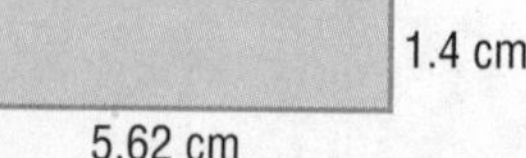

Extra Practice

Multiply.

20. 1.5 × 2.7 = 4.05

Homework Help

```
   1.5
×  2.7
   105
 + 30
  4.05
```

21. 3.1 × 0.8 = ______

22. 2.4 × 3.48 = ______

23. 5.04 × 3.2 = ______

24. 27.4 × 33.68 = ______

25. 0.451 × 0.05 = ______

26. Katelyn has a vegetable garden that measures 16.75 feet in length and 5.8 feet in width. Find the area of the garden. Justify your answer.

______

______

27. CCSS **Use Math Tools** Find examples of decimals in a newspaper or magazine, on television, or on the Internet. Write a real-world problem in which multiplies decimals.

______

______

28. CCSS **Be Precise** Find the area of the figure at the right. Justify your procedure.

______

6.1 in.
3 in.
6.9 in.
3.1 in.

29. Junnie walked for 2.5 hours at a speed of 3.2 miles per hour. Maurice walked for 1.8 hours at a speed of 4.1 miles per hour. (*Hint:* Distance equals speed times time.)

a. Who walked farther? ______

______

b. How much farther did that person walk? ______

______

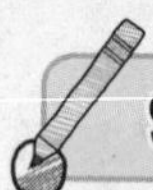

Standardized Test Practice

30. The grocery store is selling bananas for $0.35 per pound. How much will Zack pay for 3.6 pounds of bananas?

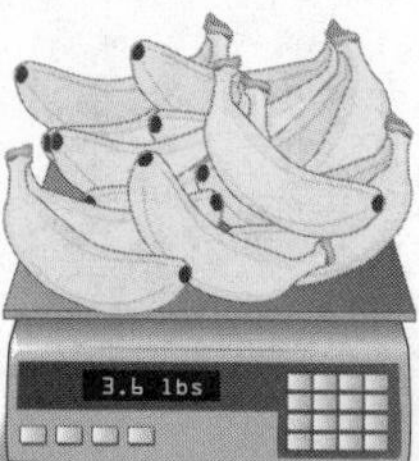

Ⓐ $1.26 Ⓒ $3.95
Ⓑ $1.40 Ⓓ $12.60

31. A turtle can walk up to 0.69 mile per hour. At this rate, how far could a turtle walk in 1.75 hours?

Ⓕ 0.1208 miles Ⓗ 2.44 miles
Ⓖ 1.2075 miles Ⓘ 12.075 miles

32. **Short Response** A soccer ball and 12 golf balls have a total mass of 1 kilogram. The mass of each golf ball is about 0.046 kilogram. What is the mass of the soccer ball?

______

33. **Short Response** Renaldo can rollerblade 9.7 miles per hour. At this rate, how far will he rollerblade in 0.75 hour?

______

CCSS Common Core Review

Divide. 5.NBT.6

34. 60 ÷ 12 = ______

35. 96 ÷ 8 = ______

36. 750 ÷ 15 = ______

37. Logan has 20 action figures. He is shipping them to a friend. He can fit 3 action figures in a box. How many boxes will he need? 5.NBT.6

______

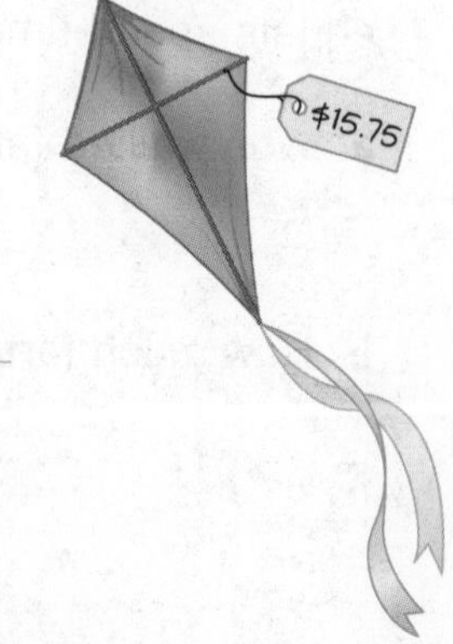

38. Three friends are dividing the cost of a kite equally. The kite costs $15.75. How much will each person pay? 5.NBT.7

______

Inquiry Lab

Multiply by Powers of Ten

HOW can number patterns be used to multiply by powers of 10?

Content Standards
6.NS.3

Mathematical Practices
1, 3, 5

Planets Each planet in our solar system orbits around the Sun at a different distance from the Sun. Mercury orbits at an average distance of 28.6 million miles. One million is 1,000,000. What is 28.6 × 1,000,000?

What do you know? ____________

What do you need to know? ____________

Investigation

Numbers like 10, 100, and 1,000 are called *powers of 10* because they can be obtained by raising 10 to a power.

Step 1 Look for a pattern. Complete the table.

Decimal		Power of 10		Product
28.6	×	0.1	=	2.86
28.6	×	1	=	28.6
28.6	×	10	=	286
28.6	×	1,000	=	28,600

Move the decimal point the ____________ number of places as the number of zeros in the power of 10.

Move the decimal point to the ____________ when multiplying by a power of 10 that is less than 1.

Move the decimal point to the ____________ when multiplying by a power of 10 that is greater than 1.

Step 2 Determine how many zeros are in 1,000,000 and move the decimal point in 28.6 the appropriate number of places.

There are ☐ zeros in 1,000,000.

28.6 million = 28.6 × 1,000,000
= 28.600000
= 28,600,000

Move the decimal point ☐ places to the right.

Analyze

Use Math Tools **Work with a partner to complete the tables.**

	Decimal		Power of 10		Product
	12.4	×	0.1	=	1.24
1.	12.4	×	0.01	=	
2.	12.4	×	0.001	=	
3.	12.4	×	0.0001	=	

	Decimal		Power of 10		Product
	1.24	×	1	=	1.24
4.	1.24	×	10	=	
5.	1.24	×	100	=	
6.	1.24	×	1,000	=	
7.	1.24	×	10,000	=	
8.	1.24	×	100,000	=	
9.	1.24	×	1,000,000	=	
10.	1.24	×	10,000,000	=	

11. **Reason Inductively** How can you find the product of a number and power of 10 without using paper and pencil or a calculator?

12. **Reason Inductively** The product of 13.6 and a power of 10 is 13,600. What is the power of 10? Explain.

Reflect

13. **Use Math Tools** Suppose you plan to purchase 10 items that each cost $4.95. Explain how to use mental math to find the cost of the 10 items.

14. **Inquiry** HOW can number patterns be used to multiply by powers of 10?

Problem-Solving Investigation
Look for a Pattern

CCSS Content Standards
6.NS.3
Mathematical Practices
1, 3, 4, 8

Case #1 Dance Party

The Student Government is organizing a spring dance. They plan to decorate with helium-filled balloons. The cost of the balloons is shown in the table.

What is the cost of 6 bags of balloons?

Number of Bags	Total Cost ($)
1	4.75
2	9.50
3	14.25
4	19.00

Understand ***What are the facts?***

The table shows the cost of the balloons. Six bags of balloons are needed.

Plan ***What is your strategy to solve this problem?***

Look for a pattern in the table. Each bag costs $4.75.

Solve ***How can you apply the strategy?***

Complete the table to find the cost of 6 bags of balloons.

Number of Bags	Total Cost ($)	
1	4.75	
2	9.50	+ 4.75
3	14.25	+ 4.75
4	19.00	+ 4.75
5		+ 4.75
6		+ 4.75

So, six bags of balloons cost $ ________.

Check ***Does the answer make sense?***

Use multiplication to check your answer. $4.75 × 6 = ________ ✓

Analyze the Strategy

Tutor

CCSS **Reason Inductively** How would the results change if the store offered a discount of $0.50 for each bag of balloons? ________________

____

Case #2 Virtual DJ

The Student Government is hiring a DJ for the spring dance. They expect the dance to last for 5 hours. The cost to hire DJ Trax is shown in the table.

How much will it cost to hire DJ Trax for the dance?

Cost to Hire DJ Trax	
Number of Hours	Total Cost ($)
1	125.50
2	251.00
3	376.50

Understand

Read the problem. What are you being asked to find?

I need to ____________________________________.

Underline key words and values in the problem. What information do you know?

The cost to hire DJ Trax is $ _________ for 1 hour, $ _________ for 2 hours, and $ _________ for 3 hours.

Is there any information that you do *not* need to know?

I do not need to know ____________________________________.

Plan

Choose a problem-solving strategy.

I will use the ____________________________________ strategy.

Solve

Use your problem-solving strategy to solve the problem.

Describe the pattern. Then complete the table.

The cost to hire DJ Trax ________________ by $ _______ for each hour.

Cost to Hire DJ Trax				
Number of Hours	1	2	3	4
Total Cost ($)	125.50	251.00	376.50	

So, it will cost $ _______ to hire DJ Trax for 5 hours.

Check

Use information from the problem to check your answer.

$ _______ × ☐ = $ _______

Collaborate Work with a small group to solve the following cases. Show your work on a separate piece of paper.

Case #3 Gaming

The table below shows the cost of a subscription to the Action Gamers Channel.

Action Gamers Channel Prices	
Number of Months	**Total Cost ($)**
1	7.95
2	15.90
3	23.85

What is the cost of a 6-month subscription?

Case #4 Number Theory

The diagram to the right is known as Pascal's Triangle.

If the pattern continues, what will the numbers in the next row be from left to right?

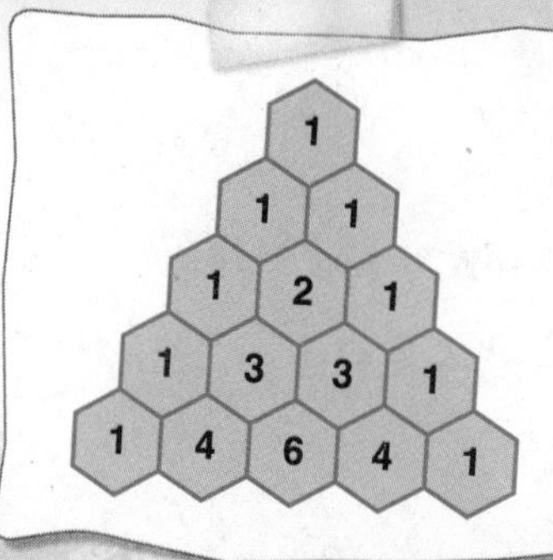

Case #5 Number Sense

Describe the pattern below. Then find the next three numbers.

3, 6.5, 11, 16.5, 23, ___, ___, ___

Circle a strategy below to solve the problem.

- Make a model.
- Make a table.
- Solve a simpler problem.
- Guess, check, and revise.

Case #6 Games

Claudio is purchasing a new gaming system. One Web site sells the system for $240 and the games for $45.99 each.

What is the total cost if Claudio purchases the system and 3 games?

Mid-Chapter Check

Vocabulary Check

1. Define *product.* Give an example of two whole number factors with a product of 9.

Skills Check and Problem Solving

Find each sum or difference. (Lesson 1)

2. 42.7 + 52.12 = ______

3. 4.7 − 3.28 = ______

4. 8.37 − 0.015 = ______

Multiply. (Lessons 3 and 4)

5. 2.3 × 5 = ______

6. 3.4 × 5.2 = ______

7. 1.2 × 0.015 = ______

8. The table shows a list of walking trails in the United States. Latisha walked the KATY Trail 6 days last week. How many miles did she walk in a week. (Lesson 3)

Location	Length of Trail (mi)
Florida Trail (FL)	4.8
Long Path (NJ)	3.3
Ohio & Erie Canal Trail (OH)	4.3
KATY Trail (MO)	5.7
Point Reyes National Seashore (CA)	5.0

9. CCSS **Be Precise** The length of a pool table is 7.1 feet and the width is 3.6 feet. Find the area of the surface of the pool table by multiplying the length by the width. (Lesson 4)

10. **Standardized Test Practice** Ashton used 12.6 gallons of gasoline to drive his car on a weekend trip. He averaged 21.5 miles per gallon. About how many miles did he travel? (Lesson 2)

    Ⓐ 20 miles
    Ⓑ 200 miles
    Ⓒ 350 miles
    Ⓓ 400 miles

Lesson 5

Divide Multi-Digit Numbers

What You'll Learn

Scan the lesson. Predict two things you will learn about dividing multi-digit numbers.

Essential Question

HOW can estimating be helpful?

Common Core State Standards

Content Standards
6.NS.2

Mathematical Practices
1, 2, 3, 4, 5, 6

Vocabulary Start-Up

When one number is divided by another, the result is called a *quotient*. The *dividend* is the number that is divided and the *divisor* is the number used to divide another number.

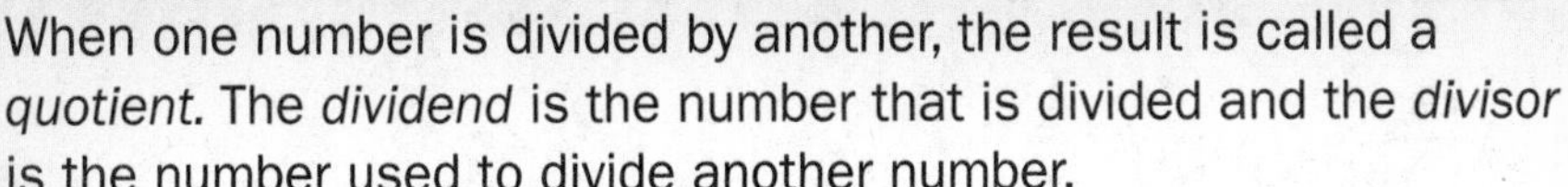

Label the division problem with the correct vocabulary term: quotient, dividend, and divisor.

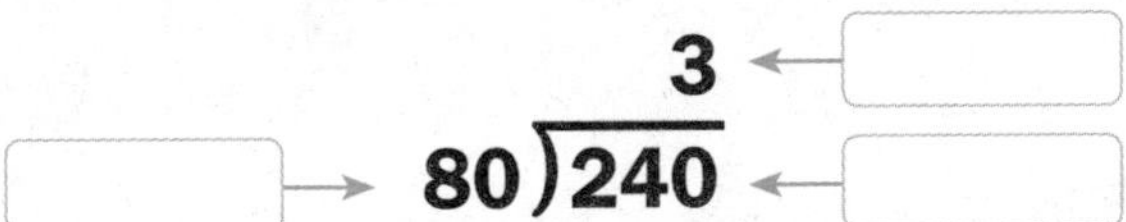

Real-World Link

Circulation When you are at rest it takes about 60 seconds for a single blood cell to travel around your body and back to your heart.

1. In 120 seconds, about how many times does a single blood cell travel around your body and back to your heart? Write the dividend, divisor, and quotient in the diagram below.

2. Camila's target heart rate should be about 200 beats per minute. Estimate the number of times Camila's heart will beat in one second if her heart is working at this rate. Explain.

Work Zone

Divide Three-Digit Dividends

In this lesson, you will divide multi-digit numbers. Use estimation to help you place the first digit in the quotient.

Examples

1. Find 351 ÷ 9.

Estimate 360 ÷ 9 = 40. So, the first digit is in the tens place.

Write 351 ÷ 9 as $9\overline{)351}$.

```
  39
9)351     Divide each place-value position from left to right.
 -27
  81
 -81
   0      Since 81 − 81 = 0, there is no remainder.
```

So, 351 ÷ 9 is 39.

Check Compare 39 to the estimate. 39 ≈ 40 ✓

2. Find $31\overline{)878}$.

Estimate 900 ÷ 30 = 30. So, the first digit is in the tens place.

```
    28 R10
31)878      Divide each place-value position from left to right.
  -62
   258
  -248
    10      Since 258 − 248 = 10 and 10 < 31, 10 is the remainder.
```

So, $31\overline{)878}$ is 28 R10.

Check 28 R10 ≈ 30 ✓

a. ____________

b. ____________

Got It? Do these problems to find out.

Find each quotient.

a. 768 ÷ 8

b. $16\overline{)318}$

Divide Four-Digit Dividends

The steps for dividing three-digit dividends and four-digit dividends are the same.

Examples

3. Find 6,493 ÷ 75.

Estimate 6,400 ÷ 80 = 80

$$\begin{array}{r} 86 \text{ R43} \\ 75\overline{)6{,}493} \\ -600 \\ \hline 493 \\ -450 \\ \hline 43 \end{array}$$

Divide each place-value position from left to right.

Check for Reasonableness 86 R43 ≈ 80 ✓

The quotient of 6,493 ÷ 75 is 86 R43.

Check Your Answer

You can check division with a remainder. Multiply the quotient by the divisor. Then add the remainder.

$$\begin{array}{r} 86 \\ \times 75 \\ \hline 6{,}450 \end{array} \qquad \begin{array}{r} 6{,}450 \\ +\ 43 \\ \hline 6{,}493 \end{array}$$

4. The average person has 1,460 dreams a year. What is the average number of dreams a person has each night?

Find 1,460 ÷ 365.

Estimate 1,600 ÷ 400 = 4

$$\begin{array}{r} 4 \\ 365\overline{)1{,}460} \\ -1{,}460 \\ \hline 0 \end{array}$$

Check for Reasonableness 4 = 4 ✓

The average number of dreams a person has each night is 4.

Got It? Do these problems to find out.

c. Find $56\overline{)4{,}321}$.

d. Find $91\overline{)8{,}465}$.

e. To promote its opening weekend, a water park gave the local middle school 1,050 free tickets. The middle school has 350 students. Each student will receive the same number of tickets. How many tickets will each student receive?

c. ________

d. ________

e. ________

Example

5. The total number of seats in a college stadium is 54,912. There are 44 sections and each section has an equal number of seats. How many seats are in each section?

Divide 54,912 by 44.

```
      1,248
44)54,912      Divide each place-value position from left to right.
  −44
    109
    −88
     211
    −176
      352
     −352
        0
```

There are 1,248 seats in each section.

Guided Practice

Find each quotient. (Examples 1–4)

1. $8\overline{)736}$ Show your work.

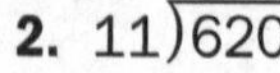

2. $11\overline{)620}$

3. $19\overline{)7{,}814}$

4. $37\overline{)3{,}511}$

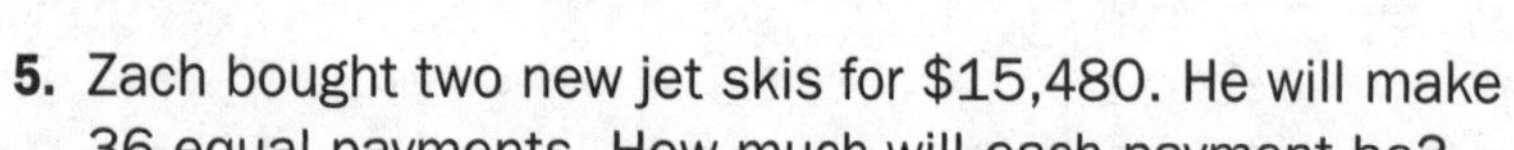

5. Zach bought two new jet skis for $15,480. He will make 36 equal payments. How much will each payment be?

(Example 5) ______

6. **Building on the Essential Question** How is estimation helpful when dividing multi-digit numbers?

______

______

______

______

Rate Yourself!

How well do you understand dividing multi-digit numbers? Circle the image that applies.

Clear

Somewhat Clear

Not So Clear

For more help, go online to access a Personal Tutor.

Independent Practice

Go online for Step-by-Step Solutions eHelp

Find each quotient. (Examples 1–3)

1. 174 ÷ 6 = ______

2. 453 ÷ 8 = ______

3. 645 ÷ 43 = ______

4. 299 ÷ 21 = ______

5. $62\overline{)8,090}$

6. $31\overline{)2,480}$

7. $34\overline{)5,780}$

8. $16\overline{)3,482}$

9. A tour bus travels 2,160 miles in 36 hours. What is the average distance the bus travels in one hour? (Example 4)

10. A charity sold 475 tickets to a dinner auction. If the charity raised $16,625 in ticket sales, what was the cost of one ticket? (Example 5)

11. A city phone book has 86 pages filled with residents' names. There are a total of 15,050 names in the book. Each page has an equal number of names on it. How many names are on each page? (Example 5)

12. CCSS **Use Math Tools** The table shows the number of servings for different size cakes at Mimi's Bakery. Suppose a high school graduation expected 2,889 guests. How many X-large sheet cakes should the school order? Explain how you solved.

Mimi's Bakery	
Sheet Cake Size	**Number of Servings**
Small	30
Medium	60
Large	90
X-Large	120

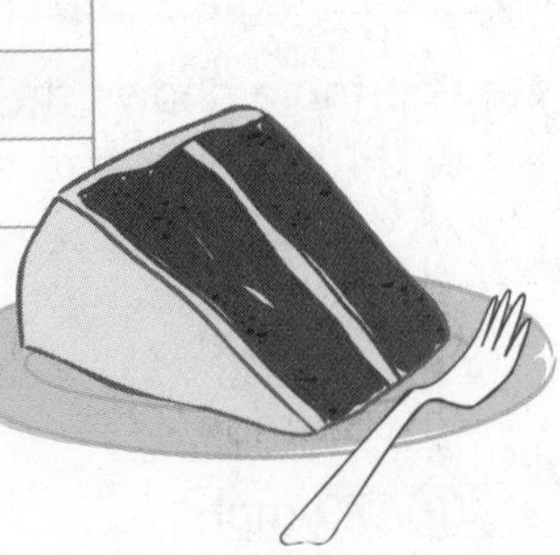

13. **Be Precise** How many 8-ounce cups can be filled from 9 gallons of juice? (*Hint:* There are 128 ounces in one gallon.)

14. **Be Precise** Water stations will be placed every 600 meters of a fifteen kilometer race. How many water stations will be needed? (*Hint:* There are 1,000 meters in one kilometer.)

H.O.T. Problems Higher Order Thinking

15. **Model with Mathematics** Write and solve a real-word problem that involves a two-digit divisor and a four-digit dividend.

16. **Persevere with Problems** If the divisor is 40, what is the least three-digit dividend that would give a remainder of 4?

17. **Justify Conclusions** Can the remainder in a division problem ever equal the divisor? Why or why not?

18. **Reason Abstractly** Use the digits 2, 4, and 8 one time each in the following problem.

☐,☐00 ÷ ☐0 = 30

Standardized Test Practice

19. The table shows the mileage Karla drove on her trip. She drove for a total of 24 hours. What is the average speed she drove?

Ⓐ 55 mph
Ⓑ 60 mph
Ⓒ 65 mph
Ⓓ 70 mph

Karla's Trip	
Day	**Miles**
1	486
2	316
3	638

Name ______________________ My Homework ______________________

Extra Practice

Find each quotient.

20. 182 ÷ 7 = 26

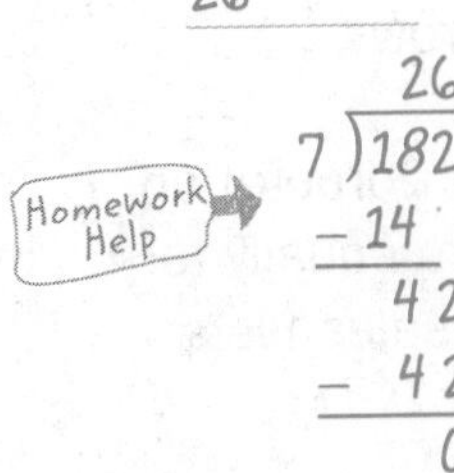

21. 345 ÷ 6 = ______

22. 792 ÷ 33 = ______

23. 811 ÷ 79 = ______

24. 44)2,876

25. 26)4,340

26. 33)9,537

27. 19)4,257

28. A city library has 9,440 nonfiction books. The librarian wants to divide the books evenly among 80 shelves. How many books will be on each shelf?

____

29. A concession stand manager ordered 20,280 souvenir cups. He wants to divide the cups evenly among the 24 concession stands. How many cups will each concession stand receive?

____

30. CCSS **Use Math Tools** The table shows the average weight of animals. How many more tons does a blue whale weigh than an African elephant? Explain how you solved. (*Hint:* There are 2,000 pounds in one ton.)

____

____

Animal Weights	
Animal	**Weight (pounds)**
African Elephant	15,000
Blue Whale	238,000
Great White Shark	5,000
Lowland Gorilla	500

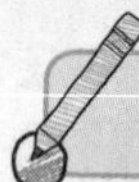

Standardized Test Practice

31. The dance team recently purchased 25 pairs of new boots for $1,350. What was the price of each pair of boots?

Ⓐ $50 Ⓒ $55

Ⓑ $54 Ⓓ $60

32. **Short Response** A toy factory assembles 19,824 toy castles over a 12-hour period of time. The same number of castles is assembled every hour. How many toy castles were assembled each hour?

____

33. The school auditorium holds 1,711 people. There are 59 seats in each row. How many rows of seats are in the auditorium?

Ⓕ 23 rows Ⓗ 29 rows

Ⓖ 25 rows Ⓘ 39 rows

34. **Short Response** The Carson Corporation distributed 58,992 sales fliers equally to 24 different cities. How many sales fliers were sent to each city?

____

Common Core Review

Divide mentally. 4.NBT.6

35. $300 \div 5 =$ ______

36. $4,800 \div 8 =$ ______

37. $4,200 \div 6 =$ ______

38. The maple tree in Logan's backyard is 58.6 feet tall. Plot 58.6 on the number line below. Then round 58.6 to the nearest whole number. 4.NF.7

____

39. There are 75 students attending a field trip. Each van will seat 8 students. How many vans will be needed? 4.NBT.6

____

40. Mr. Maxwell is shipping 80 video games. Each box will hold 12 games. How many boxes will be needed? 4.NBT.6

____

Lesson 6

Estimate Quotients

What You'll Learn

Scan the lesson. List two headings you would use to make an outline of the lesson.

-
-

Essential Question

HOW can estimating be helpful?

Vocabulary

compatible numbers

Common Core State Standards

Content Standards
6.NS.2

Mathematical Practices
1, 3, 4, 5

Vocabulary Start-Up

To determine what a compatible number is, first you must determine what compatible means. Fill in the table below.

Definition:	Example:
compatible	
What would make numbers compatible?	Non-Example:

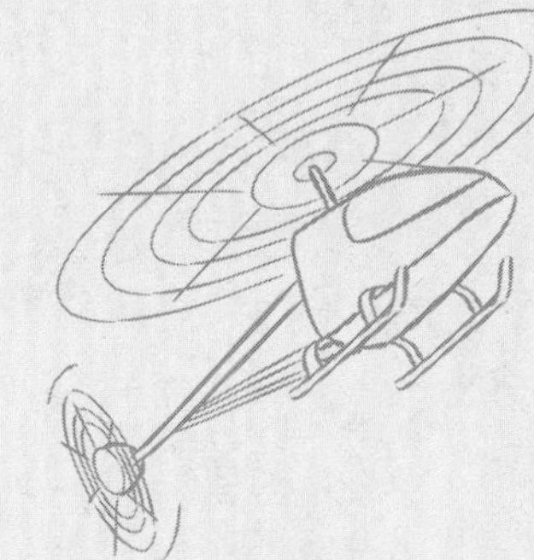

Real-World Link

Remote Control Latasha and her two sisters want to buy their little brother a remote control helicopter. The helicopter costs $28.90. They decided to split the cost equally.

1. What number that is a multiple of 3 is close to $28.90? Explain.

2. Use your answer from Exercise 1 to determine about how much each person will pay. Explain.

Work Zone

Estimate by Rounding Dividends

To estimate quotients of decimals, use rounding and compatible numbers. **Compatible numbers** are numbers that are easy to divide mentally.

Examples

1. Estimate 11.75 ÷ 3.

Round the dividend, 11.75, to a whole number.

The divisor is 3. So, round 11.75 to a whole number that is a multiple of 3.

$3\overline{)11.75} \rightarrow \overset{4}{3\overline{)12}}$ Using multiples of 3, 12 is closest to 11.75

So, 11.75 ÷ 3 is about 4.

2. The Jenkins family bought five tickets to a charity auction. The receipt shows the total cost of the tickets. Estimate the cost of each ticket. Justify your answer.

$5\overline{)61.25} \rightarrow \overset{12}{5\overline{)60}}$ Round 61.25 to 60.

Each ticket costs about $12.
Since $5 \times 12 = 60$ and $60 \approx 61.25$, the answer is reasonable.

Got It? **Do these problems to find out.**

Estimate each quotient.

a. 49.3 ÷ 7

b. $25\overline{)98.1}$

c. Suppose the Jenkins family decided to purchase 6 tickets for a total price of $64.50 using a discount. Estimate the cost of each ticket. Justify your answer.

a. ____________

b. ____________

c. ____________

Estimate by Rounding Divisors

You can also estimate quotients of decimals by rounding the divisors. Choose compatible numbers that are easy to divide mentally.

Examples

Tutor

and Reflect

How does the division fact $63 \div 9 = 7$ help you to estimate the quotient of $63 \div 8.4$? Answer below.

3. Estimate 32 ÷ 3.9.

Round the divisor, 3.9, to a whole number.

The dividend is 32. So, round 3.9 to a whole number that is a factor of 32.

$3.9\overline{)32} \rightarrow 4\overline{)32}$ with quotient 8. Round 3.9 to 4 since 32 and 4 are compatible numbers.

So, 32 ÷ 3.9 is about 8.

Check by Multiplying $3.9 \times 8 = 31.2$

$31.2 \approx 32$ ✓

4. Estimate 56 ÷ 6.8.

Round the divisor, ☐, to a whole number.

The dividend is ☐.

So, round 6.8 to a whole number that is a ________ of 56.

Round 6.8 to ☐.

$6.8\overline{)56} \rightarrow$ ☐$\overline{)56}$ with quotient ☐

So, 56 ÷ 6.8 is about ☐.

Check by Multiplying $6.8 \times$ ☐ = ☐

☐ ≈ 56 ✓

Got It? Do these problems to find out.

Estimate each quotient.

d. $54 \div 9.16$

e. $10.75\overline{)99}$

d. ________

e. ________

Example

5. **STEM A Pacific Leatherback turtle can have a mass of up to 704.4 kilograms. An Olive Ridley turtle can have a mass of up to 49.9 kilograms. About how many times heavier is the Pacific Leatherback turtle? Explain why your answer is reasonable.**

$49.9\overline{)704.4} \rightarrow 50\overline{)700}$ with quotient 14 — Round 49.9 to 50 and 704.4 to 700.

The Pacific Leatherback is about 14 times heavier than the Olive Ridley turtle.

Check for Reasonableness Since $50 \times 14 = 700$, and $700 \approx 704.4$, your answer is reasonable. ✓

Got It? **Do this problem to find out.**

f. There are approximately 250.9 million cars in the United States. Spain has approximately 25.1 million cars. About how many times more cars does the U.S. have than Spain? Explain why your answer is reasonable.

f. ______

Guided Practice

Estimate each quotient. (Examples 1, 3, and 4)

1. $25 \div 4.7 \approx$ ______

2. $40.79 \div 7 \approx$ ______

3. $38.1\overline{)984.76} \approx$ ______

4. STEM The average yearly precipitation for Gulfport, Mississippi, is 65.3 inches. About how much precipitation does the area receive each month? Explain why your answer is reasonable. (Example 2)

______

5. Mauricio bought 6.75 yards of fabric for a total of $47.50. About how much was the cost per yard? Explain why your answer is reasonable. (Example 5)

______

6. **Building on the Essential Question** When is it helpful to estimate quotients? ______

______

Rate Yourself!

How confident are you about estimating quotients? Shade the ring on the target.

For more help, go online to access a Personal Tutor.

Independent Practice

Go online for Step-by-Step Solutions 

Estimate each quotient. (Examples 1, 3, and 4)

1. $32.4 \div 3 \approx$ ______

2. $76.2 \div 18.4 \approx$ ______

3. $11.4\overline{)35.7} \approx$ ______

4. **Financial Literacy** Emily spent a total of \$38.04 on four CDs. If each CD cost the same amount, what is a reasonable amount for the cost of each CD? Explain why your answer is reasonable. (Example 2)

______

______

5 A recipe for a smoothie calls for 0.75 pound of strawberries. If Kerry has 3.15 pounds of strawberries, how many batches of the recipe can she make? (Example 5)

______

______

6. **Financial Literacy** For each handmade greeting card Jacqui sells, she makes a profit of \$0.35. In one week, she made a profit of \$42. She sells the cards for \$0.75 each.

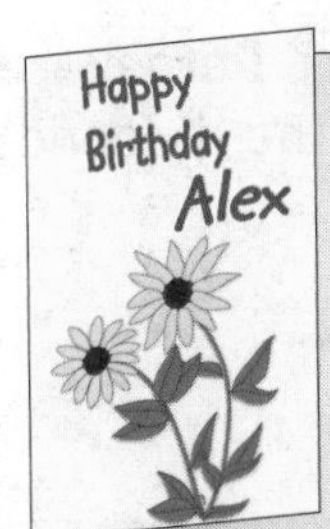

a. About how many greeting cards did Jacqui sell that week?

______

b. About how much did she earn before paying expenses?

______

7 CCSS **Justify Conclusions** The average cow produces about 53 pounds of milk per day. If one gallon of milk weighs about 8.5 pounds, estimate the number of gallons of milk a cow produces each day. Explain why your estimate is reasonable.

______

______

8. When full, a 22-gallon gas tank holds 129.8 pounds of gasoline. Estimate the weight of one gallon of gasoline. If it costs \$91.30 to fill the gas tank, estimate the cost per gallon.

______

______

9. **Use Math Tools** Use estimation and mental math to find the four missing quantities from the receipt.

Precious Pets Receipt

Qty	Description	Unit Price	Total
	Hamster cage	$35.99	$35.99
	Exercise wheel	$5.29	$10.58
	Softwood bedding	$6.29	$25.16
	Hamster food	$4.59	$36.72
		Total	$108.45

H.O.T. Problems Higher Order Thinking

10. **Model with Mathematics** Write a real-world division problem involving decimals in which you would use compatible numbers to estimate the quotient.

11. **Persevere with Problems** Determine where to place the decimal point in the dividend and divisor so that the quotient is between 23 and 25.

16023 ÷ 654

12. **Reason Inductively** Explain how you know which compatible numbers to use when estimating the quotient of a division problem involving decimals. Support your answer with an example.

Standardized Test Practice

13. Approximately 243.0 million people live in Indonesia. The population of Germany is about 82.2 million. About how many times more people live in Indonesia than in Germany?

Ⓐ about 2 times
Ⓑ about 3 times
Ⓒ about 20 times
Ⓓ about 30 times

Extra Practice

Estimate each quotient.

14. $54 \div 9.4 \approx$ ______

Homework Help

$9.4\overline{)54} \rightarrow 9\overline{)54}$ = 6

15. $45.8 \div 23.6 \approx$ ______

16. $23.3\overline{)119} \approx$ ______

17. The average annual snowfall in King Salmon, Alaska, is 45.9 inches. The snow season lasts about 7 months of the year. About how much snow does the area receive on average each month? Explain why your answer is reasonable.

18. Scientists at the zoo recently studied an anaconda that weighs 8,643.2 ounces. The average weight of the common rat is 11.8 ounces. About how many times heavier is the anaconda than the common rat? Explain why your answer is reasonable.

19. CCSS **Justify Conclusions** Aurelia would like to save $474.72 in a year to purchase a new video camera. She estimates she needs to save $40 per month. Explain why her estimate is reasonable.

20. A piggy bank containing only quarters has a mass of 850 grams when empty and 7,822 grams when filled. If a quarter weighs 5.6 grams, estimate the amount of money inside the piggy bank.

21. Melanie is making homemade stickers. She uses the recipe shown to create the glue for the stickers.

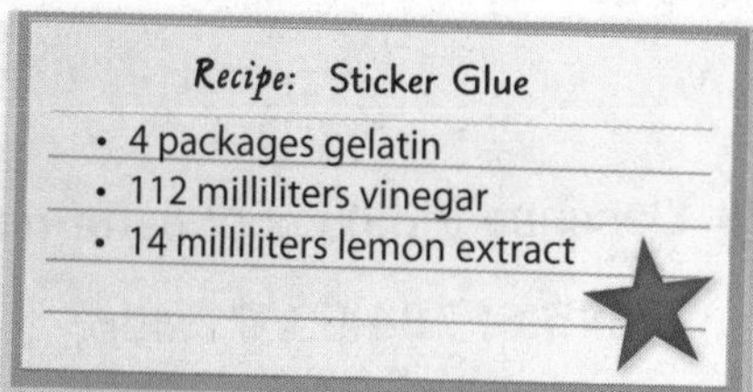

a. She has 545 milliliters of vinegar. Which is a more reasonable estimate for the number of batches she can make, 5 or 7? Explain your answer.

b. About how many times as many milliliters of vinegar are needed than lemon extract? ______

Standardized Test Practice

22. The table shows the average breakdown of body weight for a 130-pound person.

Body Part	Weight (ounces)
Water	896
Muscle	720
Skeleton	240
Head	128
Skin	96

About how many times as great is the weight of water than the weight of skin?

Ⓐ about 9 Ⓒ about 13
Ⓑ about 11 Ⓓ about 15

23. **Short Response** For a craft activity at a day care, each child will need 1.75 yards of ribbon. If there are 25 yards of ribbon available, estimate the number of children that can participate.

24. The following advertisement was in the local newspaper.

Bike Country	
26" Bike	$135.99
Folding Bike Rack	$43.95
Seat Covers	$6.59
Bike Lock	$12.89
Helmet	$29.49

The cost of a 26″ bike is equal to about how many bike locks?

Ⓕ about 7 Ⓗ about 9
Ⓖ about 8 Ⓘ about 10

25. **Short Response** Rewrite the following division problem using compatible numbers, so the quotient is a whole number.

485.87 ÷ 71.54

CCSS Common Core Review

Find each quotient. 5.NBT.7

26. $8.4 \div 10 =$ ______

27. $100\overline{)14.7} =$ ______

28. $94.5 \div 100 =$ ______

29. Describe the number pattern below. Then find the next three numbers.

7,345.6; 734.56; 73.456; 5.NBT.2 ______; ______; ______

30. The movie theater sold 825 tickets to fill 3 theaters. Each theater has an equal number of seats. How many seats are in each theater?

4.NBT.6 ______

Lesson 7

Divide Decimals by Whole Numbers

What You'll Learn

Scan the lesson. List two real-world scenarios in which you would divide decimals by whole numbers.

- ______________________________
- ______________________________

Essential Question

HOW can estimating be helpful?

Common Core State Standards

Content Standards
6.NS.3

Mathematical Practices
1, 3, 4, 5, 6

Real-World Link

Movies Charlotte, Aaron, Maddie, and Catie went to the movies and ordered snacks from the menu shown.

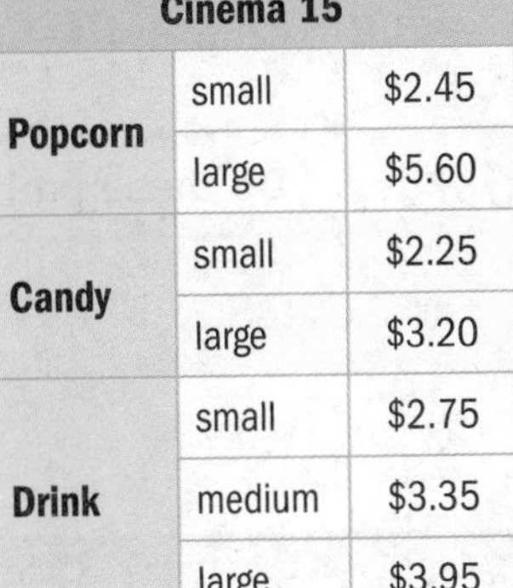

Cinema 15		
Popcorn	small	$2.45
	large	$5.60
Candy	small	$2.25
	large	$3.20
Drink	small	$2.75
	medium	$3.35
	large	$3.95

1. How much did they pay for four small popcorns?

   ______ × 4 = ______

2. What is the total cost for two small packages and one large package of candy?

   ____

3. How much do four medium drinks cost?

   ______ × 4 = ______

4. What is the total cost for Exercises 1–3?

Popcorn	$ ☐.☐
Candy	$ ☐.☐
Drinks	+ $ ☐.☐
	$ ☐.☐

5. Estimate how much each person should pay if they split the total cost evenly.

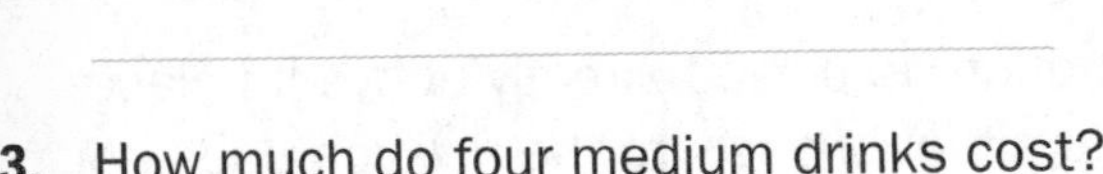

Work Zone

Divide a Decimal by a 1-Digit Number

When dividing a decimal by a whole number, divide as with whole numbers. Then place the decimal in the quotient directly above its place in the dividend.

Example

1. Find 6.8 ÷ 2. Estimate $6 \div 2 = 3$

$$\begin{array}{r} 3.4 \\ 2\overline{)6.8} \\ -6 \\ \hline 0\,8 \\ -8 \\ \hline 0 \end{array}$$

6 ones divided by 2 is 3 ones.

8 tenths divided by 2 is 4 tenths.

$6.8 \div 2 = 3.4$ Compared to the estimate, the quotient is reasonable.

Show your work.

a. ________

b. ________

c. ________

Got It? Do these problems to find out.

a. $7.5 \div 3$ **b.** $3.5 \div 7$ **c.** $9.8 \div 2$

Divide a Decimal by a 2-Digit Number

The decimal point in the quotient is placed directly above its place in the dividend. In real-world situations where the division does not result in a remainder of zero, round the quotient to a specified place.

Checking Answers
To check that the answer is correct, multiply the quotient by the divisor. In Example 2, $0.55 \times 14 = 7.7$. ✓

Example

2. Find 7.7 ÷ 14. Estimate $10 \div 10 = 1$

$$\begin{array}{r} 0.55 \\ 14\overline{)7.70} \\ -7\,0 \\ \hline 70 \\ -70 \\ \hline 0 \end{array}$$

Place the decimal point.

Annex a zero and continue dividing.

$7.7 \div 14 = 0.55$ Compared to the estimate, the quotient is reasonable.

 Do these problems to find out.

d. 9.48 ÷ 15 **e.** 3.49 ÷ 4 **f.** 55.08 ÷ 17

d. ________

e. ________

f. ________

 **Example**

3. Lin is mailing a care package to his brother. The table gives the cost for mailing packages. If Lin's care package weighs 3 pounds, how much is the cost per pound?

Weight (Pounds)	Cost ($)
1	4.80
2	5.63
3	6.74
4	7.87

To find the cost per pound, divide $6.74 by 3.

```
   2.246  ← Place the decimal point after dividing to thousandths.
3)6.740
 – 6
   07
 – 06
    14
  – 12
     20    Annex a zero and continue dividing.
   – 18
      2    The remainder will never be zero.
```

Round 2.246 to 2.25 because hundredths are the smallest denomination used in money. It costs about $2.25 per pound to mail the package.

Check Use a bar diagram and multiplication to check your work.

6.75		
2.25	**2.25**	**2.25**

2.25 × 3 = 6.75

6.75 ≈ 6.74 ✓

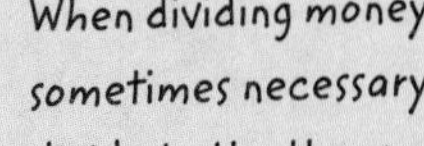

Dividing Money
When dividing money, it is sometimes necessary to divide to the thousandths place and then round to the hundredths.

Got It? Do this problem to find out.

g. Find the cost per pound of a two-pound and four-pound package.

g. ________

Example

4. Ryan and his brother are sharing the cost of a video game. The video game costs $28.60. If Ryan saved $20 to buy the game, how much does he have left after paying his share?

Step 1 Determine how much Ryan will pay.

$$\begin{array}{r} 14.30 \\ 2\overline{)28.60} \\ -2 \\ \hline 08 \\ -8 \\ \hline 0\,6 \\ -6 \\ \hline 0 \end{array}$$

Place the decimal point.

Ryan's share is $14.30.

Step 2 Determine how much Ryan will have left.

$$\begin{array}{r} \$20.00 \\ -\ \$14.30 \\ \hline \$5.70 \end{array}$$

So, Ryan has $5.70 left.

Got It? **Do this problem to find out.**

h. ______

h. Kristen and her two friends are sharing the cost of a funnel cake. The funnel cake costs $5.49. If Kristen has $2.00, how much will she have left after she pays her share?

Guided Practice

Divide. Round to the nearest tenth if necessary. (Examples 1 and 2)

1. 3.6 ÷ 4 = ______

Show your work.

2. 12.32 ÷ 22 = ______

3. 69.904 ÷ 34 = ______

4. Light travels 5.88 trillion miles in one year. How far will light travel in one month? (Examples 1 and 3) ______

5. Four dozen bagels costs $30.00. How much change will you receive if you pay for a dozen bagels with a ten-dollar bill? (Examples 2 and 4) ______

6. **Building on the Essential Question** How can estimating quotients help you to place the decimal correctly? ______

Rate Yourself!

How confident are you about dividing decimals by whole numbers? Check the box that applies.

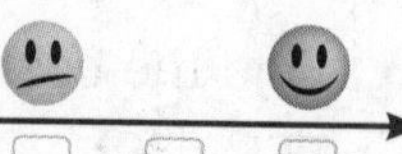

For more help, go online to access a Personal Tutor.

FOLDABLES Time to update your Foldable!

John Giustina/Iconica/Getty Images

Independent Practice

Go online for Step-by-Step Solutions 

Divide. Round to the nearest tenth if necessary. (Examples 1 and 2)

1. 39.39 ÷ 3 = ______

2. 7.24 ÷ 7 = ______

3. 118.5 ÷ 5 = ______

4. 11.4 ÷ 19 = ______

5. 55.2 ÷ 46 = ______

6. 336.752 ÷ 31 = ______

7. The Gonzalez family is taking a cruise that costs $3,082.24 for a family of four. How much does it cost per person? (Example 3)

______

8. Find the average height of the buildings shown in the table. (*Hint:* To find the average, add the values and divide by the number of values.) (Example 4)

World's Tallest Buildings (thousands of feet)				
1.667	1.483	1.483	1.451	1.381

______

9. CCSS **Be Precise** Mr. Jamison will stain the deck in his backyard. The deck has an area of 752.4 square feet. If the deck is 33 feet long, how wide is it? Justify your procedure. ______

______

10. CCSS **Be Precise** The Verrazano-Narrows Bridge in New York City is 4.26 thousand feet long and is the seventh longest suspension bridge in the world. There are 3 feet in a yard. How long is the bridge in yards? Justify your procedure. ______

______

11. The Student Council is raising money by selling bottled water at a band competition. The table shows the prices for different brands. Which brand costs the least per bottle? Explain your reasoning. ______________________

Cost of Bottled Water (20 oz bottles)		
Brand A	6-pack	$3.45
Brand B	12-pack	$5.25
Brand C	24-pack	$10.99

H.O.T. Problems Higher Order Thinking

12. CCSS **Persevere with Problems** Find each of the following quotients. Then find a pattern and explain how you can use this pattern to mentally divide 0.0096 by 3.

$844 \div 2$ $0.844 \div 2$ $84.4 \div 2$ $0.0844 \div 2$ $8.44 \div 2$ $0.00844 \div 2$

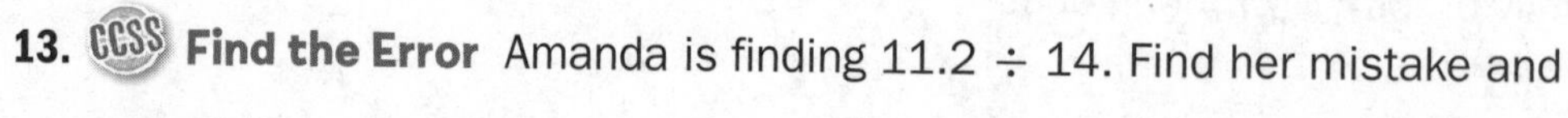

13. CCSS **Find the Error** Amanda is finding $11.2 \div 14$. Find her mistake and correct it. ______________________

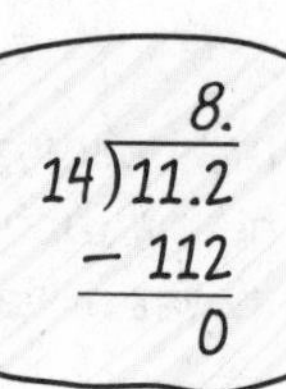

14. CCSS **Reason Inductively** Is the quotient $2.7 \div 3$ greater than or less than 1? Explain.

15. CCSS **Use Math Tools** Explain how you can use estimation to place the decimal point in the quotient $42.56 \div 22$.

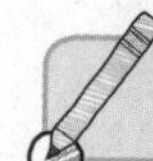

Standardized Test Practice

16. The Patterson family bought steak for a family picnic. They paid $71.94 for 6 pounds of ribeye steak. What was the price per pound?

Ⓐ $1.20 Ⓒ $10.99

Ⓑ $9.99 Ⓓ $11.99

Name ________________ My Homework ________________

Extra Practice

Divide. Round to the nearest tenth if necessary.

17. 36.8 ÷ 2 = 18.4

Homework Helper

$$\begin{array}{r} 18.4 \\ 2\overline{)36.8} \\ -2 \\ 16 \\ -16 \\ 0\,8 \\ -\,8 \\ 0 \end{array}$$

18. 124.2 ÷ 9 = ______

19. 6.271 ÷ 4 = ______

20. 10.22 ÷ 14 = ______

21. 59.84 ÷ 32 = ______

22. 751.2 ÷ 25 = ______

23. The Franklin Middle School jazz band plans to have a car wash to raise $468.75 for a new sound system. In the past, they washed an average of 125 cars at each car wash. What should they charge per car so they reach their goal?

______

24. Marcel Park is weeding the rectangular vegetable garden. The garden has an area of 599.5 square feet. If the garden is 22 feet wide, how long is the garden? Justify your procedure.

______

25. CCSS **Use Math Tools** The table shows the prices for different party toy packages from the Tomtown Toy Company. Which item costs the least per toy? Explain your reasoning. ______

Cost of Party Toy Packages		
Dominoes	6-pack	$3.98
Peg Games	12-pack	$9.99
Mini Footballs	24-pack	$17.98

Standardized Test Practice

26. The table shows the number of subscribers to several Internet providers.

Internet Provider	Subscribers (millions)
Company A	2.45
Company B	3.12
Company C	2.83

If Company B earned $119 million last month, about how much did each subscriber pay?

Ⓐ $30　Ⓑ $40　Ⓒ $50　Ⓓ $60

27. **Short Response** Tanner and three neighborhood friends are buying a basketball hoop that costs $249.84. If the cost is divided equally, how much will each person pay?

______

28. **Short Response** Marvin completed 8 rounds of a trivia game and earned 94.4 points. If he earned the same number of points each round, how many points did he earn each round?

______

CCSS Common Core Review

Add or subtract mentally. Use compensation. 5.NBT.7

29. $0.47 + 0.36 =$ ______

30. $26.5 - 9.3 =$ ______

31. $29.4 + 1.7 =$ ______

32. Use <, >, or = to compare 34.3 and 34.32. 5.NBT.3b

______

33. A king cobra has a mass of 8.845 kilograms. Round the mass to the nearest tenth kilogram. 5.NBT.4

______

34. The same king cobra is 4.237 meters long. Round the length to the nearest meter. 5.NBT.4

______

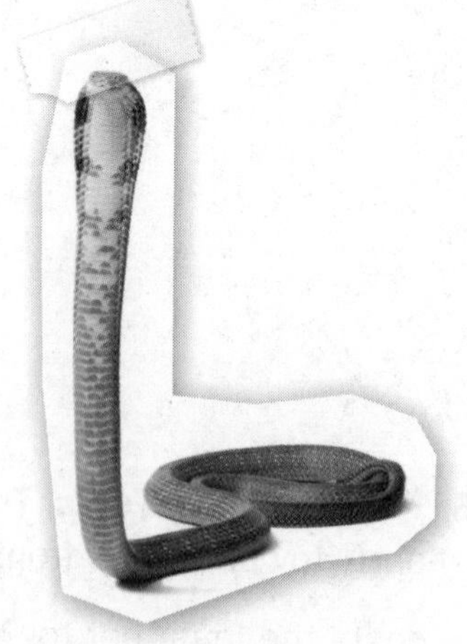

Life On White/Getty Images

Lesson 8

Divide Decimals by Decimals

What You'll Learn

Scan the lesson. Predict two things you will learn about dividing decimals by decimals.

-
-

Essential Question

HOW can estimating be helpful?

Common Core State Standards

Content Standards
6.NS.3

Mathematical Practices
1, 3, 4, 5

Real-World Link

Art An art studio has 36 gallons of acrylic paint. They separate it into 9 containers. How many gallons are in each container?

Division Problem	Quotient
36 ÷ 9	

Use the division problem to find patterns and complete the tables below. Then use these patterns to describe the dividends, divisors, and quotients in each set.

1.

Division Problem	Quotient
36 ÷ 0.9	40
36 ÷ 0.09	400
36 ÷ 0.009	4,000
36 ÷ 0.0009	

2.

Division Problem	Quotient
3.6 ÷ 9	0.4
0.36 ÷ 9	0.04
0.036 ÷ 9	0.004
0.0036 ÷ 9	

3.

Division Problem	Quotient
3.6 ÷ 0.9	4
0.36 ÷ 0.09	4
0.036 ÷ 0.009	4
0.0036 ÷	4

Work Zone

Divide by Decimals

When dividing by decimals, change the divisor into a whole number. To do this, multiply both the divisor and the dividend by the same power of 10. Then divide as with whole numbers.

Examples

1. Find 1.71 ÷ 0.9. Estimate $2 \div 1 = 2$

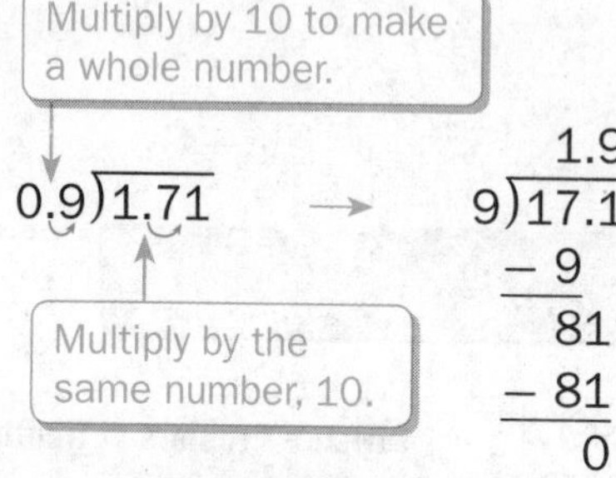

9)17.1 = 1.9
− 9
81
− 81
0

Place the decimal point.

Divide as with whole numbers.

1.71 divided by 0.9 is 1.9. Compared to the estimate, the quotient is reasonable.

Check $1.9 \times 0.9 = 1.71$ ✓

2. Find 2.64 ÷ 0.6. Estimate ☐ ÷ ☐ = ☐

0.6)2.64

Multiply 0.6 by ☐ to make a whole number.

Multiply the dividend, ☐, by the same power of 10.

0.6)2.64 Place the decimal point in the quotient.

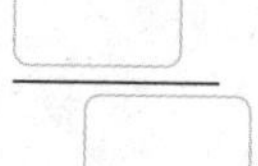

Divide as with whole numbers.

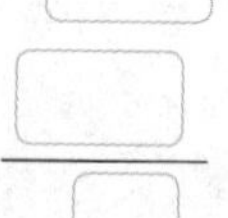

2.64 divided by 0.6 is ☐.

Compared to the estimate, is the quotient reasonable? ______

Show your work.

a. ______

b. ______

c. ______

Got It? Do these problems to find out.

a. 54.4 ÷ 1.7 **b.** 8.424 ÷ 0.36 **c.** 0.0063 ÷ 0.007

Zeros in the Quotient and Dividend

Line up the numbers by place value as you divide. Annex zeros in the quotient in order to keep digits with the correct place value. Annex zeros in the dividend to continue dividing after the decimal point.

Examples

3. Find 52 ÷ 0.4.

$0.4\overline{)52.0}$

Multiply each by 10.

$$\begin{array}{r} 130. \\ 4\overline{)520.} \\ -\,4 \\ \hline 12 \\ -\,12 \\ \hline 00 \end{array}$$

Place the decimal point.

Write a zero in the ones place of the quotient because 0 ÷ 4 = 0.

So, 52 ÷ 0.4 = 130.

4. Find 0.009 ÷ 0.18.

$0.18\overline{)0.009}$

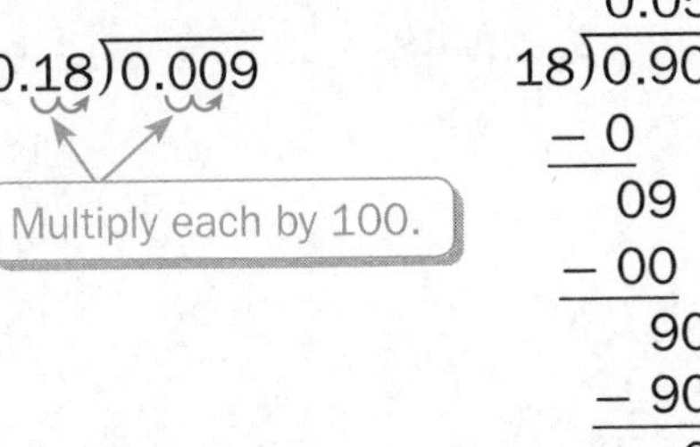

$$\begin{array}{r} 0.05 \\ 18\overline{)0.90} \\ -\,0 \\ \hline 09 \\ -\,00 \\ \hline 90 \\ -\,90 \\ \hline 0 \end{array}$$

Place the decimal point.

9 tenths divided by 18 is 0, so write a 0 in the tenths place.

Annex a 0 in the dividend and continue to divide.

So, 0.009 ÷ 0.18 is 0.05.

Checking Answers

You can always check your answer to a division problem by multiplying the quotient by the divisor.

5. Find 11.2 ÷ 0.07.

Multiply 0.07 and 11.2 by ☐.

$0.07\overline{)11.20}$ Place the decimal point in the quotient. Divide as with whole numbers.

Got It? Do these problems to find out.

Show your work.

d. 5.6 ÷ 0.014 **e.** 6.24 ÷ 200 **f.** 0.4 ÷ 25

d. ______

e. ______

f. ______

Example

6. How many times as many Internet users are there in Japan than in Spain? Round to the nearest tenth.

Internet Users in 2008 (millions)	
China	1,321.9
United States	301.1
Japan	127.4
France	63.7
Spain	40.4
Canada	33.4

Find 127.4 ÷ 40.4.

$$40.4\overline{)127.4} \rightarrow 404\overline{)1274.00}$$

```
        3.15
404)1274.00
   - 1212
       620
     - 404
      2160
    - 2020
       140
```

To the nearest tenth, 127.4 ÷ 40.4 = 3.2. So, there are about 3.2 times as many Internet users in Japan than in Spain.

Guided Practice

Divide. (Examples 1–5)

1. 3.69 ÷ 0.3 = ______

2. 0.0338 ÷ 1.3 = ______

3. 2.943 ÷ 2.7 = ______

4. Alicia bought 5.75 yards of fleece fabric to make blankets for a charity. She needs 1.85 yards of fabric for each blanket. How many blankets can Alicia make with the fabric she bought? (Example 6)

______

5. **Building on the Essential Question** When is it helpful to round the quotient to the nearest hundredth?

______

______

______

______

Name ________________ My Homework ________________

Independent Practice

Go online for Step-by-Step Solutions

Divide. (Examples 1–5)

1. 1.44 ÷ 0.4 = ______

Show your work.

2. 16.24 ÷ 0.14 = ______

3. 0.6 ÷ 0.0024 = ______

4. 96.6 ÷ 0.42 = ______

5. 13.5 ÷ 0.03 = ______

6. 0.12 ÷ 0.15 = ______

7. CCSS **Use Math Tools** The average person's *stride length,* the distance covered by one step, is approximately 2.5 feet long. How many steps would the average person take to travel 50 feet? (Example 6)

______

8. STEM Alaska has a coastline of about 6.64 thousand miles. Florida has about 1.35 thousand miles of coastline. How many times more coastline does Alaska have than Florida? Round to the nearest tenth if necessary. Justify your procedure.

______

9. CCSS **Model with Mathematics** Refer to the graphic novel frame below for Exercises a–b.

a. How many hours does Raj need to work to earn the remainder of the money he needs to buy the video game system? ______

b. Suppose Raj receives a raise for his hard work and now earns $6.25 per hour. How many hours would he need to work to earn $132?

______

10. A necklace is being made with beads that are 1.25 centimeters in diameter each. The necklace is 30 centimeters long. How many beads are needed? ______________________

11. CCSS **Use Math Tools** Use the table that shows popular sports car colors in North America.

Popular Sports Car Colors	
Color	**Portion of Responses**
Silver	0.2
Blue	0.16
Black	0.14
Red	0.09
Other	0.41

a. How many times more respondents chose silver than red? Round to the nearest tenth if necessary.

____

b. How many times more respondents chose either silver or black than red? Round to the nearest tenth if necessary.

____

H.O.T. Problems Higher Order Thinking

12. CCSS **Persevere with Problems** Find two positive decimals a and b that make the following statement true. Then find two positive decimals a and b that make the statement false.

If $a < 1$ and $b < 1$, then $a \div b < 1$.

____

____

13. CCSS **Which One Doesn't Belong?** Identify the problem that does not have the same quotient as the other three. Explain your reasoning.

$49 \div 7$	$4.9 \div 7$	$0.49 \div 0.7$	$0.049 \div 0.07$

____

____

Standardized Test Practice

14. To the nearest tenth, how many times as many people in the U.S. own dogs as own birds?

Ⓐ 6.8

Ⓑ 12.2

Ⓒ 26.6

Ⓓ 35.8

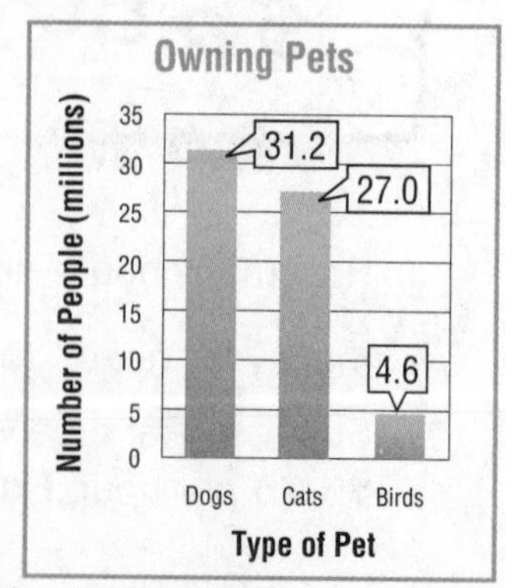

Name ______________________ My Homework ______________________

Extra Practice

Divide.

15. 0.68 ÷ 3.4 = 0.2

Homework Help: 3.4)0.68 = 0.2; −68; 0

16. 2.07 ÷ 0.9 = ______

17. 0.16728 ÷ 3.4 = ______

18. 1.08 ÷ 2.7 = ______

19. 8.4 ÷ 0.02 = ______

20. 0.242 ÷ 0.4 = ______

21. A submarine sandwich 1.5 feet long is cut into 0.25-foot pieces. How many pieces will there be?

______

22. CCSS **Use Math Tools** Find examples of decimals in a newspaper or magazine. Write a real-world problem in which you would divide decimals.

______

______

______

23. The table shows the five most populated countries in the world. How many times as many people live in China than in the United States? Round to the nearest tenth if necessary.

______

Most Populated Countries

Country	Approximate Population (billions)
China	1.325
India	1.13
United States	0.304
Indonesia	0.235
Brazil	0.19

24. CCSS **Justify Conclusions** Lake Superior, along the U.S.-Canadian border, has a maximum depth of 1.333 thousand feet. There are 5,280 feet in one mile. How deep is Lake Superior in miles? Round to the nearest hundredth if necessary. Explain your answer.

______

______

Standardized Test Practice

25. Ava paid $4.90 for 2.5 pounds of walnuts. What is the cost of one pound of walnuts?

Ⓐ $0.96

Ⓑ $1.76

Ⓒ $1.86

Ⓓ $1.96

26. Short Response Max bicycled 6.25 miles in 30.5 minutes. On average, how far did he bicycle each minute? Round to the nearest tenth.

____

27. The table shows the approximate number of people in the world who speak either Spanish or French.

Language	Speakers (billions)
Spanish	0.425
French	0.129

To the nearest tenth, how many times as many people speak Spanish as French?

Ⓕ 0.2 Ⓗ 0.3

Ⓖ 1.1 Ⓘ 3.3

28. Short Response About 24.8 million people live in Texas. About 0.6 million people live in Vermont. How many times as many people live in Texas than in Vermont? Round to the nearest tenth if necessary.

____

CCSS Common Core Review

Fill in each ◯ with $<$, $>$, or $=$ to make a true sentence. 4.NF.2

29. $\frac{2}{4}$ ◯ $\frac{5}{8}$

30. $\frac{6}{12}$ ◯ $\frac{5}{10}$

31. $\frac{1}{2}$ ◯ $\frac{6}{14}$

32. Plot the fraction $\frac{7}{12}$ on the number line. Is $\frac{7}{12}$ closer to 0, $\frac{1}{2}$, or 1?

4.NF.2 ______________________

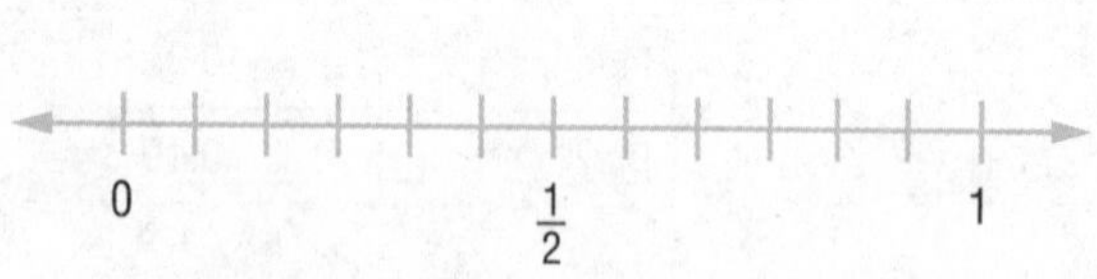

33. Lily spent $\frac{1}{6}$ of her free time practicing soccer and $\frac{5}{12}$ of her free time playing a video game. What fraction of her free time did she spend on these two activities? 5.NF.2

____

21ST CENTURY CAREER in Design

Sports Equipment Designer

Do you have a passion for sports and a strong interest in science? Are you a creative thinker who always has new ideas or better ways of doing things? If so, then you should consider a career designing sports equipment. Sports equipment designers combine creativity and engineering principles to create equipment that is cutting edge and helps improve athletic performance. They design everything from baseball bats and footballs to lacrosse protective gear and racing wheelchairs.

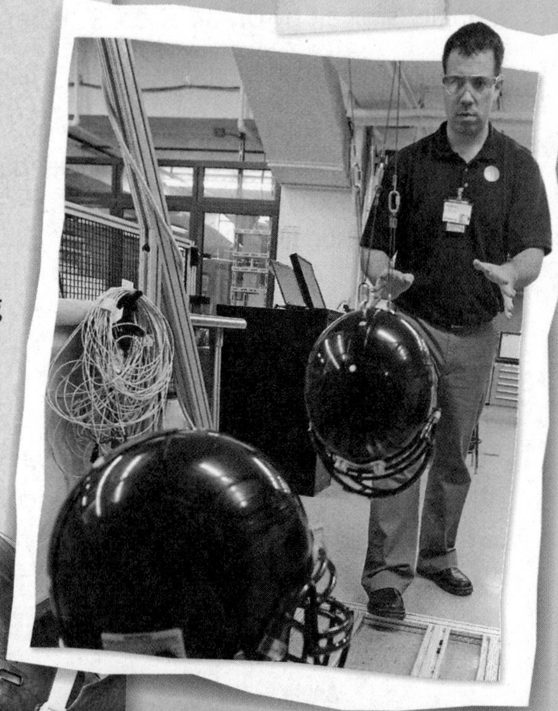

Explore college and careers at ccr.mcgraw-hill.com

Is This the Career for You?

Are you interested in a career as a sports equipment designer? Take some of the following courses in high school.

- Algebra
- Biology
- Calculus
- Computer Science
- Physics

Find out how math relates to a career in Design.

Gaining a Competitive Edge

When a punter kicks a football, the ball has both horizontal motion and vertical motion. The table shows these values when a football is kicked at 25 meters per second.

Use the information in the table to solve each problem. Assume that each football is kicked at 25 meters per second. Round to the nearest tenth if necessary.

1. The *hang time,* or time that a football is in the air, of a football that is kicked at a 27° angle is given by 0.204 × 11.3. What is the approximate hang time? ________

2. How much greater is the hang time of a football that is kicked at a 62° angle than one that is kicked at a 45° angle? Use the expressions 0.204 × 22.1 and 0.204 × 17.7.

____

3. The final distance from the punter to a football kicked at a 27° angle is approximately 22.3 × 11.3 × 0.2. What is the distance from the punter to the football?

____

4. Find the distance of a football that is kicked at an angle of 62° if the distance is found by using the expression 11.7 × 22.1 × 0.2. ________

5. The hang time of a football is about 3 seconds. Find 3 ÷ 0.204 to determine the vertical motion of the football. ________

6. A football reaches its maximum height in $y \div 9.8$ seconds. A football is kicked at a 62° angle. At the same time, another football is kicked at a 27° angle. Which reaches its maximum height first? Explain.

____

____

Punting A Football		
Angle of Kick	Horizontal Motion (m/s)	Vertical Motion (m/s)
	x	y
27°	22.3	11.3
45°	17.7	17.7
62°	11.7	22.1

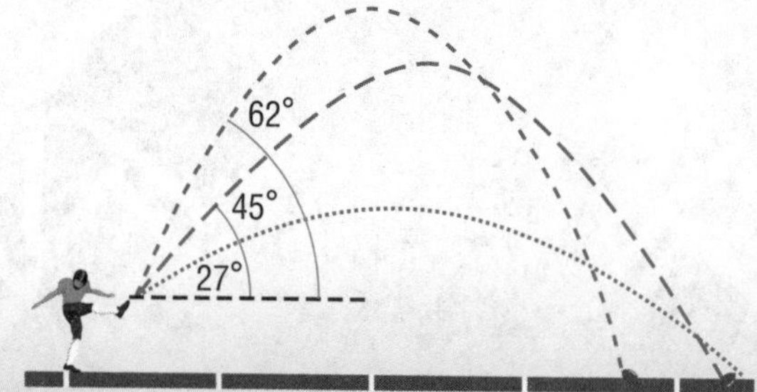

Career Project

It's time to update your career portfolio! Choose a piece of sports equipment and describe how it has changed over the past 20 years. List the reasons for the changes.

____

____

____

____

____

Suppose you are an employer hiring a sports equipment designer. What questions would you ask a potential employee?

- ________
- ________
- ________

Chapter Review

Vocabulary Check

Write the correct term for each clue in the crossword puzzle.

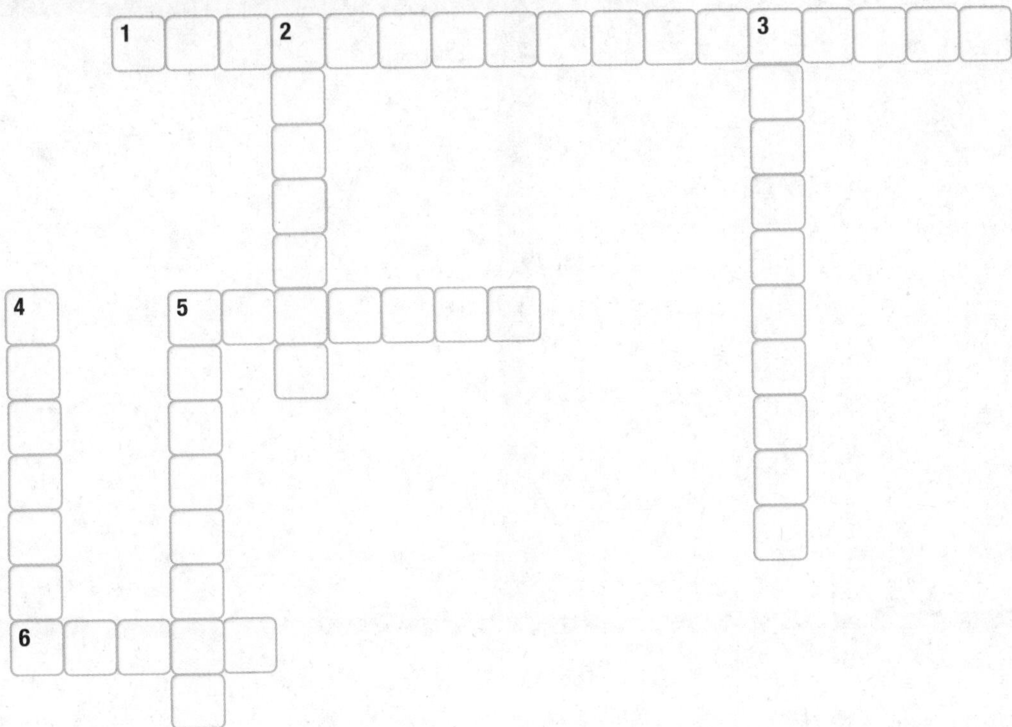

Across

1. easy to divide mentally
5. a number that has a digit in the tenths place, hundredths place, or beyond
6. to find an approximate value for a number

Down

2. the answer to a multiplication problem
3. a number with more than one digit
4. the number by which the dividend is being divided
5. a number that is being divided

Key Concept Check

Use Your FOLDABLES®

Use your Foldable to help review the chapter.

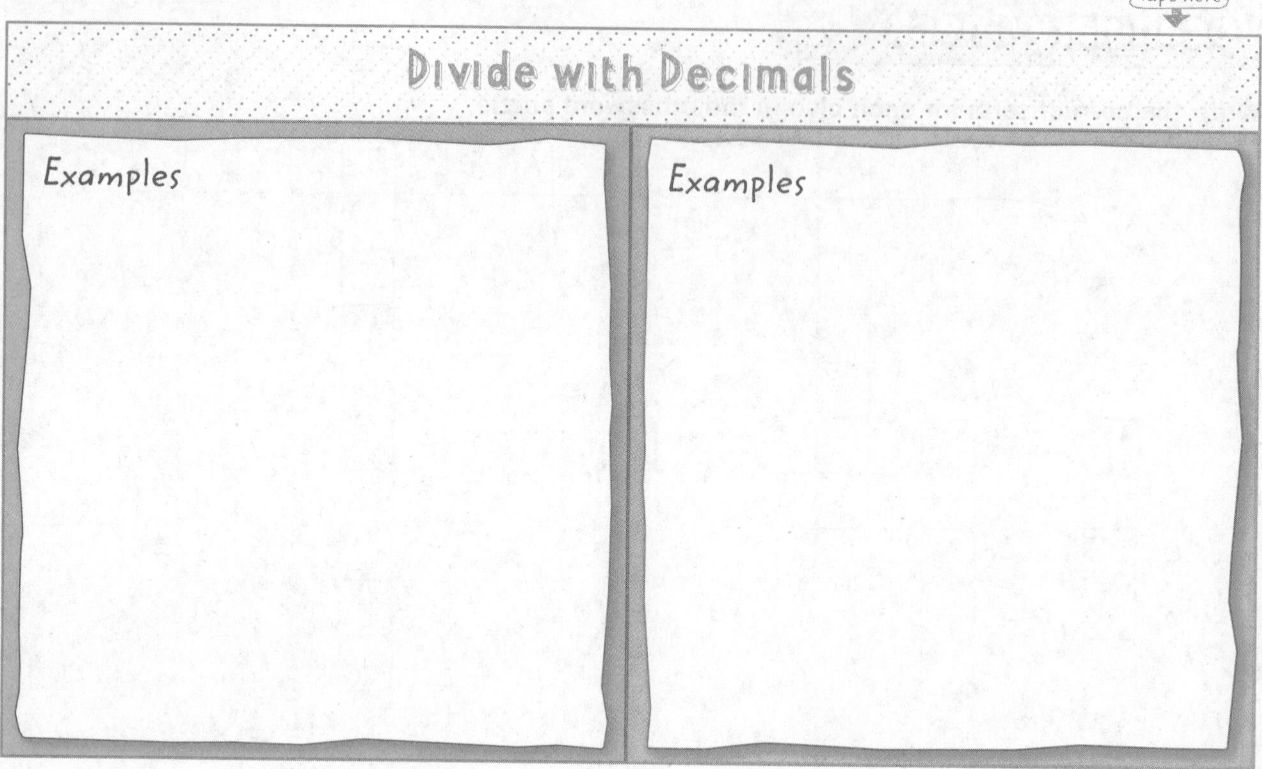

Got it?

Complete the cross number puzzle by solving the problems.

1		2		3		
		4				
	5			6		7
			8			
9					11	
			10			
12				13		

Across

1. 34.5 × 14

3. 569.6 ÷ 3.2

4. 18.5 × 40

5. 50.4 ÷ 2.4

6. 562.39 + 304.61

9. 42.5 × 116

10. 339.2 × 2.5

12. 1,584 ÷ 4.5

13. 1,218 ÷ 6

Down

1. 24.3 + 15.7

2. 21.2 × 17.5

3. 33.75 × 3.2

5. 146.53 + 92.47

7. 2,628 ÷ 36.5

8. 24 × 4.5

9. 263.4 + 199.6

11. 35.2 × 25

Problem Solving

1. A car travels 57.9 miles per hour for 3.2 hours. Estimate the number of miles driven. (Lesson 1) ______

2. A loaf of bread costs $1.79. How much would five loaves cost? (Lesson 3) ______

3. What is the area of the base of the fountain below? (Lesson 4) ______

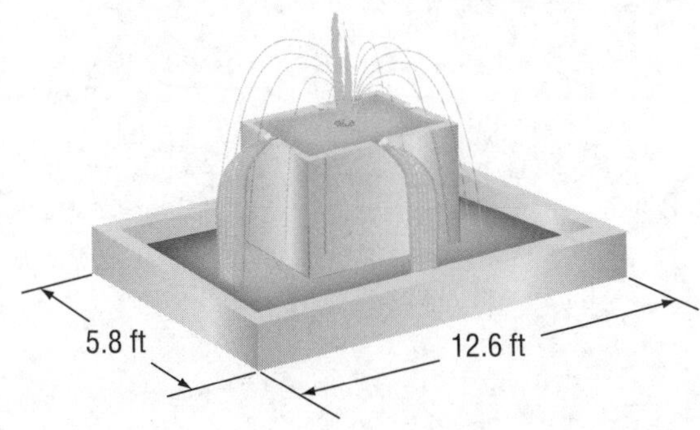

4. A marathon race is 26.2 miles long. Lacey ran the marathon in 3.6 hours. On average, how many miles did she run per hour? Round to the nearest tenth. (Lesson 8) ______

5. The speed of light is 1.86 × 100,000 miles per second. Write this number in standard form. (Lesson 4) ______

6. How many dimes are in $8,590? (Lesson 7) ______

7. **CCSS Use Math Tools** The table shows the height of members of Evan's family. His sister, Cindy, is 0.8 times his height. Which is a reasonable height for Cindy: about 4 feet, 4.5 feet, or 6 feet? Explain. (Lesson 4) ______

Family Member	Height (ft)
Evan	5.75
Grace	5.5
Jasper	6.25
Tron	5.25

Reflect

Answering the Essential Question

Use what you learned about computing with multi-digit numbers to complete the graphic organizer.

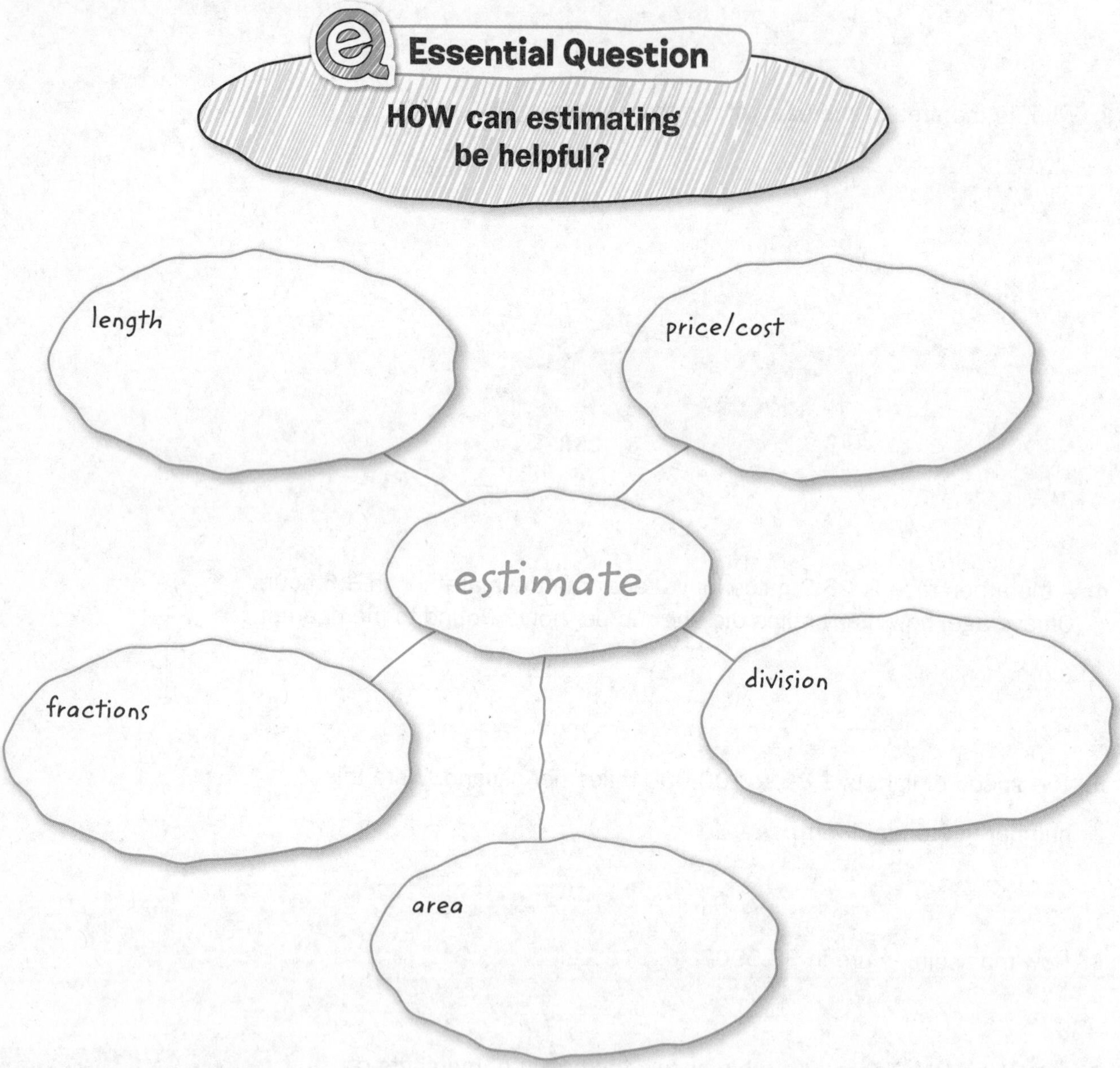

Essential Question

HOW can estimating be helpful?

estimate

length

price/cost

fractions

division

area

Answer the Essential Question. HOW can estimating be helpful?

Chapter 4
Multiply and Divide Fractions

Essential Question

WHAT does it mean to multiply and divide fractions?

Common Core State Standards

Content Standards
6.NS.1, 6.RP.3, 6.RP.3d

Mathematical Practices
1, 2, 3, 4, 5, 6, 7, 8

Math in the Real World

Construction Builders measure using fractions and convert measurements between different units.

A $12\frac{2}{3}$-foot board needs to be divided into 19-inch pieces. On the board below, make the divisions to show the number of pieces.

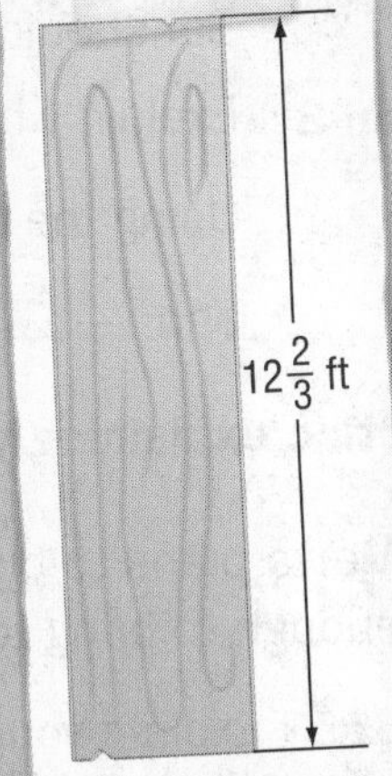

FOLDABLES® Study Organizer

1 Cut out the Foldable on page FL9 of this book.

2 Place your Foldable on page 336.

3 Use the Foldable throughout this chapter to help you learn about multiplying and dividing fractions.

What Tools Do You Need?

Vocabulary

Commutative Property
dimensional analysis
reciprocals
unit ratio

Study Skill: Writing Math

Explain Your Answer When you explain your answer, you give reasons why your answer is correct.

Sal wants to buy 5 packages of the limited edition cards. Is $20 enough for 5 packages, or does Sal need to bring $25 to the store to buy them? Explain your answer.

Football Card Prices	
Package	**Price ($)**
All-Star	3.75
Limited Edition	4.59
Deluxe	5.99

Step 1 Estimate.

5 × $4 = ________ Round down.

5 × $5 = ________ Round up.

Step 2 Answer the question.

Sal should bring ________ to the store.

Step 3 Explain why. Write your explanation in complete sentences.

Using the estimate, Sal knows that the actual cost is between ________ and ________. So ________ is not enough.

Practice explaining your answer.

1. Marta plans to buy 2 baseballs and 1 baseball glove. Is $50 enough to bring to the store or does Marta need to bring $55? Explain your answer. ________________________________

When Will You Use This?

Your Turn! You will solve this problem in the chapter.

Try the Quick Check below.
Or, take the Online Readiness Quiz.

Quick Review

CCSS **Common Core Review** 5.NF.1, 5.NF.2

Example 1

Estimate $4\frac{5}{6} + 1\frac{1}{8}$.

Think: $\frac{5}{6} > \frac{1}{2}$. So, $4\frac{4}{6}$ rounds to 5.

Think: $\frac{1}{8} < \frac{1}{2}$. So, $1\frac{1}{8}$ rounds to 1.

$$4\frac{5}{6} + 1\frac{1}{8}$$
$$5 + 1 = 6$$

So, $4\frac{5}{6} + 1\frac{1}{8}$ is about 6.

Example 2

Add $5\frac{9}{10} + 3\frac{3}{4}$. Write in simplest form.

$5\frac{9}{10} \rightarrow 5\frac{18}{20}$

$+\ 3\frac{3}{4} \rightarrow 3\frac{15}{20}$

$8\frac{33}{20}$ or $9\frac{13}{20}$

Rename the fractions using the LCD, 20.

So, $5\frac{9}{10} + 3\frac{3}{4}$ is equal to $9\frac{13}{20}$.

Quick Check

Estimate with Fractions **Estimate each sum or difference.**

1. $6\frac{7}{8} + 5\frac{1}{4} \approx$ ______
2. $3\frac{1}{7} + 8\frac{1}{9} \approx$ ______
3. $12\frac{1}{5} - 5\frac{5}{6} \approx$ ______

Add and Subtract Fractions **Add or subtract. Write in simplest form.**

4. $\frac{9}{10} + \frac{3}{5} =$ ______
5. $7\frac{2}{3} - 3\frac{1}{7} =$ ______
6. $9\frac{7}{8} - 2\frac{5}{6} =$ ______

7. Suppose a plant grew $4\frac{1}{2}$ inches one week and $2\frac{3}{8}$ inches the next week. How many inches did it grow during both weeks?

______

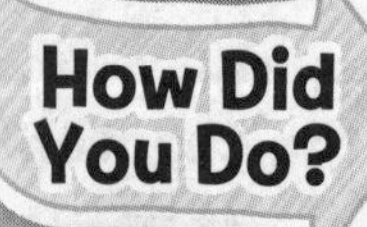

Which problems did you answer correctly in the Quick Check? Shade those exercise numbers below.

① ② ③ ④ ⑤ ⑥ ⑦

Lesson 1

Estimate Products of Fractions

What You'll Learn

Scan the lesson. List two headings you would use to make an outline of the lesson.

-

-

Essential Question

WHAT does it mean to multiply and divide fractions?

Common Core State Standards

Content Standards
Preparation for 6.NS.1

Mathematical Practices
1, 3, 4, 5

Real-World Link

Nature A wildlife preserve has 16 tigers, about $\frac{1}{3}$ of which are male. The tigers are represented by the counters below.

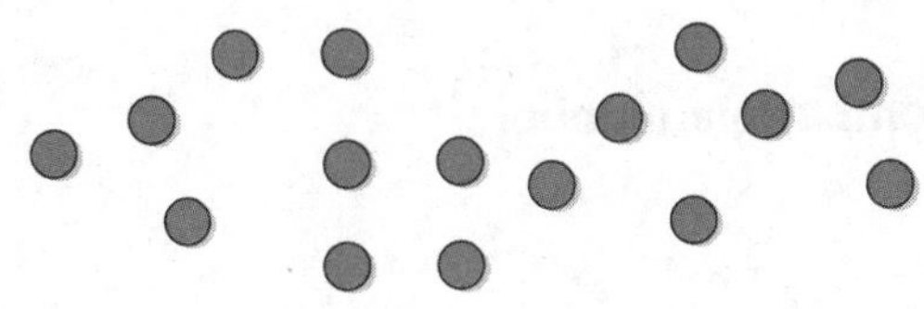

1. Can you separate the counters into three equal groups? Explain.

2. Fill in the multiples of 3 on the number line. Place a dot at 16.

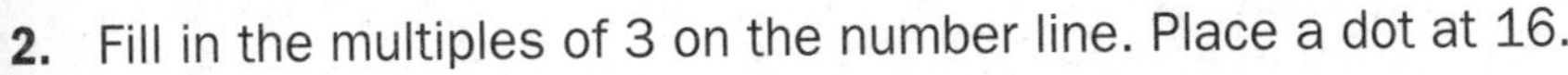

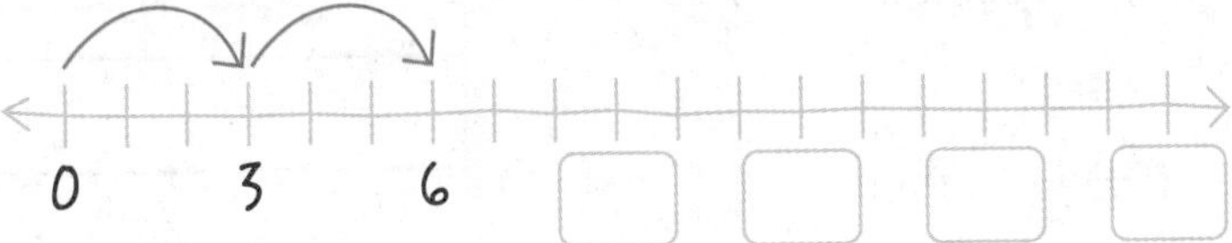

3. What multiple of 3 is closest to 16? ☐

4. Arrows jump from 0 to 3 to 6. Continue the pattern. How many jumps are from 0 to your answer in Exercise 3? ☐ jumps

5. About how many tigers in the preserve are male? Explain.

Work Zone

Estimate Using Compatible Numbers

You have already used compatible numbers to estimate quotients. You can also use compatible numbers to estimate products.

Examples

1. Estimate $\frac{1}{4} \times 13$.

Find a multiple of 4 close to 13. 12 and 4 are compatible numbers since $12 \div 4 = 3$.

Method 1 **Use a model.**

Divide the bar representing 12 into 4 sections.

3	3	3	3

(12)

Each section is $\frac{1}{4}$ of 12, or 3.

Label each section 3.

Method 2 **Use compatible numbers.**

$\frac{1}{4} \times 13 \approx \frac{1}{4} \times 12$

≈ 3

So, $\frac{1}{4} \times 13$ *is about* 3.

Compatible Numbers

Use compatible numbers to estimate the product of a fraction and a whole number. Find compatible numbers using the denominator of the fraction and the whole number.

2. Estimate $\frac{2}{5}$ of 11.

Find a multiple of 5 close to 11. 10 and 5 are compatible numbers since $10 \div 5 = 2$.

$\frac{1}{5} \times 11 \approx \frac{1}{5} \times 10$

≈ 2

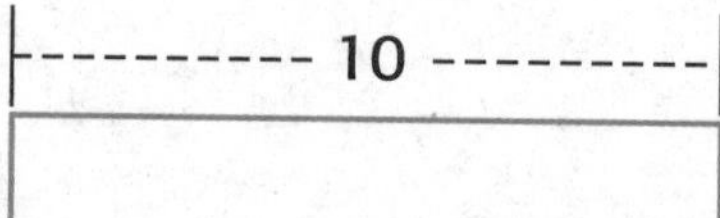

If $\frac{1}{5}$ of 10 is 2, then

$\frac{2}{5}$ of 10 is 2×2, or ☐.

So, $\frac{2}{5} \times 11$ *is about* ☐.

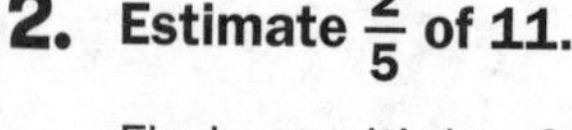

Show your work.

a. ____

b. ____

c. ____

Got It? **Do these problems to find out.**

a. $\frac{1}{5} \times 16$ **b.** $\frac{5}{6} \times 13$ **c.** $\frac{3}{4}$ of 23

Estimate by Rounding to 0, $\frac{1}{2}$, or 1

Key Concept

Words If the numerator of a fraction between 0 and 1 is almost as large as the denominator, round up. If the numerator is much smaller than the denominator, round down.

Examples $\frac{7}{8}$ rounds to 1; $\frac{1}{8}$ rounds to 0

Rounding fractions can help you find products of fraction factors.

Examples

3. Estimate $\frac{1}{3} \times \frac{7}{9}$.

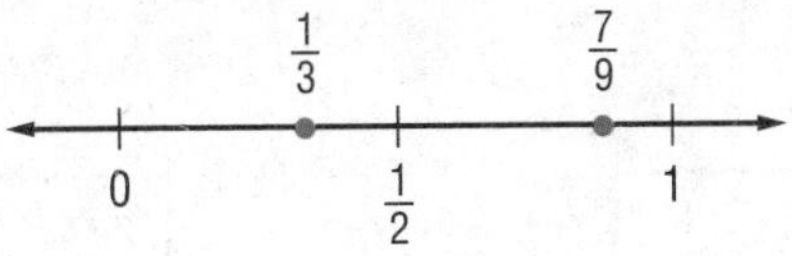

Number line: 0, $\frac{1}{3}$, $\frac{1}{2}$, $\frac{7}{9}$, 1

Dots are placed at $\frac{1}{3}$ and $\frac{7}{9}$.

Round to 0, $\frac{1}{2}$, or 1. $\frac{1}{3}$ is about $\frac{1}{2}$ and $\frac{7}{9}$ is about 1.

$\frac{1}{3} \times \frac{7}{9} \rightarrow \frac{1}{2} \times 1$

$\frac{1}{2} \times 1 = \frac{1}{2}$

So, $\frac{1}{3} \times \frac{7}{9}$ *is about* $\frac{1}{2}$.

4. Estimate $\frac{1}{9} \times \frac{4}{5}$.

Place dots on the number line at $\frac{1}{9}$ and $\frac{4}{5}$.

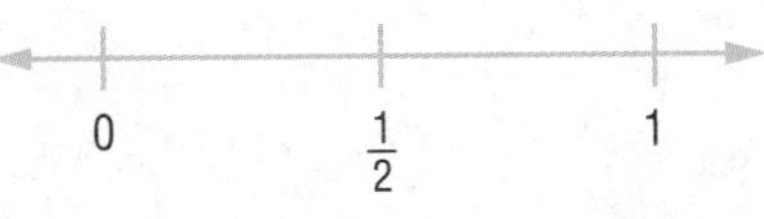

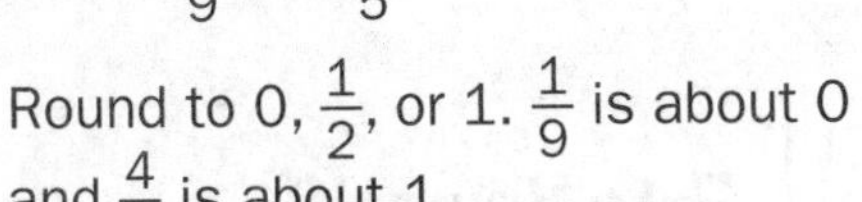

Round to 0, $\frac{1}{2}$, or 1. $\frac{1}{9}$ is about 0 and $\frac{4}{5}$ is about 1.

$\frac{1}{9} \times \frac{4}{5} \rightarrow 0 \times 1 =$ ☐

So, $\frac{1}{9} \times \frac{4}{5}$ *is about* ☐.

Got It? Do these problems to find out.

Estimate each product.

d. $\frac{5}{8} \times \frac{9}{10}$

e. $\frac{5}{6} \times \frac{9}{10}$

f. $\frac{5}{6}$ of $\frac{1}{9}$

d. ______

e. ______

f. ______

Real World Example

5. Estimate the area of the flower bed.

Round each mixed number to the nearest whole number.

$14\frac{7}{8} \times 6\frac{1}{8} \rightarrow 15 \times 6 = 90$

So, the area *is about* 90 square feet.

$6\frac{1}{8}$ ft

$14\frac{7}{8}$ ft

Guided Practice

Estimate each product. Use a bar diagram if needed. (Examples 1–4)

1. $\frac{1}{8} \times 15 \approx$ ________

2. $\frac{2}{5}$ of $26 \approx$ ________

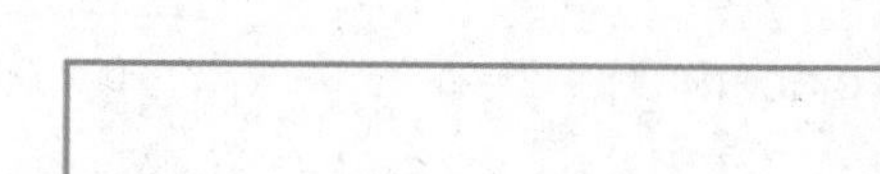

3. $\frac{1}{5} \times \frac{8}{9} \approx$ ________

4. $6\frac{2}{3} \times 4\frac{1}{5} \approx$ ________

5. A border is made of $32\frac{2}{3}$ bricks that are $1\frac{1}{6}$ feet long. About how long is the border? (Example 5)

6. A kitchen measures $24\frac{1}{6}$ feet by $9\frac{2}{3}$ feet. Estimate the area of the kitchen. (Example 5)

7. **Building on the Essential Question** Why is estimating products of fractions useful?

Rate Yourself!

How confident are you about estimating the products of fractions? Color the square that applies.

For more help, go online to access a Personal Tutor.

Name ______________________ My Homework ______________________

Independent Practice

Go online for Step-by-Step Solutions

Estimate each product. Use a bar diagram if needed. (Examples 1–4)

1. $\frac{1}{4} \times 21 \approx$ ______________

2. $\frac{5}{7}$ of $22 \approx$ ______________

3. $\frac{5}{7} \times \frac{1}{9} \approx$ ______________

4. $4\frac{1}{3} \times 2\frac{3}{4} \approx$ ______________

5. Cyrus is inviting 11 friends over for pizza. He would like to have enough pizza so each friend can have $\frac{1}{4}$ of a pizza. About how many pizzas should he order? (Example 5)

____

6. Hakeem's front porch measures $9\frac{3}{4}$ feet by 4 feet. Estimate the area of his front porch. (Example 5)

____

7. CCSS **Use Math Tools** Refer to the graphic novel frame for Exercises a–b.

a. If each bag holds $3\frac{3}{4}$ pounds, estimate how many pounds of birdseed Elisa, Luis, and Dwayne purchased. ______________

____

b. Suppose each bag costs \$14.99. Estimate the total cost of 5 bags.

____

Estimate the area of each rectangle.

8. ______

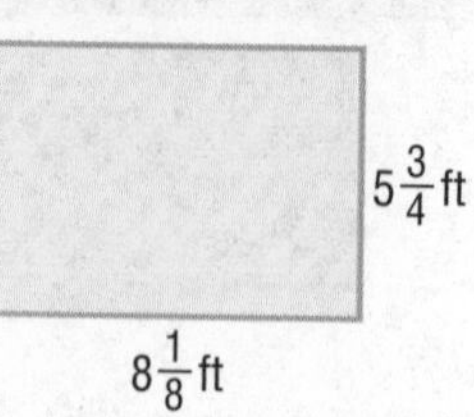

9. ______

$3\frac{1}{4}$ in.

$9\frac{5}{8}$ in.

10. **STEM** Seattle, Washington, received rain on $\frac{7}{10}$ of the days in a recent month. If this pattern continues, about how many days would it *not* rain in 90 days? ______

H.O.T. Problems Higher Order Thinking

11. **CCSS Justify Conclusions** By what fraction would you multiply $8\frac{1}{2}$ so that the product is about 5? Explain your reasoning. ______

12. **CCSS Persevere with Problems** Determine which point on the number line could be the graph of the product of the numbers graphed at *C* and *D*.

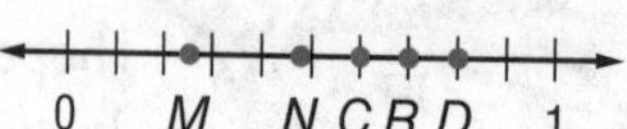

Explain your reasoning. ______

Standardized Test Practice

13. Which is the best estimate of the area of the rectangle?

Ⓐ 2 yd^2

Ⓑ 3 yd^2

Ⓒ 6 yd^2

Ⓓ 10 yd^2

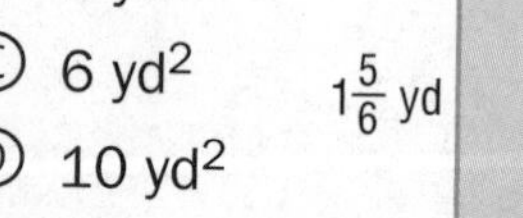

$2\frac{2}{3}$ yd

$1\frac{5}{6}$ yd

Name ____________________ My Homework ____________________

Extra Practice

Estimate each product. Use a bar diagram if needed.

14. $\frac{1}{5} \times 26 \approx$ 5

$26 \approx 25; \frac{1}{\cancel{5}_1} \times \cancel{25}^{5} = \frac{5}{1}$ or 5

15. $\frac{1}{3}$ of $41 \approx$ 14

$41 \approx 42; \frac{1}{\cancel{3}_1} \times \cancel{42}^{14} = \frac{14}{1}$ or 14

16. $\frac{1}{6}$ of $17 \approx$ ________

17. $\frac{2}{9}$ of $88 \approx$ ________

18. $\frac{2}{3} \times 10 \approx$ ________

19. $\frac{3}{8} \times 4 \approx$ ________

Estimate the area of the each rectangle.

20. ____________________

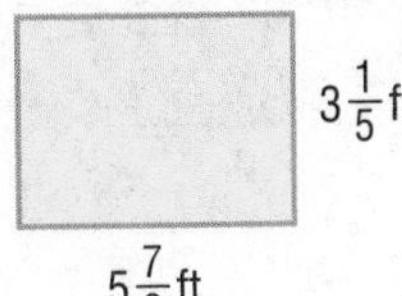

$3\frac{1}{5}$ ft

$5\frac{7}{9}$ ft

21. ____________________

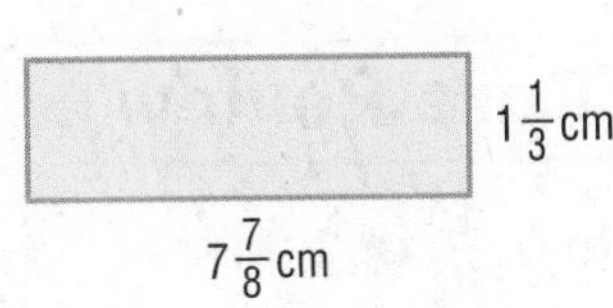

$1\frac{1}{3}$ cm

$7\frac{7}{8}$ cm

22. Tara would like to finish $\frac{2}{5}$ of her book by next Friday. If the book has 203 pages, about how many pages does she need to read? ____________________

23. Javier is organizing his movie collection. He discovers that $\frac{5}{8}$ of his movies are action movies. If he has 46 movies, about how many are action movies? ____________________

24. CCSS **Justify Conclusions** Marco has a collection of 38 state quarters. If $\frac{3}{5}$ of his quarters are dated 2005, what is the approximate value of the quarters from 2005? Explain your answer to a classmate.

____

Standardized Test Practice

25. The table shows the number of students in grades 6–8 who went to a local museum. Of these, between one half and three fourths packed their lunch. Which of the following ranges could represent the number of students who packed their lunch?

Students Visiting the Museum	
Grade	**Number of Students**
6	45
7	48
8	40

Ⓐ Less than 65

Ⓑ Between 65 and 100

Ⓒ Between 100 and 130

Ⓓ More than 130

26. According to a survey, $\frac{3}{5}$ of the students prefer outdoor activities after school. If 63 students were surveyed, about how many would prefer playing sports outdoors?

Ⓕ 24 students

Ⓖ 30 students

Ⓗ 36 students

Ⓘ 48 students

27. In Mrs. Petrocelli's class, $\frac{5}{7}$ of the students are wearing jeans today. If there are about 27 students in her class, how many students are wearing jeans today?

Ⓐ 4 students

Ⓑ 8 students

Ⓒ 20 students

Ⓓ 24 students

28. Short Response A drawing of a square room is shown. Estimate the area in square feet of tile needed to cover the floor.

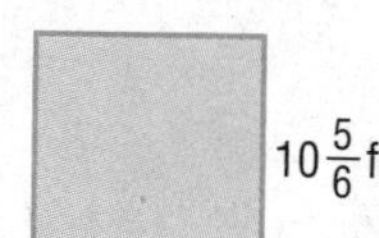

Common Core Review

Round each fraction to 0, $\frac{1}{2}$, or 1. 4.NF.2

29. $\frac{7}{12} \approx$ ______

30. $\frac{5}{6} \approx$ ______

31. $\frac{2}{11} \approx$ ______

32. $\frac{4}{9} \approx$ ______

33. Graph $\frac{3}{4}$, $\frac{1}{3}$, $\frac{1}{4}$, and $\frac{2}{3}$. 4.NF.2

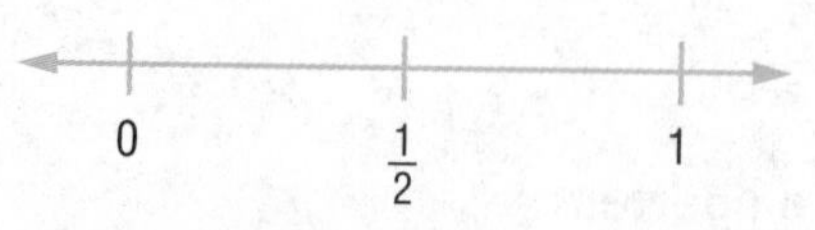

34. Graph $\frac{5}{10}$, $\frac{4}{5}$, $\frac{2}{5}$, and $\frac{7}{10}$. 4.NF.2

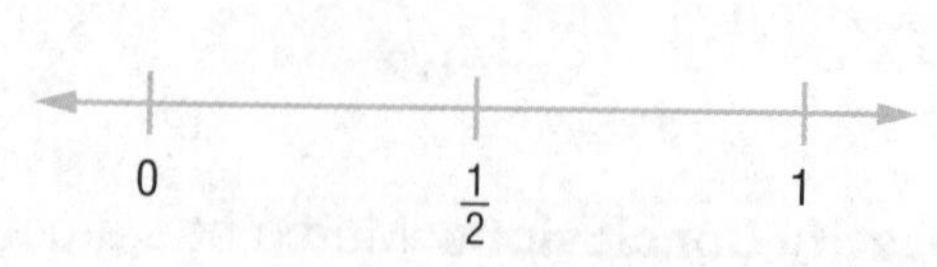

35. Jasper wants to paint one wall of his room. If the wall is 12 feet wide and 10 feet tall, what is the area of the wall? 4.MD.3 ______

Lesson 2

Multiply Fractions and Whole Numbers

What You'll Learn

Scan the lesson. Predict two things you will learn about multiplying fractions and whole numbers.

-
-

Essential Question

WHAT does it mean to multiply and divide fractions?

Vocabulary

Commutative Property

Common Core State Standards

Content Standards
Preparation for 6.NS.1

Mathematical Practices
1, 3, 4, 7

Vocabulary Start-Up

A commuter train travels back and forth but does not change the distance traveled. In mathematics, operations that follow the **Commutative Property** can be performed in any order. For example, addition and multiplication are commutative.

Draw a line to "Commutative" if the examples can be done in either order. Draw a line to "Not Commutative" if the order changes the outcome.

Commutative

Not Commutative

$12 \div 6; 6 \div 12$

tying your left shoe; tying your right shoe

$5 \times 7; 7 \times 5$

play a soccer game; change into your team uniform

$15 + 5; 5 + 15$

Real-World Link

Some morning routines can be done in any order. Sometimes, the order matters. Describe a situation when the order you perform two actions is important.

Key Concept

Multiply a Whole Number by a Fraction

Words Write the whole number as a fraction. Multiply the numerators and multiply the denominators.

Example

$5 \times \frac{3}{4} = \frac{5}{1} \times \frac{3}{4}$ Write 5 as $\frac{5}{1}$.

$= \frac{5 \times 3}{1 \times 4}$ Multiply.

$= \frac{15}{4}$ or $3\frac{3}{4}$ Simplify.

Work Zone

Example

1. **Find $2 \times \frac{2}{5}$.**

Method 1 **Use an area model.**

Shade $\frac{2}{5}$ of each of the first two columns.

A total of $\frac{4}{5}$ has been shaded.

Shade $\frac{4}{5}$ on the third column.

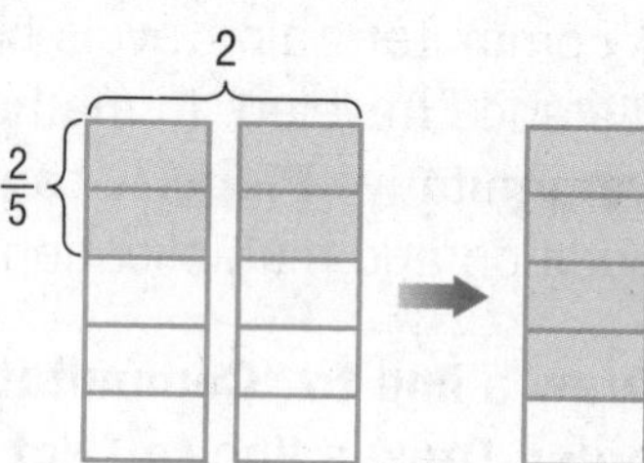

Method 2 **Use an equation.**

Estimate $2 \times \frac{1}{2} = 1$

$2 \times \frac{2}{5} = ■$

$2 \times \frac{2}{5} = \frac{2}{1} \times \frac{2}{5}$ Write 2 as $\frac{2}{1}$.

$= \frac{2 \times 2}{1 \times 5}$ Multiply.

$= \frac{4}{5}$ Simplify.

Using either method, $2 \times \frac{2}{5}$ is $\frac{4}{5}$.

Check for Reasonableness $\frac{4}{5} \approx 1$ ✓

a. ____________

b. ____________

c. ____________

Got It? **Do these problems to find out.**

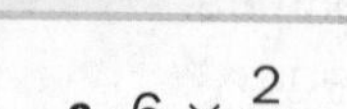

a. $6 \times \frac{2}{3}$

b. $9 \times \frac{1}{3}$

c. $4 \times \frac{1}{8}$

Multiply a Fraction by a Whole Number

When multiplying whole numbers and fractions, the order of the factors does not change the product. So, $4 \times \frac{3}{5} = \frac{3}{5} \times 4$. This is an example of the Commutative Property.

Examples

Tutor

2. **Find $\frac{3}{5} \times 4$.**

Method 1 **Use an area model.**

Shade $\frac{3}{5}$ of each of the 4 columns.

A total of $\frac{12}{5}$ or $2\frac{2}{5}$ has been shaded.

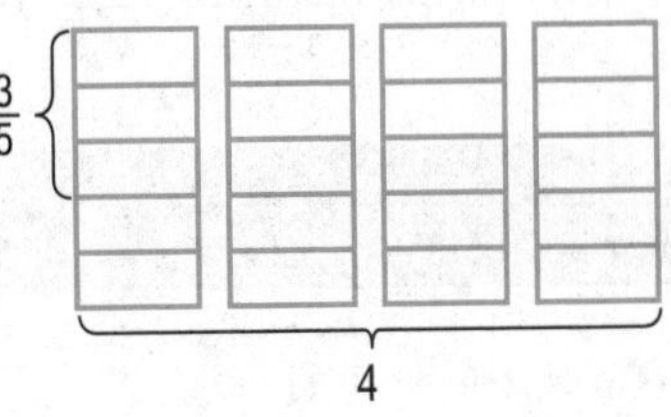

Method 2 **Use an equation.** Estimate $\frac{1}{2} \times 4 = 2$

$\frac{3}{5} \times 4 = \blacksquare$

$\frac{3}{5} \times 4 = \frac{3}{5} \times \frac{4}{1}$ Write 4 as $\frac{4}{1}$.

$= \frac{3 \times 4}{5 \times 1}$ Multiply.

$= \frac{12}{5}$ Simplify. Compare to the estimate.

$= 2\frac{2}{5}$

Using either method, $\frac{3}{5} \times 4$ is $2\frac{2}{5}$.

Renaming
To rename an improper fraction as a mixed number, divide the numerator by the denominator. Write the remainder as a fraction with the divisor as the denominator.

3. **Find $\frac{1}{4} \times 5$.**

Estimate $\frac{1}{4} \times 4 = 1$

$\frac{1}{4} \times 5 = \frac{1}{4} \times \frac{5}{1}$ Write 5 as $\frac{5}{1}$.

$= \frac{1 \times 5}{4 \times 1}$ Multiply.

$= \frac{5}{4}$ or $1\frac{1}{4}$ Simplify.

Check for Reasonableness $1\frac{1}{4} \approx 1$ ✓

Got It? **Do these problems to find out.**

d. $\frac{1}{2} \times 3$

e. $\frac{2}{5} \times 4$

f. $\frac{3}{4} \times 5$

d. ______

e. ______

f. ______

Example

4. A sloth spends $\frac{4}{5}$ of its life asleep. If a sloth lives to be 28 years old, how many years does it spend asleep?

Find $\frac{4}{5}$ of 28. — Estimate $\frac{4}{5}$ of 30 is 24.

$\frac{4}{5} \times 28 = \frac{4}{5} \times \frac{28}{1}$ — Write 28 as $\frac{28}{1}$.

$= \frac{4 \times 28}{5 \times 1}$ — Multiply.

$= \frac{112}{5}$ or $22\frac{2}{5}$ — Simplify. Compare to the estimate.

A sloth spends $22\frac{2}{5}$ years of its life asleep.

Guided Practice

Multiply. Write in simplest form. (Examples 1–3)

1. $10 \times \frac{4}{5} =$ ______

2. $2 \times \frac{3}{4} =$ ______

3. $\frac{3}{8} \times 11 =$ ______

4. $\frac{3}{7} \times 9 =$ ______

5. A cat spends $\frac{2}{3}$ of its life asleep. If a cat lives to be 15 years old, how many years did it spend asleep? (Example 4) ______

6. **Building on the Essential Question** How is the process used to multiply a fraction and a whole number similar to the process used to multiply two whole numbers?

______

______

______

______

Rate Yourself!

Are you ready to move on? Shade the section that applies.

YES ? NO

For more help, go online to access a Personal Tutor.

Tutor

FOLDABLES *Time to update your Foldable!*

Independent Practice

Go online for Step-by-Step Solutions  eHelp

Multiply. Write in simplest form. (Examples 1–3)

1. $20 \times \frac{3}{4} =$ ______

2. $14 \times \frac{2}{7} =$ ______

3. $10 \times \frac{1}{5} =$ ______

4. $\frac{3}{4} \times 6 =$ ______

5. $\frac{2}{5} \times 11 =$ ______

6. $\frac{1}{4} \times 6 =$ ______

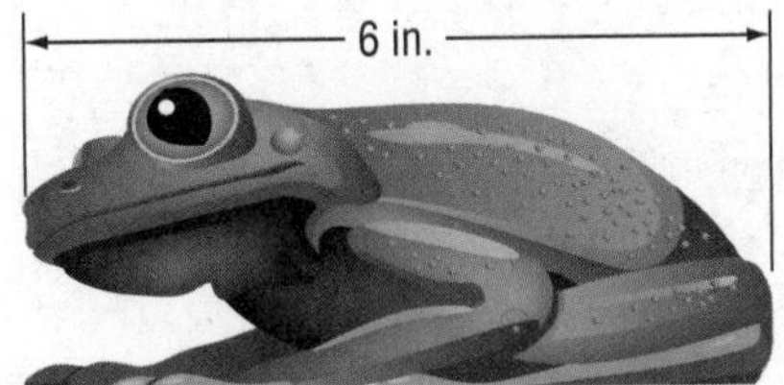

7. **STEM** The male Cuban tree frog is about $\frac{2}{5}$ the size of the female Cuban tree frog. The average size of the female Cuban tree frog is shown at the right. What is the size of the male Cuban tree frog? (Example 4) ______

8. The Mississippi River is the second longest river in the United States, second only to the Missouri River. The Mississippi River is about $\frac{23}{25}$ the length of the Missouri River. If the Missouri River is 2,540 miles long, how long is the Mississippi River? (Example 4)

______

______

9. One evening, $\frac{2}{3}$ of Mrs. Thorne's students watched a reality television show. Of Mrs. Lombardo's students, $\frac{4}{5}$ watched the same reality show. Which teacher had more students that watched the reality show? Explain.

Teacher	Number of Students
Mrs. Thorne	36
Mrs. Lombardo	30
Mr. Hollern	28

______

______

10. CCSS **Persevere with Problems** The table shows where sixth grade students at Sharonton Middle School attended fifth grade. There are 156 sixth grade students. How many more students attended Sharonton Elementary than Deacon Elementary?

School	Fraction of Students
Sharonton Elementary	$\frac{1}{2}$
Deacon Elementary	$\frac{1}{4}$
Banyon Elementary	$\frac{1}{6}$
New Students	$\frac{1}{12}$

______

11. **CCSS Persevere with Problems** Students at Marzo Middle School were recently surveyed. The results reported $\frac{1}{4}$ of sixth grade students, $\frac{3}{10}$ of seventh grade students, and $\frac{2}{7}$ of eighth grade students plan a career in STEM. In which grade do the most students plan to have careers in STEM?

Grade	Total Students
6	152
7	160
8	147

H.O.T. Problems Higher Order Thinking

12. **CCSS Identify Structure** Write a problem involving the multiplication of a fraction and a whole number with a product that is between 8 and 10.

13. **CCSS Find the Error** Noah is finding $\frac{3}{4}$ of 8. Find his mistake and correct it.

14. **CCSS Persevere with Problems** Use the digits 2, 3, and 5 to create a fraction and a whole number with a product greater than 2.

Standardized Test Practice

15. Jenny made five loaves of banana bread that had $\frac{1}{4}$ cup of oil in each loaf. How many cups of oil were used in all?

Ⓐ 5 Ⓒ $1\frac{1}{4}$

Ⓑ 4 Ⓓ $\frac{3}{4}$

Extra Practice

Multiply. Write in simplest form.

16. $12 \times \frac{1}{3} =$ 4

$\frac{12}{1} \times \frac{1}{3} = \frac{4}{1}$ or 4

17. $18 \times \frac{1}{3} =$ 6

$\frac{18}{1} \times \frac{1}{3} = \frac{6}{1}$ or 6

18. $8 \times \frac{1}{4} =$ ______

19. $\frac{1}{5} \times 7 =$ ______

20. $\frac{3}{7} \times 8 =$ ______

21. $\frac{5}{6} \times 15 =$ ______

22. For a singing contest in which 42,000 votes were cast, the winner received $\frac{3}{5}$ of the votes. How many votes did the winner *not* receive?

______

23. STEM In a recent year, the weather was partly cloudy $\frac{2}{5}$ of the days. Assuming there are 365 days in a year, how many days were partly cloudy?

______

24. CCSS **Model with Mathematics** Write a real-world problem that involves multiplying a fraction and a whole number. Solve the problem and use estimation to check for reasonableness.

______

______

Standardized Test Practice

25. Leonard used $\frac{2}{7}$ of his paycheck to pay his cell phone bill. How much was Leonard's cell phone bill?

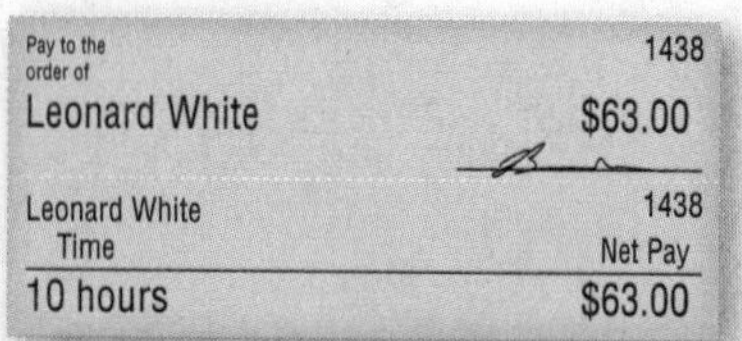

Ⓐ $12
Ⓑ $18
Ⓒ $27
Ⓓ $36

26. There are 150 students in the band and 90 students in the chorus. One half of the band members and $\frac{4}{5}$ of the chorus members participated in a charity concert. How many more band members than chorus members participated in the concert?

Ⓕ 3
Ⓖ 18
Ⓗ 27
Ⓘ 72

27. **Short Response** It takes $\frac{3}{8}$ yards of fabric to make a blanket. How many yards of fabric will it take to make 16 blankets?

______

______

28. David made 10 batches of muffins. He used $\frac{2}{3}$ cup of milk in each batch. How much milk did David use?

Ⓐ 6 cups
Ⓑ $6\frac{1}{3}$ cups
Ⓒ $6\frac{2}{3}$ cups
Ⓓ 20 cups

CCSS Common Core Review

Multiply. 5.NBT.5.

29. 22 × 13 = ______

30. 18 × 11 = ______

31. 17 × 9 = ______

32. Hayley's guitar lesson lasts $\frac{3}{4}$ hour. How many minutes does Hayley spend at her guitar lesson? Use the clock to help you find your answer.
5.NF.6 ______

______

33. Miguel has one foot of string. He cuts the string into fourths. How many inches is each piece of string? 5.MD.1 ______

______

Lesson 3

Multiply Fractions

What You'll Learn

Scan the lesson. List two real-world scenarios in which you multiply fractions.

- ______________________________
- ______________________________

Essential Question

WHAT does it mean to multiply and divide fractions?

Common Core State Standards

Content Standards
Preparation for 6.NS.1

Mathematical Practices
1, 3, 4, 5, 7

Real-World Link

Reptiles A chameleon's body is about $\frac{1}{2}$ the length of its tongue. A certain chameleon has a tongue that is $\frac{2}{3}$ foot long.

$\frac{2}{3}$ ft

$\frac{1}{2}$ size of tongue

Use an area model to show $\frac{1}{2}$ of $\frac{2}{3}$ or $\frac{1}{2} \times \frac{2}{3}$.

1. Divide the rectangle into 2 rows. Then divide it into 3 columns.
2. Shade a rectangle that is $\frac{1}{2}$ unit wide by $\frac{2}{3}$ unit long.

3. Refer to the model. The section that was shaded represents $\frac{1}{2} \times \frac{2}{3}$. What fraction represents $\frac{1}{2} \times \frac{2}{3}$? ______
4. What is the relationship between the numerators and denominators of the factors and the numerator and denominator of the product?

____

____

____

Key Concept

Multiply Fractions

Work Zone

Words Multiply the numerators and multiply the denominators.

Models

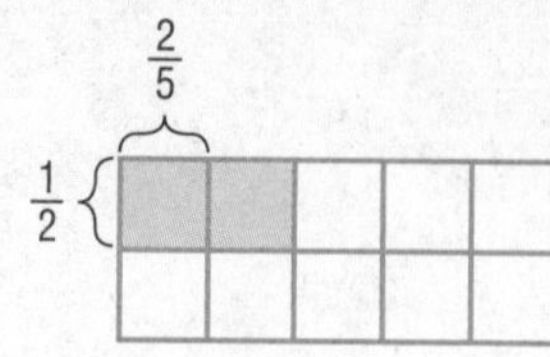

Numbers $\frac{2}{5} \times \frac{1}{2} = \frac{2 \times 1}{5 \times 2}$

Symbols $\frac{a}{b} \times \frac{c}{d} = \frac{a \times c}{b \times d}$, where b and d are not 0.

Example

1. Find $\frac{1}{3} \times \frac{1}{4}$. Write in simplest form.

Method 1 **Use a model.**

Divide the rectangle into 4 rows. Then divide the rectangle into 3 columns.

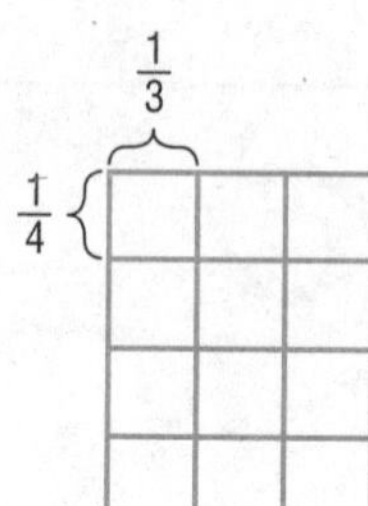

Shade a section that is $\frac{1}{4}$ unit wide by $\frac{1}{3}$ unit long.

The section that is shaded represents $\frac{1}{4} \times \frac{1}{3}$, or $\frac{1}{12}$.

Method 2 **Use an equation.**

$\frac{1}{3} \times \frac{1}{4} = \blacksquare$

$\frac{1}{3} \times \frac{1}{4} = \frac{1 \times 1}{3 \times 4}$ Multiply the numerators. Multiply the denominators.

$= \frac{1}{12}$ Simplify.

So, $\frac{1}{3} \times \frac{1}{4}$ is $\frac{1}{12}$.

a. ________

b. ________

c. ________

Got It? **Do these problems to find out.**

a. $\frac{1}{2} \times \frac{3}{5}$ **b.** $\frac{1}{3} \times \frac{3}{4}$ **c.** $\frac{2}{3} \times \frac{5}{6}$

Simplify Before Multiplying

If the numerators and the denominators have a common factor you can simplify *before* you multiply. Remember that factors are two or more numbers that are multiplied together to form a product.

$$\frac{\overset{1}{\cancel{2}}}{3} \times \frac{5}{\underset{3}{\cancel{6}}} = \frac{1}{3} \times \frac{5}{3}$$ Think: $2 \div 2 = 1$ Think: $6 \div 2 = 3$

$$= \frac{5}{9}$$

Examples

2. **Find $\frac{3}{4} \times \frac{5}{6}$.**

Estimate $\frac{1}{2} \times 1 = \frac{1}{2}$

$$\frac{3}{4} \times \frac{5}{6} = \frac{\overset{1}{\cancel{3}} \times 5}{4 \times \underset{2}{\cancel{6}}}$$ Divide both the numerator and the denominator by 3.

$$= \frac{5}{8}$$ Simplify. Compare to the estimate.

Check for reasonableness $\frac{1}{2} \approx \frac{5}{8}$ ✓

3. **Find $\frac{4}{9} \times 18$.**

Estimate $\frac{1}{2} \times 18 = 9$

$$\frac{4}{9} \times 18 = \frac{4}{9} \times \frac{18}{1}$$ Write 18 as a fraction with a denominator of 1.

$$= \frac{4 \times \overset{2}{\cancel{18}}}{\underset{1}{\cancel{9}} \times 1}$$ Divide both the numerator and the denominator by 9.

$$= \frac{8}{1} \text{ or } 8$$ Simplify. Compare to the estimate.

Check for reasonableness $9 \approx 8$ ✓

Simplifying
When multiplying fractions, it is easier to find the answer if you simplify before multiplying.

Got It? Do these problems to find out.

d. $\frac{3}{4} \times \frac{4}{9}$ **e.** $\frac{5}{6} \times \frac{9}{10}$ **f.** $\frac{3}{5} \times 10$

d. ________

e. ________

f. ________

Example

4. Frank had $\frac{1}{2}$ of the lawn left to mow. On Saturday, he mowed $\frac{2}{3}$ of what was left. What fraction of the entire lawn did Frank mow on Saturday?

$$\frac{1}{2} \times \frac{2}{3} = \frac{1 \times \overset{1}{\cancel{2}}}{\underset{1}{\cancel{2}} \times 3}$$ Divide both the numerator and denominator by 2.

$$= \frac{1}{3}$$ Simplify.

So, Frank mowed $\frac{1}{3}$ of the lawn on Saturday.

Guided Practice

Multiply. Write in simplest form. (Examples 1–3)

1. $\frac{1}{8} \times \frac{1}{2} =$ ______
2. $\frac{2}{3} \times \frac{4}{5} =$ ______
3. $\frac{4}{5} \times 10 =$ ______

4. $\frac{3}{4} \times 12 =$ ______
5. $\frac{3}{10} \times \frac{5}{6} =$ ______
6. $\frac{3}{5} \times \frac{5}{6} =$ ______

7. Rick has $\frac{1}{2}$ of a footlong sub left from yesterday. He ate $\frac{1}{3}$ of the leftover sandwich as a snack. What fraction of the entire sandwich did he eat as a snack? (Example 4)

8. **Building on the Essential Question** If two positive fractions are less than 1, why is their product also less than 1?

Rate Yourself!

Are you ready to move on? Shade the section that applies.

I have a few questions. | I'm ready to move on. | I have a lot of questions.

For more help, go online to access a Personal Tutor.

FOLDABLES *Time to update your Foldable!*

Name ______________________ My Homework ______________________

Independent Practice

Go online for Step-by-Step Solutions

Multiply. Write in simplest form. (Examples 1–3)

1. $\frac{1}{3} \times \frac{2}{5} =$ ______

2. $\frac{3}{4} \times \frac{5}{8} =$ ______

3. $\frac{2}{3} \times 4 =$ ______

4. $\frac{5}{6} \times 15 =$ ______

5. $\frac{2}{3} \times \frac{1}{4} =$ ______

6. $\frac{4}{9} \times \frac{3}{8} =$ ______

7. **Financial Literacy** Juanita spent $\frac{3}{4}$ of her allowance at the mall. Of the money spent at the mall, $\frac{1}{2}$ was spent on new earphones. What part of her allowance did Juanita spend on earphones? (Example 4)

____

8. A paint store has 35 gallons of paint in storage, $\frac{2}{5}$ of which are for outdoor use. The others are for indoor use. If each gallon costs $22, what is the total cost of the indoor paint in storage?

____

9. Homeroom 101 and Homeroom 102 share a hallway bulletin board. If Homeroom 101 uses $\frac{3}{5}$ of their half to display artwork, what fraction of the bulletin board is used to display Homeroom 101's artwork?

____

10. CCSS **Use Math Tools** Mr. Williams' physical education class lasts for $\frac{7}{8}$ hour.

a. How many minutes are spent warming up and cooling down?

____

b. How many minutes are *not* spent on instruction? Explain.

____

Part of $\frac{7}{8}$-hour Class	
playing game	$\frac{1}{2}$
instruction	$\frac{1}{5}$
warm-up and cool-down	$\frac{3}{10}$

11.  **Multiple Representations** Use the bar diagram.

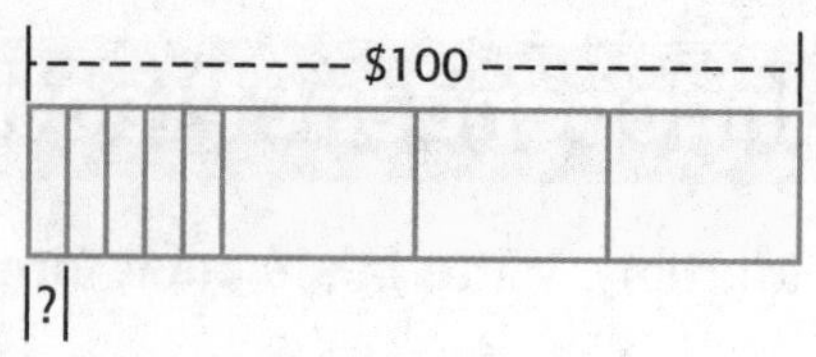

a. **Words** Write a real-world problem represented by the bar diagram.

b. **Models** Draw an area model to represent the situation.

c. **Words** Explain how you would solve your problem.

H.O.T. Problems Higher Order Thinking

12.  **Reason Inductively** State whether each statement is *true* or *false*. If the statement is *false*, provide a counterexample.

a. The product of two fractions that are each between 0 and 1 is also between 0 and 1.

b. The product of a mixed number between 4 and 5 and a fraction between 0 and 1 is always less than 4.

c. The product of two mixed numbers that are each between 4 and 5 is between 16 and 25.

13. **Identify Structure** If the product of two positive fractions *a* and *b* is $\frac{15}{56}$, find three pairs of possible values for *a* and *b*.

14. **Persevere with Problems** Justify why $\frac{a}{b} \times \frac{b}{c} \times \frac{c}{d} \times \frac{d}{e}$ is equal to $\frac{a}{e}$ when *b*, *c*, *d*, and *e* are not zero.

Standardized Test Practice

15. In a recent survey, $\frac{5}{8}$ of pet owners stated that they allow their pet to go outside. Of these, $\frac{1}{3}$ allow their pet outside without supervision. Which expression gives the fraction of the pet owners surveyed that allow their pet outside without supervision?

Ⓐ $\frac{5}{8} + \frac{1}{3}$

Ⓑ $\frac{5}{8} - \frac{1}{3}$

Ⓒ $\frac{5}{8} \times \frac{1}{3}$

Ⓓ $\frac{5}{8} \div \frac{1}{3}$

Extra Practice

Multiply. Write in simplest form.

16. $\frac{1}{8} \times \frac{3}{4} = \frac{3}{32}$

$\frac{1 \times 3}{8 \times 4} = \frac{3}{32}$

Homework Help

17. $\frac{2}{5} \times \frac{3}{7} = \frac{6}{35}$

$\frac{2 \times 3}{5 \times 7} = \frac{6}{35}$

18. $\frac{3}{4} \times 2 =$ ______

19. $\frac{3}{8} \times 11 =$ ______

20. $\frac{3}{5} \times \frac{5}{7} =$ ______

21. $\frac{2}{5} \times \frac{5}{6} =$ ______

22. The bleachers at a football game are $\frac{7}{8}$ full, and $\frac{1}{2}$ of the fans in the bleachers are rooting for the home team. What fraction of the bleachers are filled with home-team fans? Justify your procedure.

______

______

23. The table shows the fraction of the votes that each candidate received. If 230 students voted, how many students voted for each candidate? ______

______

______

Candidate	Fraction of Votes
Nyemi	$\frac{3}{5}$
Luke	$\frac{3}{10}$
Natalie	$\frac{1}{10}$

24. CCSS **Model with Mathematics** Alberto rode $\frac{5}{8}$ of the water rides at a water park. His sister, Reina, rode half of the rides that Alberto rode. What fraction of the water rides did Reina *not* ride? Support your answer with a model. ______

______

25. CCSS **Justify Conclusions** Lee is making chocolate chip cookies and the recipe calls for $\frac{3}{4}$ cup of chocolate chips. If she wants to make $\frac{2}{3}$ of the recipe, what fraction of a cup of chocolate chips will she need? Explain.

______

______

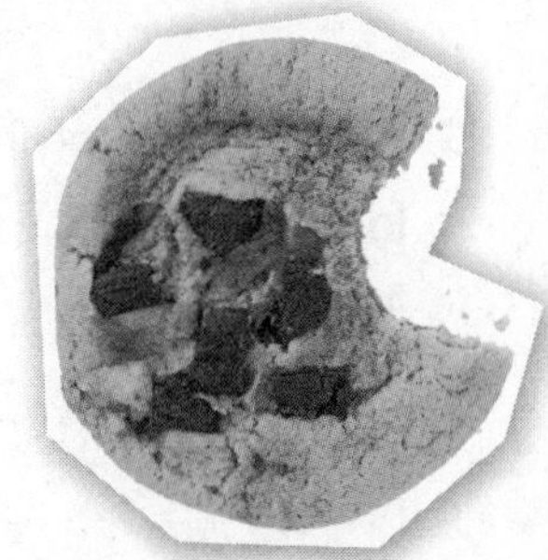

Standardized Test Practice

26. Scott is taking a $\frac{3}{4}$-hour dance class twice a week for 8 weeks. How many hours will Scott have spent in dance class at the end of the 8 weeks?

Ⓐ 6 hours

Ⓑ 8 hours

Ⓒ 12 hours

Ⓓ 16 hours

27. Amanda is stringing beads to make an anklet. The beads are $\frac{1}{4}$-inch wide. The anklet has a string of 16 beads so far. How long is the string of beads?

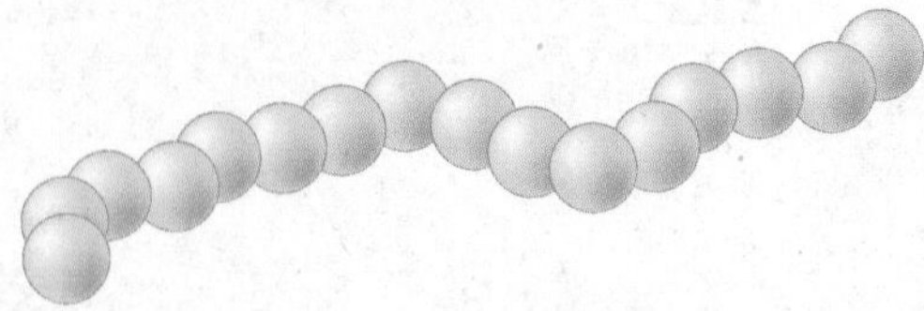

Ⓕ 8 inches

Ⓖ $6\frac{1}{2}$ inches

Ⓗ 4 inches

Ⓘ $3\frac{3}{4}$ inches

28. **Short Response** Four fifths of Terrence's text messages are to his friends. One half of those messages are to his friend Bianca. What fraction of Terrence's text messages are to Bianca?

Common Core Review

Multiply. 5.NBT.5

29. $12 \times 6 \times 9 =$ ______

30. $5 \times 22 \times 3 =$ ______

31. $15 \times 8 \times 11 =$ ______

32. Elise planted a row of flowers in an area with the dimensions shown at the right. What is the area of her flower garden? 5.NF.4b

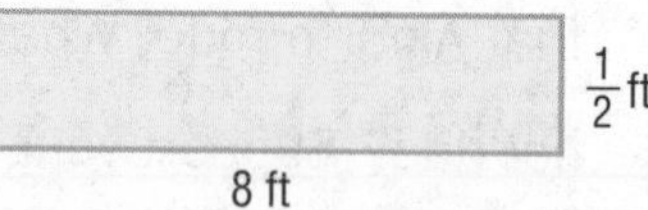

33. Without multiplying, determine whether the product of $5 \times \frac{4}{5}$ is located on the number line at point *A*, *B*, or *C*. Explain your reasoning. 5.NF.5b

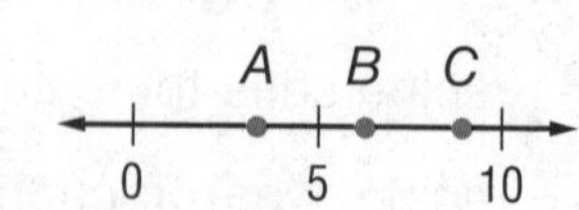

Lesson 4

Multiply Mixed Numbers

What You'll Learn

Scan the lesson. List two headings you would use to make an outline of the lesson.

- ______________________________
- ______________________________

Essential Question

WHAT does it mean to multiply and divide fractions?

Common Core State Standards

Content Standards
Preparation for 6.NS.1

Mathematical Practices
1, 3, 4, 5

Real-World Link

Watch

Animals The eyeball of an Atlantic Giant Squid is about 12 times as large as the average human eyeball. The average human eyeball is $1\frac{1}{4}$ inches across. Use a bar diagram to compare the average size of a human eyeball to the average size of a Atlantic Giant Squid's eyeball.

squid eyeball

$1\frac{1}{4}$

human eyeball

1. Use the diagram above to compare the average size of the Atlantic Giant Squid's eyeball to the average size of the human eyeball. Use repeated addition.

2. Write a multiplication expression that shows the size of the Atlantic Squid's eyeball. ______________

3. Write the multiplication expression from Exercise 2 using improper fractions. Multiply to find the size of the squid's eyeball.

____

Work Zone

Multiply a Fraction and a Mixed Number

To multiply a fraction and a mixed number, first write the mixed number as an improper fraction. Remember that when mixed numbers are written as improper fractions, the denominator does not change. Then multiply as with fractions.

$$2\frac{1}{2} \times \frac{1}{4} = \frac{5}{2} \times \frac{1}{4}$$
$$= \frac{5 \times 1}{2 \times 4}$$
$$= \frac{5}{8}$$

Examples

Tutor

1. Find $\frac{1}{3} \times 1\frac{3}{4}$. Write in simplest form.

Estimate Use compatible numbers. $\frac{1}{2} \times 2 = 1$

$\frac{1}{3} \times 1\frac{3}{4} = \frac{1}{3} \times \frac{7}{4}$ Write $1\frac{3}{4}$ as $\frac{7}{4}$.

$= \frac{1 \times 7}{3 \times 4}$ Multiply.

$= \frac{7}{12}$ Simplify. Compare to the estimate.

2. Find $5\frac{1}{2} \times \frac{1}{3}$. Write in simplest form.

Estimate $\square \times \frac{\square}{\square} = \square$

$5\frac{1}{2} \times \frac{1}{3} = \frac{\square}{\square} \times \frac{1}{3}$ Write $5\frac{1}{2}$ as an improper fraction.

$= \frac{\square \times \square}{\square \times \square}$ Multiply.

$= \frac{\square}{\square}$ or $1\frac{\square}{\square}$ Simplify.

Check for Reasonableness $1\frac{5}{6} \approx 2$ ✓

Show your work.

a. ____________

b. ____________

Got It? Do these problems to find out.

a. $\frac{2}{3} \times 2\frac{1}{2}$

b. $\frac{3}{8} \times 3\frac{1}{3}$

Multiply Mixed Numbers

To multiply two mixed numbers, write each mixed number as an improper fraction. Use the greatest common factor, or GCF, to simplify.

Examples

Tutor

3. Find $1\frac{7}{8} \times 3\frac{1}{3}$. Write in simplest form.

$1\frac{7}{8} \times 3\frac{1}{3} = \frac{15}{8} \times \frac{10}{3}$ — Write $1\frac{7}{8}$ as $\frac{15}{8}$. Write $3\frac{1}{3}$ as $\frac{10}{3}$.

$= \frac{\overset{5}{\cancel{15}}}{\underset{4}{\cancel{8}}} \times \frac{\overset{5}{\cancel{10}}}{\underset{1}{\cancel{3}}}$ — Divide 15 and 3 by their GCF, 3. Then divide 10 and 8 by their GCF, 2.

$= \frac{25}{4}$ — Multiply the numerators and multiply the denominators.

$= 6\frac{1}{4}$ — Simplify.

4. The Hoover Dam contains $4\frac{1}{2}$ million cubic yards of concrete. The Grand Coulee Dam, in Washington state, contains $2\frac{2}{3}$ times as much concrete. How much concrete does it contain?

Estimate $4 \times 3 = 12$

$4\frac{1}{2} \times 2\frac{2}{3} = \frac{9}{2} \times \frac{8}{3}$ — Write the mixed numbers as improper fractions.

$= \frac{\overset{3}{\cancel{9}}}{\underset{1}{\cancel{2}}} \times \frac{\overset{4}{\cancel{8}}}{\underset{1}{\cancel{3}}}$ — Divide 9 and 3 by their GCF, 3. Then divide 8 and 2 by their GCF, 2.

$= \frac{3}{1} \times \frac{4}{1}$ — Multiply the numerators and multiply the denominators.

$= \frac{12}{1}$ or 12 — Simplify.

There are 12 million cubic yards of concrete in the Grand Coulee Dam.

Check for Reasonableness $12 = 12$ ✓

STOP and Reflect

Is the product of two mixed numbers greater than or less than both the factors? Explain below.

Got It? Do this problem to find out.

c. Mr. Wilkins is laying bricks to make a rectangular patio. The area he is covering with bricks is $15\frac{1}{2}$ feet by $9\frac{3}{4}$ feet. What is the area of the patio?

c. ______________

Example

5. **Mr. Conrad's pecan pie recipe calls for $1\frac{3}{4}$ cups of pecans. He plans to make 8 pies for his family reunion. How many cups of pecans will Mr. Conrad need?**

Estimate $2 \times 8 = 16$

$1\frac{3}{4} \times 8 = \frac{7}{4} \times \frac{8}{1}$ Write the mixed number as an improper fraction. Write the whole number as a fraction with a denominator of 1.

$= \frac{7}{\cancel{4}_1} \times \frac{\cancel{8}^2}{1}$ Divide 8 and 4 by their GCF, 4.

$= \frac{7}{1} \times \frac{2}{1}$ Multiply the numerators and multiply the denominators.

$= \frac{14}{1}$ or 14 Simplify.

Check for Reasonableness $14 \approx 16$ ✓

Mr. Conrad will need 14 cups of pecans.

Guided Practice

Multiply. Write in simplest form. (Examples 1–3)

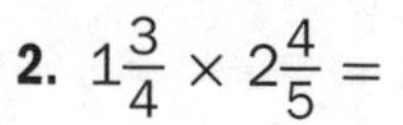

1. $\frac{1}{2} \times 2\frac{3}{8} =$ ______

2. $1\frac{3}{4} \times 2\frac{4}{5} =$ ______

3. $1\frac{2}{3} \times 2\frac{4}{7} =$ ______

4. Melanie is training for a track meet. She ran $2\frac{1}{4}$ miles 5 times this week. How far did Melanie run this week?

(Examples 4 and 5) ______

5. **Building on the Essential Question** How do you multiply mixed numbers? ______

Rate Yourself!

☐ I understand how to multiply mixed numbers.

Great! You're ready to move on!

☐ I still have some questions about multiplying mixed numbers.

No Problem! Go online to access a Personal Tutor.

FOLDABLES Time to update your Foldable!

Name ______________________ My Homework ______________________

Independent Practice

Go online for Step-by-Step Solutions

Multiply. Write in simplest form. (Examples 1–3)

1. $\frac{1}{2} \times 2\frac{1}{3} =$ ______

2. $1\frac{7}{8} \times \frac{4}{5} =$ ______

3. $\frac{7}{8} \times 3\frac{1}{4} =$ ______

4. $1\frac{2}{3} \times 1\frac{1}{4} =$ ______

5. $3\frac{3}{4} \times 2\frac{2}{5} =$ ______

6. $6\frac{2}{3} \times 3\frac{3}{10} =$ ______

7. A carp can travel at a speed of $3\frac{7}{10}$ miles per hour. At this rate, how far can a carp travel in $2\frac{1}{2}$ hours? (Example 4)

______

8. Juliette is making fruit salad. She purchased $9\frac{2}{3}$ ounces each of 6 different fruits. How many ounces of fruit did she purchase? (Example 5)

______

9. A waffle recipe calls for $2\frac{1}{4}$ cups of flour. If Chun wants to make $1\frac{1}{2}$ times the recipe, how much flour does he need? (Example 4)

______

10. CCSS **Model with Mathematics** Use the formula $d = rt$ to find the distance d a long-distance runner can run at a rate r of $9\frac{1}{2}$ miles per hour for time t of $1\frac{3}{4}$ hours.

______

11. STEM Earth is about $92\frac{9}{10}$ million miles from the Sun. Use the table shown.

a. How far is Venus from the Sun? ______

b. How far is Mars from the Sun? ______

c. How far is Jupiter from the Sun? ______

d. How far is Saturn from the Sun? ______

Planet	Approximate Number of Times as Far from the Sun as Earth
Venus	$\frac{3}{4}$
Mars	$1\frac{1}{2}$
Jupiter	$5\frac{1}{4}$
Saturn	$9\frac{1}{2}$

Multiply. Write in simplest form.

12. $\frac{3}{4} \times 2\frac{1}{2} \times \frac{4}{5} =$ ______

13. $\frac{1}{7} \times 5\frac{5}{6} \times 1\frac{1}{4} =$ ______

H.O.T. Problems Higher Order Thinking

14. **Persevere with Problems** Analyze each product in the table.

First Factor		Second Factor		Product
$\frac{1}{2}$	×	$\frac{3}{4}$	=	$\frac{3}{8}$
1	×	$\frac{3}{4}$	=	$\frac{3}{4}$
$\frac{3}{2}$	×	$\frac{3}{4}$	=	$\frac{9}{8}$

a. Why is the first product less than $\frac{3}{4}$?

b. Why is the second product equal to $\frac{3}{4}$?

c. Why is the third product greater than $\frac{3}{4}$?

15. **Use Math Tools** Without multiplying, determine whether the product of $2\frac{1}{2} \times \frac{2}{3}$ is located on the number line at point *A*, *B*, or *C*. Explain your reasoning.

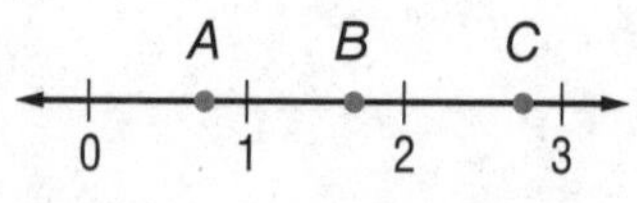

Standardized Test Practice

16. Which number when multiplied by $\frac{3}{4}$ gives a product between $\frac{3}{4}$ and 1?

Ⓐ 0

Ⓑ $\frac{1}{4}$

Ⓒ $\frac{3}{4}$

Ⓓ $1\frac{1}{4}$

Name ______ My Homework ______

Extra Practice

Multiply. Write in simplest form.

17. $\frac{3}{4} \times 2\frac{5}{6} =$ $2\frac{1}{8}$

$\frac{3}{4} \times 2\frac{5}{6} = \frac{\cancel{3}^{1}}{4} \times \frac{17}{\cancel{6}_{2}}$

Homework Help

$= \frac{1 \times 17}{4 \times 2}$

$= \frac{17}{8}$ or $2\frac{1}{8}$

18. $1\frac{4}{5} \times \frac{5}{6} =$ $1\frac{1}{2}$

$1\frac{4}{5} \times \frac{5}{6} = \frac{\cancel{9}^{3}}{\cancel{5}_{1}} \times \frac{\cancel{5}^{1}}{\cancel{6}_{2}}$

$= \frac{3 \times 1}{1 \times 2}$

$= \frac{3}{2}$ or $1\frac{1}{2}$

19. $\frac{3}{10} \times 2\frac{5}{6} =$ ______

20. $3\frac{1}{5} \times 3\frac{1}{6} =$ ______

21. $4\frac{1}{2} \times 2\frac{5}{6} =$ ______

22. $3\frac{3}{5} \times 5\frac{5}{12} =$ ______

23. CCSS **Model with Mathematics** A reproduction of Claude Monet's *Water-Lilies* has dimensions $34\frac{1}{2}$ inches by $36\frac{1}{2}$ inches. Find the area of the painting.

______

24. A photograph is $5\frac{1}{3}$ inches wide. It is being enlarged to 3 times its original size. What is the width of the enlarged photograph?

______

25. Jalisa is making bracelets with leather bands. Each bracelet uses $7\frac{3}{4}$ inches of leather banding. She plans to make 4 bracelets. How many inches of leather banding will she need?

______

26. CCSS **Use Math Tools** Find examples of mixed numbers in a newspaper or magazine, on television, or on the Internet. Write a real-world problem in which you would multiply mixed numbers.

______

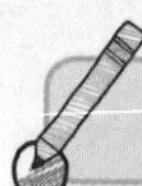

Standardized Test Practice

27. Davis runs at a speed of $4\frac{3}{4}$ miles per hour. At this rate, how far can he run in $3\frac{1}{2}$ hours?

Ⓐ $4\frac{1}{3}$
Ⓑ $4\frac{1}{2}$
Ⓒ $12\frac{3}{8}$
Ⓓ $16\frac{5}{8}$

28. **Short Response** Ben is taking guitar classes three times a week for 8 weeks. Each class will last $1\frac{3}{4}$ hours. How many hours will Ben have spent in guitar classes in 8 weeks?

______

29. Ally's picture frame is shown. What is the area of Ally's picture frame?

Ⓕ $8\frac{1}{4}$ in²
Ⓖ $63\frac{1}{16}$ in²
Ⓗ $67\frac{1}{16}$ in²
Ⓘ $268\frac{1}{4}$ in²

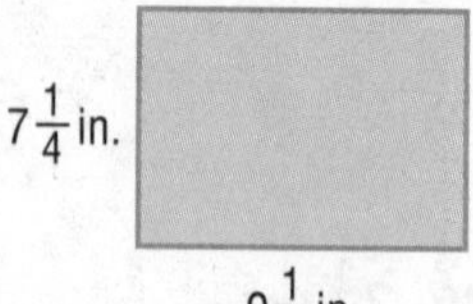

CCSS Common Core Review

Find each equivalent measurement. 5.MD.1

30. 1 foot = ______ inches

31. 1 gallon = ______ quarts

32. 1 yard = ______ feet

33. 1 cup = ______ fluid ounces

34. 24 inches = ______ feet

35. 9 feet = ______ yards

36. Leah's younger sister measures $\frac{41}{12}$ feet tall. Rewrite this as a mixed number. 5.NF.3

______

37. Graph the points (1, 3), (2, 6), and (3, 9) on the coordinate plane. 5.G.2

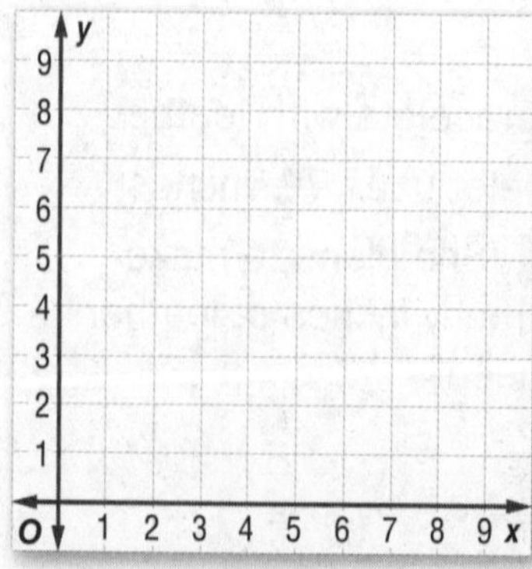

Lesson 5

Convert Measurement Units

What You'll Learn

Scan the lesson. List two headings you would use to make an outline of the lesson.

- ______________________________________
- ______________________________________

Essential Question

WHAT does it mean to multiply and divide fractions?

Vocabulary

unit ratio
dimensional analysis

Common Core State Standards

Content Standards
6.RP.3, 6.RP.3d

Mathematical Practices
1, 3, 4, 6

Real-World Link

Watch

Animals The table shows the approximate weights in tons of several large land animals. One ton is equivalent to 2,000 pounds. You can use a ratio table to convert each weight from tons to pounds.

Animal	Weight (T)
Grizzly bear	1
White rhinoceros	4
Hippopotamus	5
African elephant	8

1. Complete the ratio table. The first two ratios are done for you. To produce equivalent ratios, multiply the quantities in each row by the same number.

Tons	1	4	5	8
Pounds	2,000	8,000		

2. Use the coordinate plane shown.
   a. Graph the ordered pairs (tons, pounds) from the table on the coordinate plane.
   b. Label the horizontal axis *Weight in Tons.*
   c. Label the vertical axis *Weight in Pounds.*
   d. Connect the points and describe the graph.

____

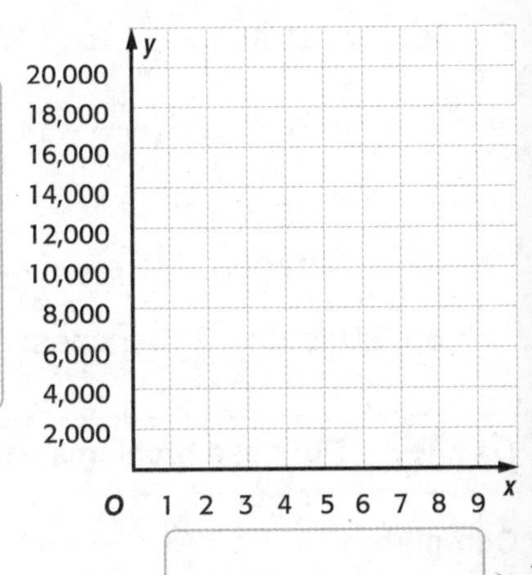

Work Zone

Convert Larger Units to Smaller Units

Each relationship in the table can be written as a ratio. For example, you know that 1 yard = 3 feet. You can use the ratio $\frac{3 \text{ ft}}{1 \text{ yd}}$ to convert from yards to feet.

Customary Conversions			
Type of Measure	**Larger Unit**	**⟶**	**Smaller Unit**
Length	1 foot (ft) 1 yard (yd) 1 mile (mi)	= = =	12 inches (in.) 3 feet 5,280 feet
Weight	1 pound (lb) 1 ton (T)	= =	16 ounces (oz) 2,000 pounds
Capacity	1 cup (c) 1 pint (pt) 1 quart (qt) 1 gallon (gal)	= = = =	8 fluid ounces (fl oz) 2 cups 2 pints 4 quarts

Like a unit rate, a **unit ratio** is one in which the denominator is 1 unit. So, the ratio $\frac{3 \text{ ft}}{1 \text{ yd}}$ is a unit ratio.

Dimensional analysis is the process of including units of measurement as factors when you compute.

Multiplying by 1

The ratio $\frac{3 \text{ ft}}{1 \text{ yd}}$ is equivalent to 1 because the numerator and the denominator represent the same amount.

Example

1. Convert 20 feet to inches.

Since 1 foot = 12 inches, the unit ratio is $\frac{12 \text{ in.}}{1 \text{ ft}}$.

$20 \text{ ft} = 20 \text{ ft} \times \frac{12 \text{ in.}}{1 \text{ ft}}$ — Multiply by $\frac{12 \text{ in.}}{1 \text{ ft}}$.

$= 20 \cancel{\text{ft}} \times \frac{12 \text{ in.}}{1 \cancel{\text{ft}}}$ — Divide out common units, leaving the desired unit, inches.

$= 20 \times 12 \text{ in.}$ — Multiply.

$= 240 \text{ in.}$

So, 20 feet = 240 inches.

a. ____________

b. ____________

c. ____________

Got It? Do these problems to find out.

Complete.

a. 36 yd = ■ ft **b.** $\frac{3}{4}$ T = ■ lb **c.** $1\frac{1}{2}$ qt = ■ pt

Example

2. Marco mixes $\frac{1}{4}$ cup of fertilizer with soil before planting each bulb. How many fluid ounces of fertilizer does he use per bulb?

$\frac{1}{4}\text{ c} = \frac{1}{4}\not{c} \times \frac{8\text{ fl oz}}{1\not{c}}$ — Since 1 cup = 8 fluid ounces, multiply by $\frac{8\text{ fl oz}}{1\text{ c}}$. Then, divide out common units.

$= \frac{1}{4} \times 8\text{ fl oz}$ — Multiply.

$= 2\text{ fl oz}$

So, 2 fluid ounces of fertilizer are used per bulb.

Got It? Do this problem to find out.

d. Jen runs $\frac{1}{8}$ of a mile before tennis practice. How many feet does she run before practice?

d. ______

Convert Smaller Units to Larger Units

Remember that the ratios $\frac{1\text{ yd}}{3\text{ ft}}$ and $\frac{3\text{ ft}}{1\text{ yd}}$ are equivalent. To convert from smaller units to larger units, choose the ratio that allows you to divide out the common units.

Example: $12\text{ ft} \times \frac{1\text{ yd}}{3\text{ ft}}$ ✓

Example

3. Convert 15 quarts to gallons.

Since 1 gallon = 4 quarts, and quarts are smaller units than gallons, use the ratio $\frac{1\text{ gal}}{4\text{ qt}}$.

$15\text{ qt} = 15\text{ qt} \times \frac{1\text{ gal}}{4\text{ qt}}$ — Multiply by $\frac{1\text{ gal}}{4\text{ qt}}$.

$= 15\not{qt} \times \frac{1\text{ gal}}{4\not{qt}}$ — Divide out common units, leaving the desired unit, gallons.

$= 15 \times \frac{1}{4}\text{ gal}$ — Multiplying 15 by $\frac{1}{4}$ is the same as dividing 15 by 4.

$= 3.75\text{ gal}$

Got It? Do these problems to find out.

e. 2,640 ft = ■ mi **f.** 100 oz = ■ lb **g.** 3 c = ■ pt

e. ______

f. ______

g. ______

Example

4. Umeka needs $4\frac{1}{2}$ feet of fabric to make a costume for a play. How many yards of fabric does she need?

$4\frac{1}{2}$ ft $= 4\frac{1}{2}$ ft $\times \frac{1 \text{ yd}}{3 \text{ ft}}$ — Since 1 yard = 3 feet, multiply by $\frac{1 \text{ yd}}{3 \text{ ft}}$. Then, divide out common units.

$= \frac{\overset{3}{\cancel{9}}}{2} \times \frac{1}{\underset{1}{\cancel{3}}}$ yd — Write $4\frac{1}{2}$ as an improper fraction. Then divide out common factors.

$= \frac{3}{2}$ yd or $1\frac{1}{2}$ yd — Multiply.

So, Umeka needs $1\frac{1}{2}$ yards of fabric.

Guided Practice

Complete. (Examples 1 and 3)

1. $5\frac{1}{3}$ yd = ______ ft

2. $4\frac{1}{2}$ pt = ______ c

3. 12 qt = ______ gal

4. 28 in. = ______ ft

5. A large grouper can weigh $\frac{1}{3}$ ton. How much does a large grouper weigh to the nearest pound? (Example 2) ______

6. The world's narrowest electric vehicle is about 35 inches wide. How wide is this vehicle to the nearest foot? (Example 4) ______

7.  **Building on the Essential Question** How can you use ratios to convert units of measurement?

______

______

______

______

Rate Yourself!

Are you ready to move on? Shade the section that applies.

YES ? NO

For more help, go online to access a Personal Tutor.

Name ______________________ My Homework ______________________

Independent Practice

Go online for Step-by-Step Solutions

Complete. (Examples 1 and 3)

1. 18 ft = ______ yd

2. 2 lb = ______ oz

3. 6.5 c = ______ fl oz

4. 2 mi = ______ ft

5. 5,000 lb = ______ T

6. $2\frac{3}{4}$ qt = ______ pt

7. One of the largest pumpkins ever grown weighed about $\frac{3}{4}$ ton. How many pounds did the pumpkin weigh? (Example 2)

____

8. A 40-foot power boat is for sale by owner. How long is the boat to the nearest yard? (Example 4)

____

9. A 3-pound pork loin can be cut into 10 pork chops of equal weight. How many ounces is each pork chop? ______________________

10. CCSS **Model with Mathematics** Will a 2-quart pitcher hold the entire recipe of citrus punch given at the right? Explain your reasoning.

____

____

____

____

11. **Multiple Representations** Use the graph at the right.

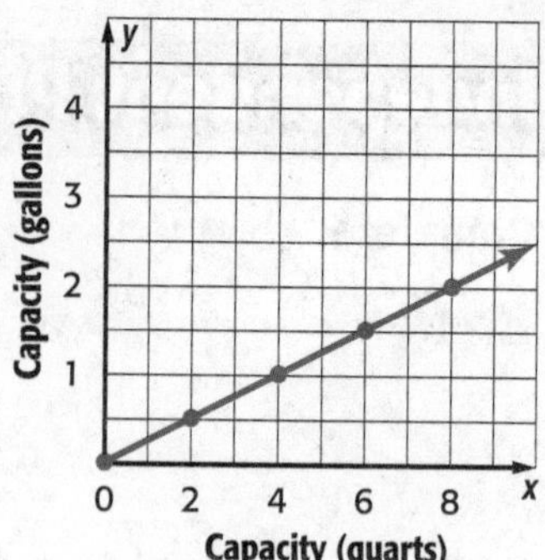

a. **Numbers** What does an ordered pair from this graph represent? ______

b. **Measurement** Use the graph to find the capacity in quarts of a 2.5-gallon container. Explain your reasoning.

______

H.O.T. Problems Higher Order Thinking

12. **Model with Mathematics** Write a real-world problem in which you would need to convert pints to cups. ______

Persevere with Problems **Fill in each ◯ with <, >, or = to make a true sentence. Justify your answers.**

13. 16 in. ◯ $1\frac{1}{2}$ ft

14. $8\frac{3}{4}$ gal ◯ 32 qt

15. **Persevere with Problems** Give two different measurements that are equivalent to $2\frac{1}{2}$ quarts. ______

Standardized Test Practice

16. Which of the following situations is represented by the graph?

Ⓐ conversion of inches to yards

Ⓑ conversion of feet to inches

Ⓒ conversion of miles to feet

Ⓓ conversion of yards to feet

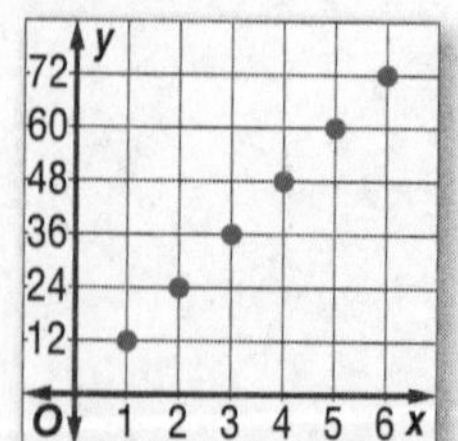

Name ______________________ My Homework ______________________

Extra Practice

Complete.

17. 72 oz = $4\frac{1}{2}$ lb

$72 \text{ oz} = 72 \text{ oz} \times \frac{1 \text{ lb}}{16 \text{ oz}}$

$= \frac{72}{1} \times \frac{1}{16} \text{ lb}$

$= \frac{9}{2} \text{ or } 4\frac{1}{2}$

18. 4 gal = 16 qt

$4 \text{ gal} = 4 \text{ gal} \times \frac{4 \text{ qt}}{1 \text{ gal}}$

$= \frac{4}{1} \times \frac{4}{1} \text{ qt}$

$= \frac{16}{1} \text{ or } 16 \text{ qt}$

19. 3 c = ________ fl oz

20. $1\frac{1}{4}$ mi = ________ ft

21. 13 c = ________ pt

22. $3\frac{3}{8}$ T = ________ lb

23. Speed skiing takes place on a course that is $\frac{2}{3}$ mile long. How many feet long is the course?

____

24. A total of 35 pints of blood were collected at a local blood drive. How many quarts of blood were collected?

____

25. STEM On Monday, it snowed a total of 15 inches. On Tuesday and Wednesday, it snowed an additional $4\frac{1}{2}$ inches and $6\frac{3}{4}$ inches, respectively. A weather forecaster says that over the last three days, it snowed over $2\frac{1}{2}$ feet. Is this a valid claim? Justify your answer.

____

____

CCSS **Be Precise Complete the following statements.**

26. If 16 c = 1 gal, then $1\frac{1}{4}$ gal = ________ c.

27. If 1,760 yd = 1 mi, then 880 yd = ________ mi.

28. If 36 in. = 1 yd, then $2\frac{1}{3}$ yd = ________ in.

Standardized Test Practice

29. Which relationship between the given units of measure is true?

Ⓐ One foot is $\frac{1}{12}$ of an inch.

Ⓑ One yard is $\frac{1}{3}$ of a foot.

Ⓒ One yard is $\frac{1}{3}$ of a mile.

Ⓓ One inch is $\frac{1}{12}$ of a foot.

30. How many cups of milk are shown?

Ⓕ $\frac{3}{4}$ c

Ⓖ $1\frac{1}{4}$ c

Ⓗ $2\frac{1}{2}$ c

Ⓘ 10 c

31. Which of the following lists the measurements below in order from least to greatest?

88 inches, $7\frac{1}{2}$ feet, $2\frac{1}{3}$ yards

Ⓐ 88 inches, $7\frac{1}{2}$ feet, $2\frac{1}{3}$ yards

Ⓑ $7\frac{1}{2}$ feet, 88 inches, $2\frac{1}{3}$ yards

Ⓒ $2\frac{1}{3}$ yards, $7\frac{1}{2}$ feet, 88 inches

Ⓓ $2\frac{1}{3}$ yards, 88 inches, $7\frac{1}{2}$ feet

32. **Short Response** How many quarts are equal to 15 pints? ______

Common Core Review

Divide. 5.NBT.6

33. 156 ÷ 4 = ______

34. 212 ÷ 8 = ______

35. 90 ÷ 12 = ______

36. David baked 78 cookies for a bake sale. He set aside 12 cookies to share with volunteers. The remaining cookies are bagged with 3 cookies in each bag. How many bags does David need? 4.OA.3 ______

37. Refer to the diagram of a living room. The doorway will not have a baseboard. How many feet of baseboard are needed to go around the room? 4.MD.3 ______

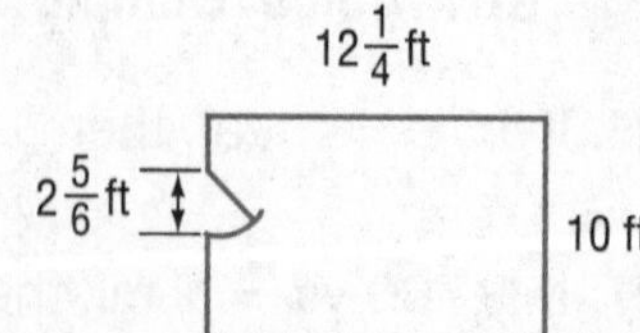

Problem-Solving Investigation
Draw a Diagram

Content Standards
6.NS.1
Mathematical Practices
1, 3, 4

Case #1 Traction Action

Manuel and his friends celebrated his birthday at the FunTimes game center. He spent $\frac{4}{7}$ of his money at the fun center on go-karts and now he has $15 left.

How much money did he spend on go-karts?

1 **Understand** *What are the facts?*

You know that Manuel spent $\frac{4}{7}$ of his money on go-karts. You need to determine how much money he spent on go-karts.

2 **Plan** *What is your strategy to solve this problem?*

He spent a fraction of his money. Draw a bar diagram.

3 **Solve** *How can you apply the strategy?*

Complete the bar diagram using information from the problem. Fill in the missing numbers to show the value of each section.

				5	5	5

Amount Spent on Go-Karts | $15

$\$\square \div \square = \$\square$

So, each section represents $ ☐.

Manuel spent 4 × $ ☐ or $ ☐ on go-karts.

4 **Check** *Does the answer make sense?*

Four sevenths of $35 is 4 × ☐ or $ ☐. ✓

Analyze the Strategy

Justify Conclusions Suppose Manuel had $9 left. How much money did he start with? Explain.

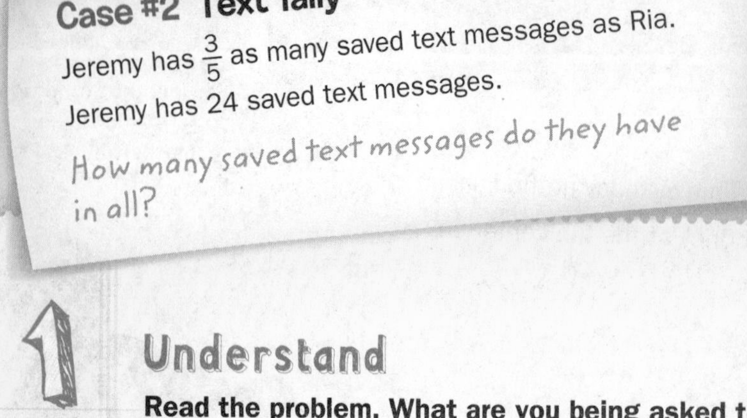

Case #2 Text Tally

Jeremy has $\frac{3}{5}$ as many saved text messages as Ria.

Jeremy has 24 saved text messages.

How many saved text messages do they have in all?

1 Understand

Read the problem. What are you being asked to find?

I need to find ______________________.

Underline key words and values in the problem. What information do you know?

Jeremy has $\frac{\square}{\square}$ as many saved texts as Ria.

Jeremy has $\square$ saved texts.

Plan

Choose a problem-solving strategy.

I will use the ______________________ strategy.

Solve

Solve the problem using your problem-solving strategy.

- Divide the tape diagram for Jeremy into 3 equal sections and the diagram for Ria into 5 equal sections.
- Jeremy has 24 messages. Fill in the boxes.

Jeremy [ ]

Ria [ ]

Ria has $\square \times \square = \square$ saved texts.

So, Jeremy and Ria have $\square + \square$, or $\square$ saved texts in all.

Check

Use information from the problem to check your answer.

$\frac{3}{5} \times \square =$ ______

Collaborate Work with a small group to solve the following cases. Show your work on a separate piece of paper.

Case #3 Internet

Francesca spent 45 minutes on the Internet yesterday.

If this is $\frac{3}{4}$ of the time she spent on the computer, how long did she spend on the computer, but not on the Internet?

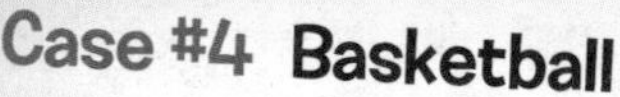

Case #4 Basketball

Mieko practiced shooting a basketball for $\frac{7}{10}$ of her total practice time. During the other time, she practiced dribbling.

If she practiced dribbling for 18 minutes, how many minutes did she practice shooting?

Case #5 Vacation

Of Joseph's vacation pictures, $\frac{4}{9}$ were of his family. The remaining photos were of famous landmarks.

If 45 photos were of landmarks, how many were of his family?

Circle a strategy below to solve the problem.

- Look for a pattern.
- Solve a simpler problem.
- Act it out.
- Make a list.

Case #6 Fruit

Use the table that shows the prices of different amounts of mixed fruit at the grocery store.

How much will 13 pounds of fruit cost?

Pounds	Cost ($)
2	4.50
4	9.00
6	13.50
8	18.00

Mid-Chapter Check

Vocabulary Check

1. CCSS **Be Precise** Define *Commutative Property*. Provide an example of an operation that is commutative. Provide an example of an operation which is not commutative. (Lesson 2)

Skills Check and Problem Solving

Multiply. Write in simplest form. (Lessons 1–4)

2. $8 \times \frac{2}{5} =$ ______

3. $\frac{7}{8} \times \frac{2}{3} =$ ______

4. $4\frac{3}{4} \times 2\frac{1}{8} =$ ______

5. A new shirt costs $14.99. If the shirt is on sale for $\frac{1}{5}$ off its price, about how much would you save? (Lesson 1) ______

6. CCSS **Justify Conclusions** Corey needs 24 boards that are $47\frac{1}{2}$ inches long. (Lesson 5)

   a. How many feet of boards should he buy? Explain. ______

   b. If you can only buy 8-foot boards, how many should he buy? Explain.

7. **Standardized Test Practice** What is the area of the picture and frame shown? (Lesson 4)

Ⓐ $84\frac{7}{12}$ square inches

Ⓑ $83\frac{5}{6}$ square inches

Ⓒ $82\frac{1}{2}$ square inches

Ⓓ $77\frac{1}{6}$ square inches

Inquiry Lab

Divide Whole Numbers by Fractions

HOW can a bar diagram help you understand what it means to divide fractions?

Content Standards
6.NS.1

Mathematical Practices
1, 3, 4

Set Design Juan is building a set for the school musical. He has a 3-foot board that he needs to equally divide into $\frac{1}{2}$-foot pieces. How many pieces will he have after he cuts the board?

What do you know? ______________________________

____

____

What do you need to find? ______________________________

____

Investigation 1

Step 1 Draw a model that represents the length of the board. Draw lines to separate the board into thirds. Each third represents one foot.

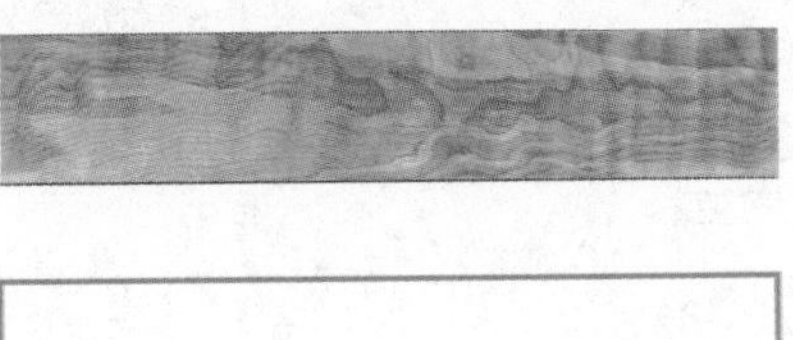

Step 2 Divide each foot into halves.

Step 3 Determine how many groups of $\frac{1}{2}$ are in 3. Circle the groups that are the size of the divisor $\frac{1}{2}$.

There are $\square$ groups of $\frac{\square}{\square}$. So, $3 \div \frac{1}{2} = \square$.

Check by multiplying: $\square \times \frac{\square}{\square} = \square$ ✓

Investigation 2

Find $4 \div \frac{2}{3}$.

Step 1 The model represents 4.

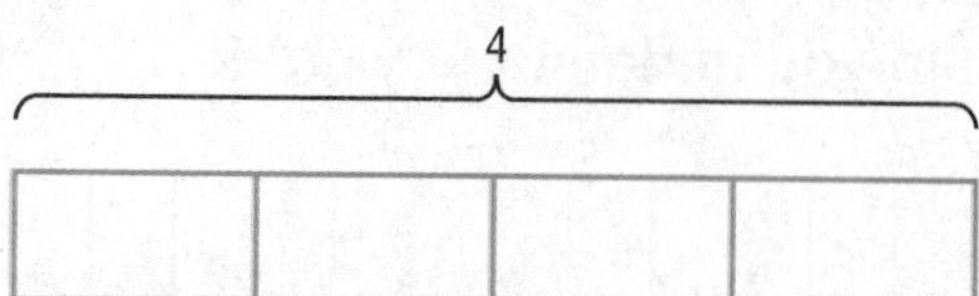

Step 2 Divide each whole into thirds.

Step 3 Circle groups of $\frac{2}{3}$ on the model. Think: How many groups of $\frac{2}{3}$ are in 4?

There are ☐ groups of $\frac{2}{3}$. So, $4 \div \frac{2}{3} =$ ☐.

Check by multiplying: ☐ $\times \frac{☐}{☐} =$ ☐ ✓

Investigation 3

Find $2 \div \frac{3}{4}$.

Step 1 The model represents 2.

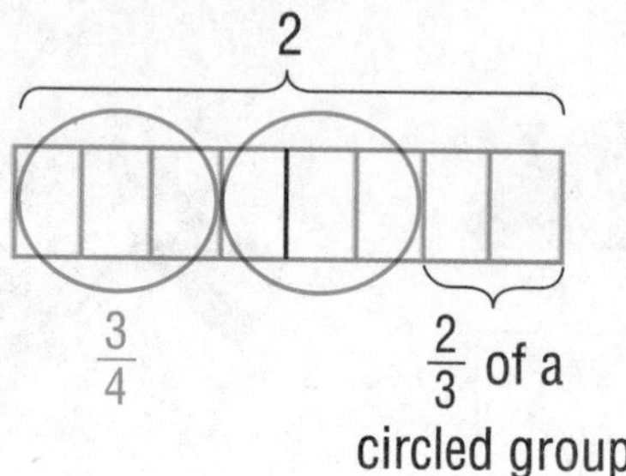

Step 2 Divide each whole into ________.

Step 3 Determine how many groups of $\frac{☐}{☐}$ are in ☐.

Circle groups of $\frac{☐}{☐}$ on the model.

There are ☐ groups of $\frac{3}{4}$ and $\frac{2}{3}$ of a group left over. So, $2 \div \frac{3}{4} =$ ☐$\frac{☐}{☐}$.

Check by multiplying: ☐$\frac{☐}{☐} \times \frac{☐}{☐} =$ ☐ ✓

CCSS Model with Mathematics **Work with a partner. Draw a diagram to find each quotient.**

1. $3 \div \frac{1}{3} =$ ________

2. $2 \div \frac{1}{4} =$ ________

3. $6 \div \frac{2}{3} =$ ________

4. $4 \div \frac{1}{2} =$ ________

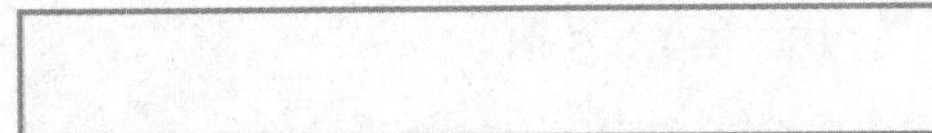

5. $3 \div \frac{3}{4} =$ ________

6. $4 \div \frac{3}{4} =$ ________

7. $5 \div \frac{2}{3} =$ ________

8. $2 \div \frac{4}{5} =$ ________

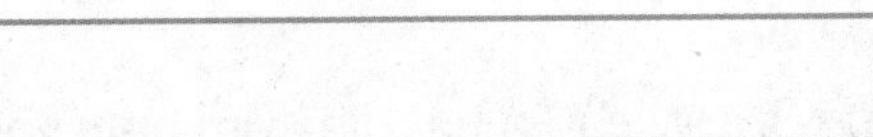

Analyze

Mikayla is modifying the recipe at the right. Use multiplication to check Mikayla's work. The first one is done for you.

	Cups of Hamburger Used	Number of Servings	Check by Multiplying	Is she correct?
	2	$2 \div \frac{2}{3} = 3$	$3 \times \frac{2}{3} = 2$	Yes
9.	3	$3 \div \frac{2}{3} = 4$		
10.	4	$4 \div \frac{2}{3} = 6$		
11.	5	$5 \div \frac{2}{3} = 7\frac{1}{2}$		
12.	6	$6 \div \frac{2}{3} = 9$		
13.	7	$7 \div \frac{2}{3} = 10\frac{2}{3}$		

14. CCSS **Reason Inductively** Compare the quotients to each of the factors in the table above. In $8 \div \frac{2}{3}$, will the quotient be greater than, less than, or equal to 8? Explain.

Reflect

15. CCSS **Model with Mathematics** Write a story context that involves $4 \div \frac{4}{5}$. Solve the problem and multiply to check your answer.

16. CCSS **Justify Conclusions** Write a real-world problem that involves the division of a whole number by a fraction. Then solve. Justify your procedure.

17. Inquiry HOW can a bar diagram help you understand what it means to divide fractions?

Divide Whole Numbers by Fractions

What You'll Learn

Scan the lesson. Predict two things you will learn about dividing whole numbers by fractions.

-
-

Essential Question

WHAT does it mean to multiply and divide fractions?

Vocabulary

reciprocals

Common Core State Standards

Content Standards
6.NS.1

Mathematical Practices
1, 3, 4, 5

Vocabulary Start-Up

Any two numbers with a product of 1 are called **reciprocals**.

Complete the table below by finding the reciprocal of $\frac{2}{3}$. Use the *guess, check,* and *revise* strategy. The first one is done for you.

Number	Product	Reciprocal
$\frac{1}{2}$	$\frac{1}{2} \times 2 = 1$	2
$\frac{2}{3}$	$\frac{2}{3} \times \frac{\square}{\square} = 1$	$\frac{\square}{\square}$

Describe the relationship between the numerator and the denominator of a number and its reciprocal.

Real-World Link

Another name for reciprocal is *multiplicative inverse.* What are some words in everyday language that are similar to reciprocal or inverse?

Pilots can fly in an *inverted* position, or upside down. How can you use the everyday meaning of *invert* to help you remember the mathematical meaning of multiplicative inverse, or reciprocal?

Work Zone

Find Reciprocals

Dividing 3 by $\frac{1}{2}$ gives the same result as multiplying 3 by 2, which is the reciprocal of $\frac{1}{2}$. Any two numbers with a product of 1 are called reciprocals.

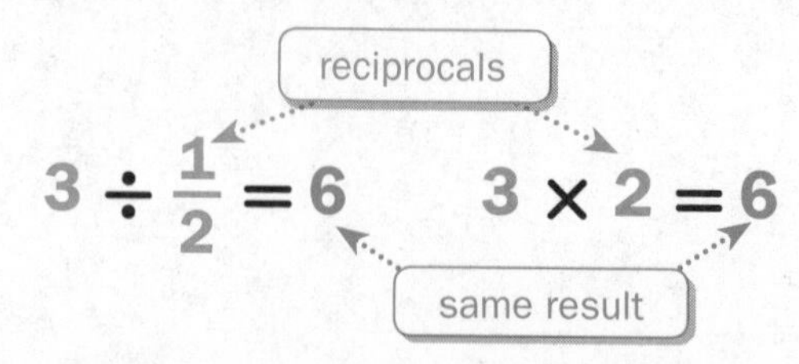

reciprocals

$$3 \div \frac{1}{2} = 6 \qquad 3 \times 2 = 6$$

same result

Examples

Tutor

1. Find the reciprocal of $\frac{2}{3}$.

Since $\frac{2}{3} \times \frac{3}{2} = 1$, the reciprocal of $\frac{2}{3}$ is $\frac{3}{2}$.

2. Find the reciprocal of $\frac{1}{8}$.

Since $\frac{1}{8} \times \frac{8}{1} = 1$, the reciprocal of $\frac{1}{8}$ is $\frac{8}{1}$ or 8.

3. Find the reciprocal of 5.

Write the whole number as a fraction.

$5 = \frac{\square}{1}$

Find the missing factor.

$\frac{\square}{\square} \times \frac{\square}{\square} = 1$

The reciprocal of 5 is $\frac{\square}{\square}$.

Reciprocals
The examples suggest that you "invert" the fraction to find the reciprocal. That is, switch the numerator and denominator. You can use reciprocals to divide fractions.

Got It? **Do these problems to find out.**

Find the reciprocal of each number.

a. $\frac{3}{5}$ **b.** $\frac{1}{3}$ **c.** 11

a. ____________

b. ____________

c. ____________

Divide by a Fraction

Key Concept

Words To divide a whole number by a fraction, multiply by its reciprocal.

Example $5 \div \frac{2}{3} = \frac{5}{1} \times \frac{3}{2}$

The division expression $5 \div \frac{2}{3}$ is read as 5 *divided by two thirds*. You need to find how many two thirds are in 5.

Examples

4. Find $2 \div \frac{1}{3}$. Write in simplest form.

Method 1 **Use a model.**

Model the dividend, 2.

Divide each whole into thirds.

Think: How many thirds are in 2?

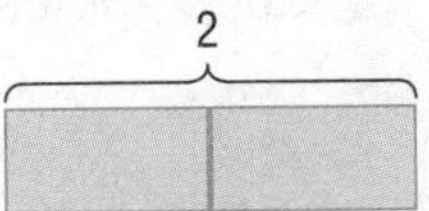

Count the total number of sections.
There are 6 sections.

Method 2 **Use an equation.**

$2 \div \frac{1}{3} = \frac{2}{1} \times \frac{3}{1}$ Multiply by the reciprocal of $\frac{1}{3}$.

$= \frac{6}{1}$ or 6 Multiply the numerators. Multiply the denominators.

5. Find $7 \div \frac{2}{3}$. Write in simplest form.

$7 \div \frac{2}{3} = \frac{\square}{\square} \times \frac{\square}{\square}$ Write the whole number as a fraction. Multiply by the reciprocal of $\frac{2}{3}$.

$\frac{\square}{\square} \times \frac{\square}{\square} = \frac{\square}{\square}$ Multiply the numerators. Multiply the denominators.

$\frac{\square}{\square} = \square\frac{\square}{\square}$ Simplify.

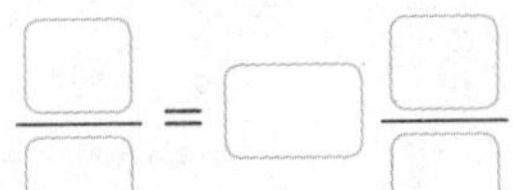

Got It? Do these problems to find out.

d. $6 \div \frac{1}{3}$ **e.** $5 \div \frac{2}{3}$ **f.** $4 \div \frac{3}{4}$

Show your work.

d. ______

e. ______

f. ______

Example

6. At summer camp, the duration of a field hockey game is $\frac{3}{4}$ hour. The camp counselors have set aside 6 hours for field hockey games. How many games can be played?

Divide 6 by three fourths.

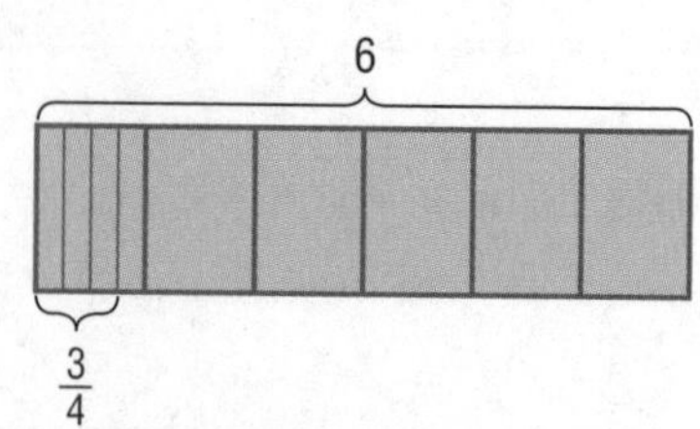

$6 \div \frac{3}{4} = \frac{6}{1} \times \frac{4}{3}$ Multiply by the reciprocal.

$= \frac{\overset{2}{\cancel{6}}}{1} \times \frac{4}{\underset{1}{\cancel{3}}}$ Divide 3 and 6 by the GCF, 3.

$= \frac{8}{1}$ or 8 Simplify.

So, 8 games can be played.

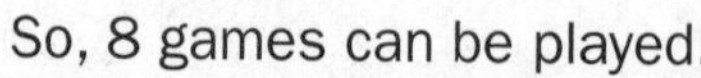

Guided Practice

Find the reciprocal of each number. (Examples 1–3)

1. $\frac{2}{3}$ ______

2. $\frac{1}{7}$ ______

3. 4 ______

Show your work.

Divide. Write in simplest form. (Examples 4 and 5)

4. $2 \div \frac{1}{3} =$ ______

5. $2 \div \frac{4}{5} =$ ______

6. $5 \div \frac{2}{7} =$ ______

7. A neighborhood development that is 4 acres is to be divided into $\frac{2}{3}$-acre lots. How many lots can be created? (Example 6) ______

8. **Building on the Essential Question** Why does a whole number divided by a fraction less than one have a quotient greater than the whole number dividend?

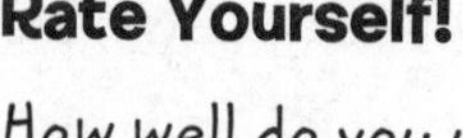

Rate Yourself!

How well do you understand dividing whole numbers by fractions? Circle the image that applies.

Clear | Somewhat Clear | Not So Clear

For more help, go online to access a Personal Tutor.

FOLDABLES Time to update your Foldable!

Name ______________________ My Homework ______________________

Independent Practice

Go online for Step-by-Step Solutions 

Find the reciprocal of each number. (Examples 1–3)

1. $\frac{3}{5}$ ______

2. $\frac{1}{4}$ ______

3. 1 ______

Divide. Write in simplest form. (Examples 4 and 5)

4. $3 \div \frac{3}{4} =$ ______

5. $5 \div \frac{3}{4} =$ ______

6. $8 \div \frac{4}{7} =$ ______

7. $6 \div \frac{3}{5} =$ ______

8. $2 \div \frac{5}{8} =$ ______

9 $4 \div \frac{8}{9} =$ ______

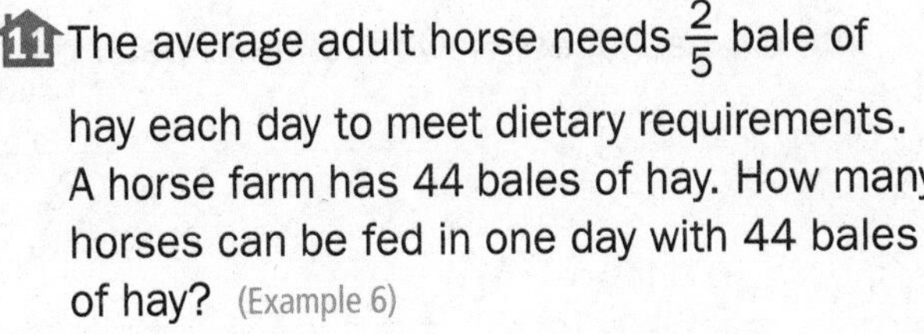

10. Jamar has an 8-foot-long piece of wood that he wants to cut to build a step stool for his tree house. If each piece is going to be $\frac{5}{6}$ foot long, what is the greatest number of pieces he will be able to use? (Example 6)

______

11 The average adult horse needs $\frac{2}{5}$ bale of hay each day to meet dietary requirements. A horse farm has 44 bales of hay. How many horses can be fed in one day with 44 bales of hay? (Example 6)

______

12. CCSS **Justify Conclusions** Ethan ordered 4 sub sandwiches for a party. Each $\frac{1}{2}$ sandwich is one serving. Does he have enough to serve 7 friends? How much is leftover or how much more is needed? Explain. ______

13. Chelsea has four hours of free time on Saturday. She would like to spend no more than $\frac{2}{3}$ of an hour on each activity. How many activities can she do during that time? Justify your procedure.

14.  **Model with Mathematics** Find an example of dividing a whole number by a fraction in a newspaper or on the Internet. Write a real-world problem in which you would divide a whole number by a fraction.

H.O.T. Problems Higher Order Thinking

15. **Find the Error** Daniella is solving $\frac{8}{9} \div 4$. Find her mistake and correct it.

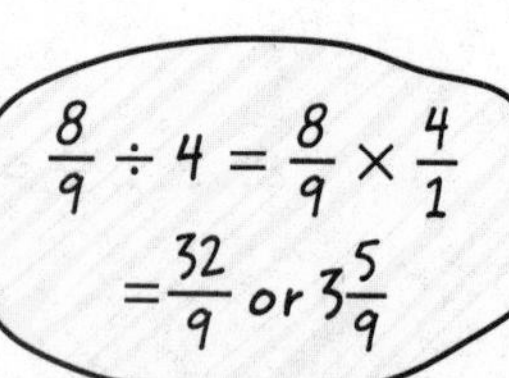

16. 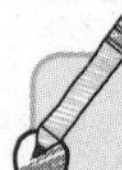 **Persevere with Problems** The Snack Shack is making a batch of trail mix. They use $9\frac{1}{3}$ pounds of granola, $9\frac{1}{3}$ pounds of mixed nuts, and $9\frac{1}{3}$ pounds of yogurt raisins to make the trail mix. They divide the mixture into 14 packages. How much is in each package? Explain.

Standardized Test Practice

17. How many $\frac{3}{4}$-pound bags of peanuts can be made from the bag of peanuts shown?

Ⓐ $3\frac{3}{4}$

Ⓑ $4\frac{1}{4}$

Ⓒ $5\frac{3}{4}$

Ⓓ $6\frac{2}{3}$

Name ____________________ My Homework ____________________

Extra Practice

Find the reciprocal of each number.

18. $\frac{1}{10}$ *10*

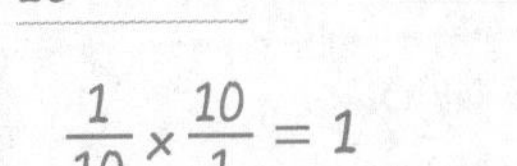

$\frac{1}{10} \times \frac{10}{1} = 1$

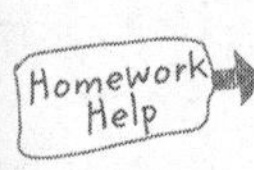

The reciprocal is $\frac{10}{1}$ or 10.

19. $\frac{7}{9}$ ______

20. 8 ______

Divide. Write in simplest form.

21. $2 \div \frac{3}{5} =$ $3\frac{1}{3}$

$2 \div \frac{3}{5} = \frac{2}{1} \times \frac{5}{3}$

$= \frac{10}{3}$ or $3\frac{1}{3}$

22. $5 \div \frac{5}{6} =$ ______

23. $3 \div \frac{5}{6} =$ ______

24. $10 \div \frac{5}{6} =$ ______

25. $4 \div \frac{5}{9} =$ ______

26. $6 \div \frac{2}{3} =$ ______

27. Turner has 6 pounds of pasta. Each time he makes dinner he uses $\frac{3}{4}$ pound of pasta. How many dinners can he make?

______

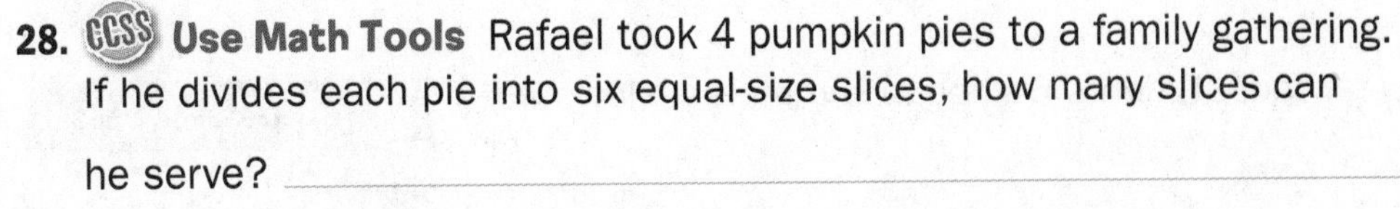

28. CCSS **Use Math Tools** Rafael took 4 pumpkin pies to a family gathering. If he divides each pie into six equal-size slices, how many slices can he serve? ______

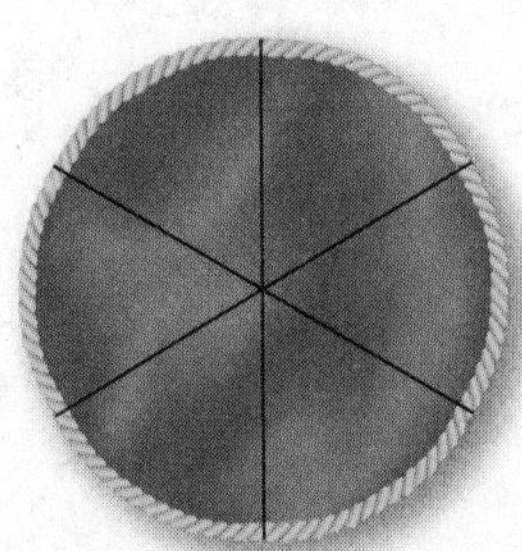

Standardized Test Practice

29. Derreck has \$4 to play video games at the mall. Each game costs a quarter to play. Which choice is *not* a correct method for determining the total number of games he can play?

Ⓐ Take the number of dollars he has and multiply it by 0.25.

Ⓑ Take the number of dollars he has and divide it by 0.25.

Ⓒ Take the number of dollars he has and multiply it by 4.

Ⓓ Take the number of dollars he has and divide it by $\frac{1}{4}$.

30. Lenora is following the recipe. How many batches of the recipe can she make if she has 5 cups of vegetable oil?

Salad Dressing
$\frac{1}{3}$ cup vegetable oil
$\frac{1}{4}$ cup vinegar
salt and pepper
to taste

Ⓕ $1\frac{2}{3}$

Ⓖ 3

Ⓗ 15

Ⓘ 20

31. **Short Response** Jayden has a 10 pound bag of flour. He needs to separate the flour into $\frac{3}{5}$-pound bags. How many bags can he make? Explain your reasoning. ____________________

____

Common Core Review

Find an equivalent fraction. 5.NF.5b

32. $\frac{2}{3} = \frac{\square}{9}$

33. $\frac{3}{5} = \frac{\square}{20}$

34. $\frac{1}{4} = \frac{\square}{24}$

35. $\frac{5}{6} = \frac{\square}{18}$

36. $\frac{3}{4} = \frac{\square}{32}$

37. $\frac{4}{7} = \frac{\square}{28}$

38. The table shows how far four students walked in 5 minutes. How far did they walk together? 5.NF.1 ____________________

	Distance (miles)
April	$\frac{3}{4}$
Ping	$\frac{1}{2}$
Hannah	$\frac{2}{3}$
Raj	$\frac{3}{4}$

39. Sixty people can receive a piece of pizza if 5 pizzas are purchased. How many people can receive a slice of pizza if 7 pizzas are purchased? 4.OA.3 ____________________

Inquiry Lab
Divide Fractions

HOW can using models help you divide one fraction by another fraction?

Content Standards
6.NS.1

Mathematical Practices
1, 3, 4

Candy Toby bought $\frac{8}{9}$ pound of mixed candy from the grocery store. He wants to divide the candy into $\frac{2}{9}$-pound bags. How many bags can Toby make?

What do you know? ______________________________

____

What do you need to find? ______________________________

____

Investigation 1

To solve the problem use the division sentence $\frac{8}{9} \div \frac{2}{9}$. This shows how many

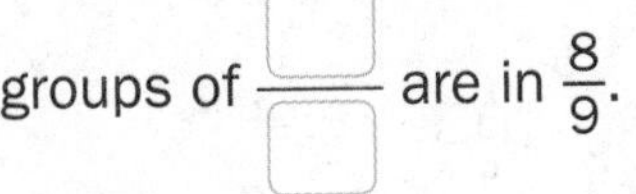

groups of $\frac{\square}{\square}$ are in $\frac{8}{9}$.

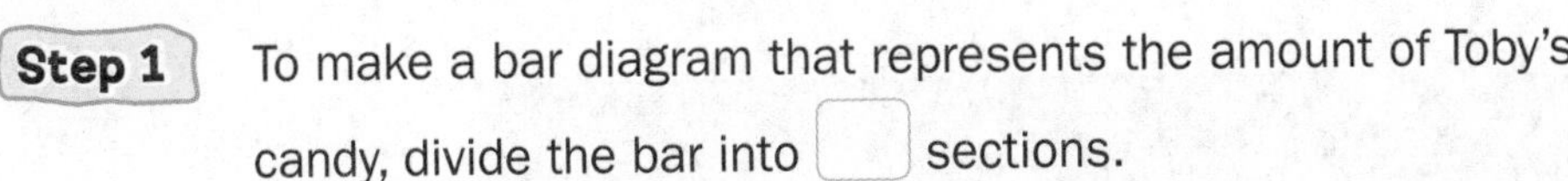

Step 1 To make a bar diagram that represents the amount of Toby's candy, divide the bar into $\square$ sections.

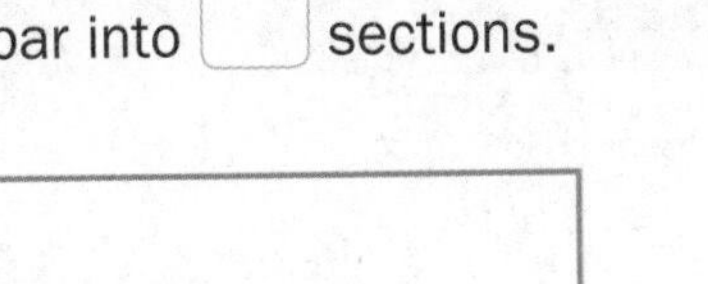

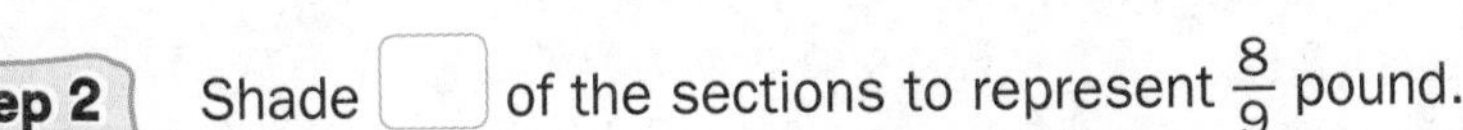

Step 2 Shade $\square$ of the sections to represent $\frac{8}{9}$ pound.

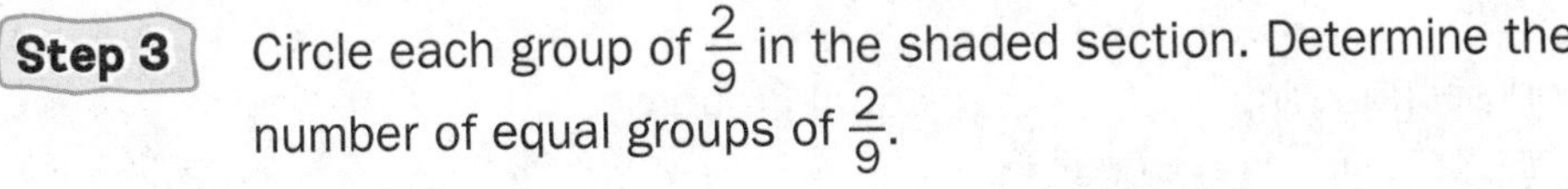

Step 3 Circle each group of $\frac{2}{9}$ in the shaded section. Determine the number of equal groups of $\frac{2}{9}$.

There are ________ groups of $\frac{2}{9}$ in $\frac{8}{9}$. So, $\frac{8}{9} \div \frac{2}{9} = \square$.

Toby can make ________ bags of candy that each have $\frac{\square}{\square}$ pound.

Investigation 2

Draw a diagram to find $\frac{3}{4} \div \frac{3}{8}$.

Step 1 Rename so the fractions have common denominators. Since 8 is a multiple of 4, rename the fraction $\frac{3}{4}$ as $\frac{\square}{\square}$.

Step 2 Draw a diagram with ☐ sections and shade ☐ of the sections to represent $\frac{6}{8}$ pound.

Step 3 Circle each group of $\frac{3}{8}$ in the shaded section. Determine the number of equal groups of $\frac{3}{8}$.

There are ______ groups of $\frac{3}{8}$ in $\frac{6}{8}$.

So, $\frac{3}{4} \div \frac{3}{8} = \square$.

Investigation 3

Draw a diagram to find $\frac{2}{3} \div 2$.

Step 1 Draw a diagram and shade the sections to represent $\frac{\square}{\square}$.

Step 2 Divide the shaded sections into ______ equal groups.

Step 3 Write the fraction that names each group. $\frac{\square}{\square}$

So, $\frac{2}{3} \div 2 = \frac{\square}{\square}$.

Model with Mathematics **Work with a partner. Draw a diagram to find each quotient.**

1. $\frac{6}{7} \div \frac{2}{7} =$ ______

2. $\frac{4}{5} \div \frac{2}{5} =$ ______

3. $\frac{6}{7} \div \frac{3}{7} =$ ______

4. $\frac{8}{10} \div \frac{2}{5} =$ ______

5. $\frac{3}{4} \div \frac{1}{2} =$ ______

6. $\frac{5}{6} \div \frac{2}{3} =$ ______

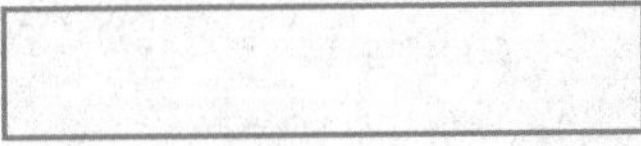

7. $\frac{4}{7} \div 2 =$ ______

8. $\frac{12}{13} \div 3 =$ ______

Analyze

Work with a partner to complete the table. The first one is done for you.

	Division Expression	Quotient	Multiplication Sentence
	$\frac{4}{5} \div \frac{1}{5}$	4	$\frac{1}{5} \times 4 = \frac{4}{5}$
9.	$\frac{8}{9} \div 8$		
10.			$\frac{3}{4} \times \frac{1}{2} = \frac{3}{8}$
11.	$\frac{6}{8} \div \frac{2}{8}$		
12.			$\frac{3}{7} \times 1 = \frac{3}{7}$
13.	$\frac{10}{11} \div 5$		
14.			$\frac{5}{9} \times 1 = \frac{5}{9}$

15. CCSS **Reason Inductively** Use the table to compare the value of the divisor and the dividend to the value of the quotient. When is the quotient greater than 1?

16. CCSS **Make a Conjecture** Some quotients in the table are less than 1. Use the table to write a rule about when the quotient of two fractions will be less than 1.

Reflect

17. CCSS **Model with Mathematics** Write a story context that involves $\frac{6}{8} \div \frac{2}{8}$. Solve the problem and multiply to check your answer.

18. Inquiry HOW can using models help you divide one fraction by another fraction?

Lesson 7

Divide Fractions

What You'll Learn

Scan the lesson. List two real-world scenarios in which you would divide fractions.

-
-

Essential Question

WHAT does it mean to multiply and divide fractions?

Common Core State Standards

Content Standards
6.NS.1

Mathematical Practices
1, 2, 3, 4, 5, 7, 8

Real-World Link

Murals Three students are painting an art mural. The art mural is half painted.

1. Use the picture at the bottom of the page. Divide the painted area into 3 equal parts.

2. Place an X over each part of the painted area. This represents the part each student has painted. Then divide the unpainted area into the same number of parts.

3. What fraction of the whole mural has each student painted? $\frac{\square}{\square}$

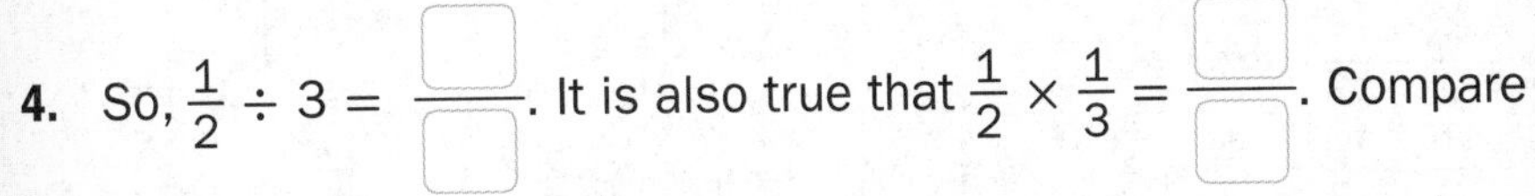

4. So, $\frac{1}{2} \div 3 = \frac{\square}{\square}$. It is also true that $\frac{1}{2} \times \frac{1}{3} = \frac{\square}{\square}$. Compare and contrast the division problem and the multiplication problem.

Key Concept Divide by a Fraction

Words To divide by a fraction, multiply by its reciprocal.

Example

Numbers $\frac{5}{6} \div \frac{2}{3} = \frac{5}{6} \times \frac{3}{2}$

Algebra $\frac{a}{b} \div \frac{c}{d} = \frac{a}{b} \times \frac{d}{c}$, where b, c, and $d \neq 0$

Work Zone

Greatest Common Factor (GCF)
The GCF of two or more numbers is the product of all common prime factors.
$4 = 2 \times 2$
$8 = 2 \times 2 \times 2$
The GCF is 2×2 or 4.

Example

1. **Find $\frac{1}{2} \div \frac{1}{3}$. Write in simplest form.**

Method 1 **Use a model.**

Model the dividend, $\frac{1}{2}$.

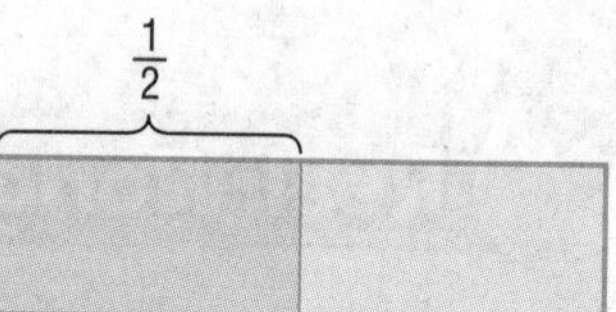

Divide the whole into thirds.

$\frac{1}{2} \div \frac{1}{3}$ means how many thirds are in $\frac{1}{2}$.

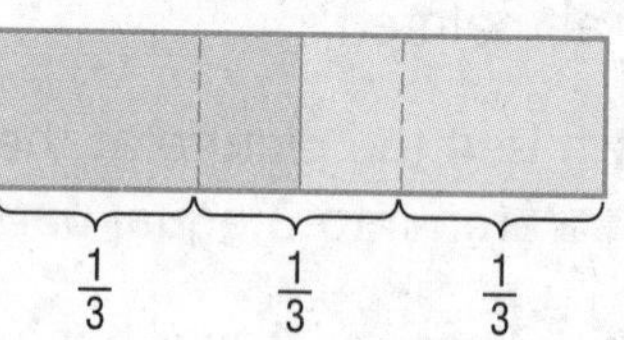

Count the number of $\frac{1}{3}$ sections that are in $\frac{1}{2}$.

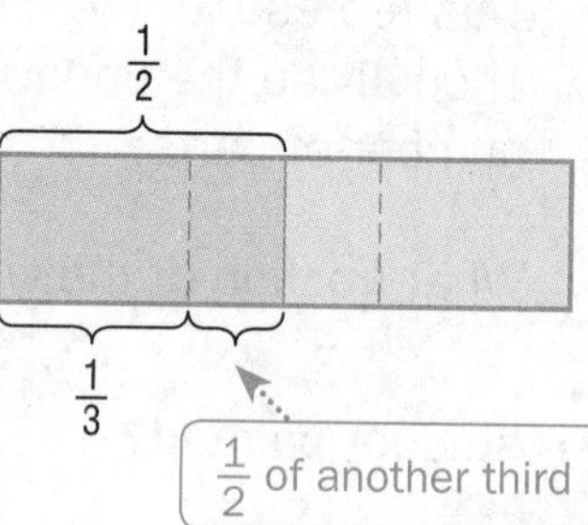

Method 2 **Use an equation.**

$\frac{1}{2} \div \frac{1}{3} = \blacksquare$

$\frac{1}{2} \div \frac{1}{3} = \frac{1}{2} \times \frac{3}{1}$ Multiply by the reciprocal, $\frac{3}{1}$.

$= \frac{3}{2}$ or $1\frac{1}{2}$ Multiply the numerators. Multiply the denominators.

So, $\frac{1}{2} \div \frac{1}{3} = 1\frac{1}{2}$.

Check by multiplying: $\frac{3}{2} \times \frac{1}{3} = \frac{1}{2}$ ✓

Show your work.

a. ______

b. ______

c. ______

Got It? **Do these problems to find out.**

a. $\frac{1}{4} \div \frac{3}{8}$ **b.** $\frac{2}{3} \div \frac{3}{8}$ **c.** $\frac{5}{6} \div \frac{1}{3}$

Example

2. Write a story context for $\frac{2}{3} \div \frac{1}{6}$. Use a model to solve.

Mariska has $\frac{2}{3}$-pound of sunflower seeds. Each day, she feeds the cardinals in her yard $\frac{1}{6}$ pound of seeds. For how many days will she be able to feed the cardinals?

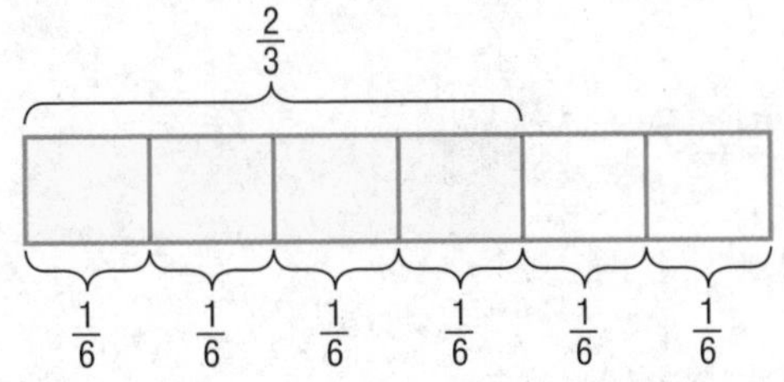

Model $\frac{2}{3}$. The whole is divided into six sections. Count the number of $\frac{1}{6}$ sections.

So, Mariska can feed the cardinals for 4 days.

Got It? Do this problem to find out.

d. Write a story context for $\frac{3}{4} \div \frac{1}{8}$. Use a model to solve.

d. ____________

Divide a Fraction by a Whole Number

When you divide a fraction by a whole number, rewrite the whole number as a fraction. Then divide as with fractions.

Example

3. Find $\frac{5}{7} \div 10$. Write in simplest form.

$\frac{5}{7} \div 10 = \frac{5}{7} \div \frac{10}{1}$ Write the whole number as a fraction with a denominator of 1.

$= \frac{5}{7} \times \frac{1}{10}$ Multiply by the reciprocal.

$= \frac{\overset{1}{\cancel{5}} \times 1}{7 \times \underset{2}{\cancel{10}}}$ Divide 5 and 10 by their GCF, 5.

$= \frac{1}{14}$ Multiply the numerators. Multiply the denominators.

Got It? Do these problems to find out.

e. $\frac{8}{9} \div 4$ **f.** $\frac{4}{5} \div 8$ **g.** $\frac{12}{13} \div 4$

e. ____________

f. ____________

g. ____________

Example

4. Ramón is making party favors. He is dividing $\frac{3}{4}$ pound of almonds into 12 packages. Write and solve an equation to find how many pounds of almonds are in each package.

To find the number of pounds in each package, solve the equation $\frac{3}{4} \div 12 = \blacksquare$.

$\frac{3}{4} \div 12 = \frac{3}{4} \times \frac{1}{12}$ — Multiply by the reciprocal, $\frac{1}{12}$.

$= \frac{\overset{1}{\cancel{3}} \times 1}{4 \times \underset{4}{\cancel{12}}}$ — Divide 3 and 12 by their GCF, 3.

$= \frac{1}{16}$ — Multiply the numerators. Multiply the denominators.

There will be $\frac{1}{16}$ pound of almonds in each package.

Guided Practice

Divide. Write in simplest form. Check by multiplying. (Examples 1 and 3)

1. $\frac{1}{4} \div \frac{1}{2} =$ ______

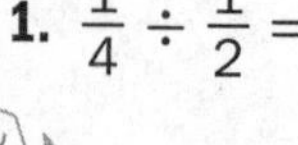

2. $\frac{5}{6} \div \frac{2}{3} =$ ______

3. $\frac{1}{8} \div 3 =$ ______

4. Write a story context for $\frac{2}{3} \div \frac{5}{6}$. Use a model to solve. (Example 2)

5. A neighborhood garden that is $\frac{2}{3}$ of an acre is to be divided into 4 equal-size sections. Write and solve an equation to find the size of each section. (Example 4)

6. **Building on the Essential Question** How is the process used to divide fractions similar to the process used to multiply fractions?

Rate Yourself!

How confident are you about dividing fractions? Shade the ring on the target.

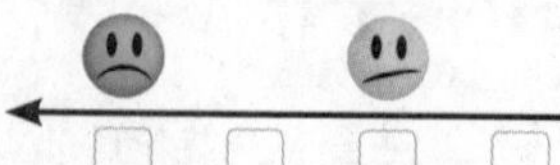

For more help, go online to access a Personal Tutor.

FOLDABLES Time to update your Foldable!

Independent Practice

Go online for Step-by-Step Solutions 

Divide. Write in simplest form. Check by multiplying. (Examples 1 and 3)

1. $\frac{1}{8} \div \frac{1}{2} =$ ______

2. $\frac{3}{4} \div \frac{2}{3} =$ ______

3. $\frac{3}{4} \div 9 =$ ______

4. $\frac{1}{6} \div \frac{4}{7} =$ ______

5. $\frac{1}{3} \div 8 =$ ______

6. $\frac{1}{3} \div \frac{5}{6} =$ ______

7. Write a story context for $\frac{5}{6} \div \frac{1}{12}$. Use a model to solve. (Example 2)

____

____

Write and solve an equation. (Example 4)

8. A piece of licorice is to be cut into 10 equal-size pieces. If the length of the piece of licorice is $\frac{2}{3}$ yard, how long will each piece of licorice be?

____

____

9. CCSS **Use Math Tools** To tie-dye one T-shirt, $\frac{3}{8}$ cup of dye is needed. The table shows the number of cups of each color of dye in Mr. Galvez's art class. How many T-shirts can be made using only orange dye?

____

____

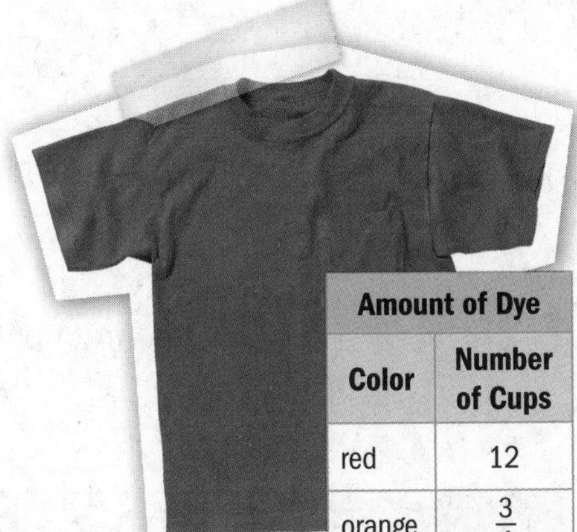

Amount of Dye	
Color	**Number of Cups**
red	12
orange	$\frac{3}{4}$

10. CCSS **Reason Abstractly** Carlota has $\frac{3}{4}$ ton of mulch she is going to divide evenly among 5 flower beds. How much mulch will each flower bed contain?

____

11. **Reason Abstractly** Complete the Venn diagram to compare and contrast the division and multiplication problems.

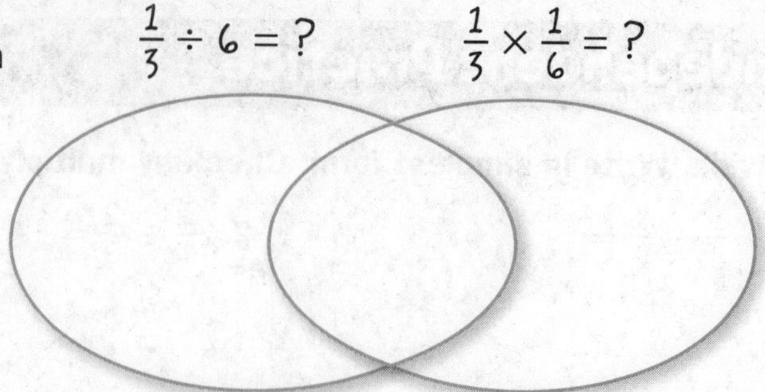

H.O.T. Problems Higher Order Thinking

12. **Identify Structure** Find two positive fractions with a quotient of $\frac{5}{6}$.

____

13. **Identify Repeated Reasoning** Is the quotient $\frac{2}{3} \div \frac{1}{2}$ greater than or less than 1? Is the quotient of $\frac{1}{2} \div \frac{2}{3}$ greater than or less than 1? Explain your reasoning. _______________

____

14. **Persevere with Problems** Complete the steps to demonstrate why you multiply by the reciprocal when dividing fractions. Find $\frac{1}{4} \div \frac{3}{8}$.

Step 1 Rewrite it as $\frac{\frac{1}{4}}{\frac{3}{8}}$.

Step 2 Multiply the numerator and the denominator by the reciprocal of $\frac{\square}{\square}$. $\frac{\frac{1}{4}}{\frac{3}{8}} = \frac{\frac{1}{4} \times \frac{\square}{\square}}{\frac{3}{8} \times \frac{\square}{\square}}$

Step 3 Simplify the denominator.

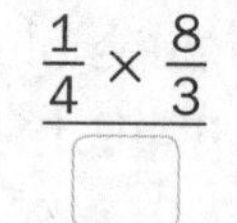

$\frac{\frac{1}{4} \times \frac{8}{3}}{\square}$

Step 4 Simplify the fraction.

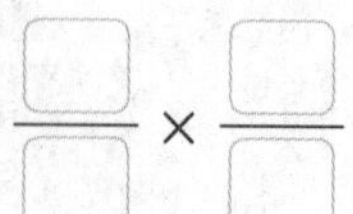

$\frac{\square}{\square} \times \frac{\square}{\square}$

Standardized Test Practice

15. In cooking, 1 drop is equal to $\frac{1}{6}$ of a dash. If a recipe calls for $\frac{2}{3}$ of a dash, which expression would give the number of drops that are needed?

Ⓐ $\frac{1}{6} + \frac{2}{3}$ Ⓑ $\frac{1}{6} \times \frac{2}{3}$ Ⓒ $\frac{1}{6} - \frac{2}{3}$ Ⓓ $\frac{2}{3} \div \frac{1}{6}$

Name ________________ My Homework ________________

Extra Practice

Divide. Write in simplest form. Check by multiplying.

16. $\frac{1}{2} \div \frac{2}{3} =$ $\frac{3}{4}$

$\frac{1}{2} \div \frac{2}{3} = \frac{1}{2} \times \frac{3}{2}$

$= \frac{3}{4}$

Homework Help

$\frac{3}{4} \times \frac{2}{3} = \frac{6}{12}$ or $\frac{1}{2}$ ✓

17. $\frac{1}{5} \div 4 =$ $\frac{1}{20}$

$\frac{1}{5} \div 4 = \frac{1}{5} \times \frac{1}{4}$

$= \frac{1}{20}$

$\frac{1}{20} \times \frac{4}{1} = \frac{4}{20}$ or $\frac{1}{5}$ ✓

18. $\frac{2}{5} \div \frac{3}{4} =$ ______

19. $\frac{2}{7} \div 2 =$ ______

20. $\frac{1}{5} \div \frac{5}{7} =$ ______

21. $\frac{1}{4} \div \frac{3}{5} =$ ______

22. Write a story context for $\frac{1}{4} \div \frac{1}{8}$. Use a model to solve.

______

______

Write and solve an equation.

23. A relay race is $\frac{1}{10}$ kilometer long. Four athletes will run an equal distance to complete the relay. How far does each athlete run?

______

______

24. Jalisa is using $\frac{5}{6}$ yard of ribbon to make bows for her party favors. Jalisa needs to make 6 bows. What is the length of the ribbon used for each bow?

______

______

25. Reaner Recycling shreds $\frac{7}{8}$ ton of aluminum each day. The machines can shred $\frac{1}{24}$ ton aluminum per cycle. How many cycles will be needed to shred the aluminum?

______

______

26. CCSS **Reason Abstractly** Reaner Recycling collected $\frac{7}{4}$ ton of aluminum last Saturday. If $\frac{7}{8}$ ton of aluminum can be shredded each day, how many days will it take to process what was collected on Saturday?

______

Standardized Test Practice

27. Which of the following numbers, when divided by $\frac{1}{2}$, gives a result less than $\frac{1}{2}$?

Ⓐ $\frac{2}{8}$ Ⓒ $\frac{2}{3}$

Ⓑ $\frac{7}{12}$ Ⓓ $\frac{5}{24}$

28. Short Response You have 60 CD cases that you would like to store on the shelf shown. If each CD case is $\frac{3}{8}$ inch wide, is there enough room on the shelf for the CD cases? Explain why or why not.

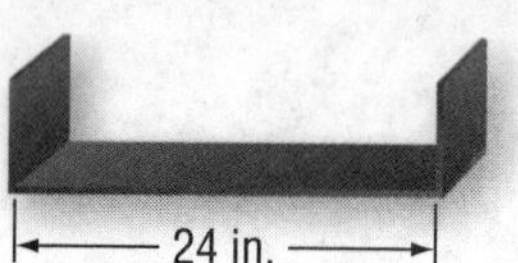

29. The city park service is delivering $\frac{3}{4}$ ton of mulch to 15 parks. Each park will receive an equal amount of mulch. How much mulch does each park receive?

Ⓕ 20 tons Ⓗ $\frac{3}{40}$ ton

Ⓖ $13\frac{1}{3}$ tons Ⓘ $\frac{1}{20}$ ton

30. After a baking contest, $\frac{2}{3}$ of a pie remained. If 8 people get slices of the remainder, how much of the pie does each person get?

Ⓐ $\frac{16}{3}$ Ⓒ $\frac{1}{8}$

Ⓑ $\frac{2}{3}$ Ⓓ $\frac{1}{12}$

CCSS Common Core Review

Find the greatest common factor of each pair of numbers. 4.OA.4

31. 4 and 8 ______

32. 6 and 3 ______

33. 12 and 8 ______

34. 6 and 8 ______

35. 12 and 16 ______

36. 9 and 15 ______

37. The set department has a 5 foot board. They cut 2 segments that are $1\frac{1}{2}$ foot each. How much of the board is left? 4.NF.3d

38. The Sanchez family is building the dog pen shown. What is the area of the dog pen? 4.MD.3

8 ft

4 ft

Lesson 8

Divide Mixed Numbers

What You'll Learn

Scan the lesson. Predict two things you will learn about dividing mixed numbers.

- ______
- ______

Essential Question

WHAT does it mean to multiply and divide fractions?

Common Core State Standards

Content Standards
6.NS.1

Mathematical Practices
1, 2, 3, 4, 6, 7

Real-World Link

Watch

Extreme Geography The deepest point in Earth's oceans is the Mariana Trench, which is located $6\frac{4}{5}$ miles beneath the ocean's surface. The average depth of Earth's oceans is $2\frac{1}{2}$ miles. By contrast, the highest elevation of Earth is Mt. Everest, which is about $5\frac{1}{2}$ miles high.

1. Write a division expression to find how many times as deep the Mariana Trench is than the average depth of the ocean.

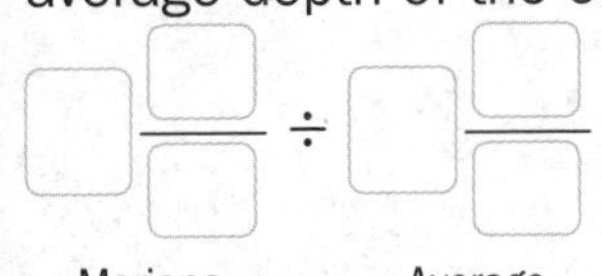

Mariana Trench ÷ Average Ocean Depth

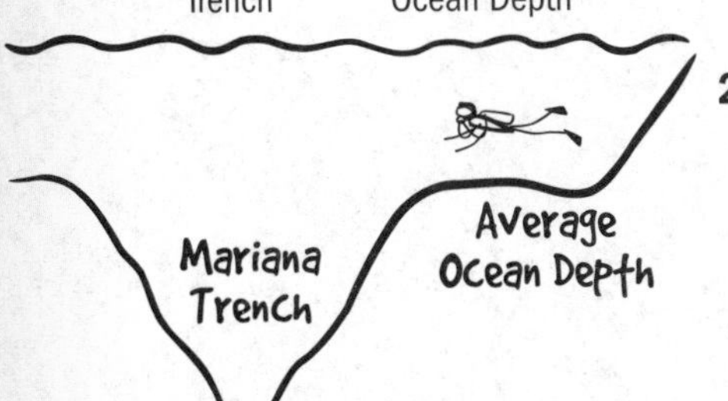

2. Write a division expression to find how many times as tall Mt. Everest is than the average depth of the ocean.

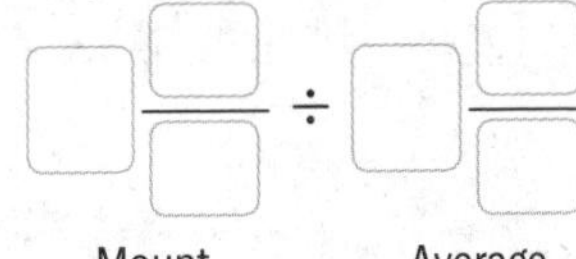

Mount Everest ÷ Average Ocean Depth

3. Rewrite the mixed number $6\frac{4}{5}$ as an improper fraction. $\frac{\square}{\square}$

4. Rewrite the mixed number $5\frac{1}{2}$ as an improper fraction. $\frac{\square}{\square}$

Work Zone

Divide a Mixed Number by a Fraction

Dividing mixed numbers is similar to dividing fractions. To divide mixed numbers, write the mixed numbers as improper fractions and then divide as with fractions.

Examples

Check by Multiplying

You can check the answer of Example 1 by multiplying the quotient by the divisor.

$\frac{\overset{7}{\cancel{35}}}{\underset{4}{\cancel{8}}} \times \frac{\overset{1}{\cancel{2}}}{\underset{1}{\cancel{5}}} = \frac{7}{4}$, or $1\frac{3}{4}$

1. Find $1\frac{3}{4} \div \frac{2}{5}$.

Estimate $2 \div \frac{1}{2} = 4$

$1\frac{3}{4} \div \frac{2}{5} = \frac{7}{4} \div \frac{2}{5}$ — Write the mixed number as an improper fraction.

$= \frac{7}{4} \times \frac{5}{2}$ — Multiply by the reciprocal.

$= \frac{35}{8}$ or $4\frac{3}{8}$ — Simplify.

Check for Reasonableness $4\frac{3}{8} \approx 4$ ✓

2. Find $3\frac{3}{4} \div \frac{4}{5}$.

Estimate ☐ ÷ ☐ = ☐

$3\frac{3}{4} \div \frac{4}{5} = \frac{☐}{☐} \div \frac{4}{5}$ — Write the mixed number as an improper fraction.

$= \frac{☐}{☐} \times \frac{☐}{☐}$ — Multiply by the reciprocal.

$= \frac{☐}{☐}$ or ☐$\frac{☐}{☐}$ — Simplify.

Check for Reasonableness ☐$\frac{☐}{☐} \approx$ ☐ ✓

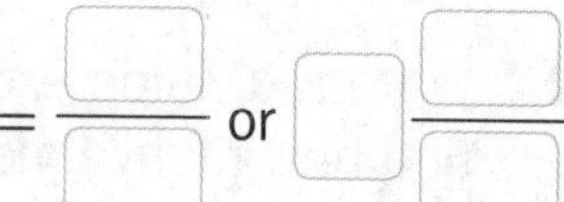

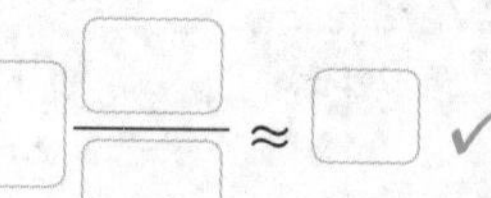

Show your work.

a. ______

b. ______

c. ______

Got It? Do these problems to find out.

Divide. Write in simplest form. Check by multiplying.

a. $2\frac{3}{8} \div \frac{1}{4}$ **b.** $2\frac{1}{2} \div \frac{3}{7}$ **c.** $5\frac{5}{8} \div \frac{3}{4}$

Divide by a Mixed Number

To divide a mixed number by another mixed number, change both mixed numbers to improper fractions. Remember to simplify before you multiply.

Examples

3. Find $5\frac{1}{2} \div 2\frac{1}{2}$.

Estimate $6 \div 3 = 2$

$5\frac{1}{2} \div 2\frac{1}{2} = \frac{11}{2} \div \frac{5}{2}$ Write mixed numbers as improper fractions.

$= \frac{11}{2} \times \frac{2}{5}$ Multiply by the reciprocal.

$= \frac{11}{\cancel{2}_1} \times \frac{\cancel{2}^1}{5}$ Divide 2 and 2 by the GCF, 2.

$= \frac{11}{5}$ or $2\frac{1}{5}$ Simplify. Compare to the estimate.

4. Find $4\frac{2}{3} \div 1\frac{3}{4}$.

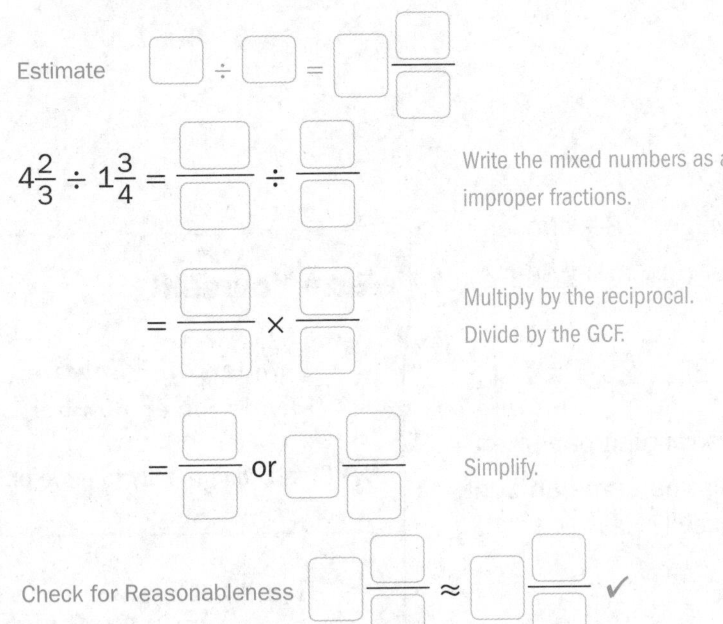

Estimate ☐ ÷ ☐ = ☐ $\frac{☐}{☐}$

$4\frac{2}{3} \div 1\frac{3}{4} = \frac{☐}{☐} \div \frac{☐}{☐}$ Write the mixed numbers as an improper fractions.

$= \frac{☐}{☐} \times \frac{☐}{☐}$ Multiply by the reciprocal. Divide by the GCF.

$= \frac{☐}{☐}$ or ☐ $\frac{☐}{☐}$ Simplify.

Check for Reasonableness ☐ $\frac{☐}{☐} \approx$ ☐ $\frac{☐}{☐}$ ✓

STOP and Reflect

How is dividing two mixed numbers similar to dividing two fractions?

Got It? Do these problems to find out.

d. $4\frac{1}{5} \div 2\frac{1}{3}$ **e.** $8 \div 2\frac{1}{2}$ **f.** $1\frac{5}{9} \div 2\frac{1}{3}$

d. ________

e. ________

f. ________

Example

5. The average adult male Giant Panda weighs about $1\frac{1}{5}$ times as much as the average adult female. If the average weight of a male Giant Panda is 330 pounds, how much does the average female Giant Panda weigh?

To find the average weight, solve the equation $330 \div 1\frac{1}{5} = \blacksquare$.

$330 \div 1\frac{1}{5} = \frac{330}{1} \div \frac{6}{5}$ Write the mixed number as an improper fraction.

$= \frac{330}{1} \times \frac{5}{6}$ Multiply by the reciprocal.

$= \frac{\overset{55}{\cancel{330}}}{1} \times \frac{5}{\underset{1}{\cancel{6}}}$ Divide 330 and 6 by their GCF, 6.

$= \frac{275}{1}$ or 275 Simplify.

So, the average female Giant Panda weighs about 275 pounds.

Guided Practice

Divide. Write in simplest form. Check by multiplying. (Examples 1–4)

1. $3\frac{1}{2} \div \frac{1}{2} =$ ______

2. $2\frac{2}{3} \div 1\frac{1}{6} =$ ______

3. $6\frac{2}{3} \div 2\frac{6}{7} =$ ______

4. A box of snack-size cracker packs weighs $28\frac{1}{2}$ ounces. Each snack pack weighs $4\frac{3}{4}$ ounces. How many snack packs are in the box? (Example 5)

______

5. The soccer team has $16\frac{1}{2}$ boxes of wrapping paper left to sell. If each of the 12 players sells the same amount, how many boxes should each player sell? (Example 5)

______

6. **Building on the Essential Question** How do you divide mixed numbers? ______

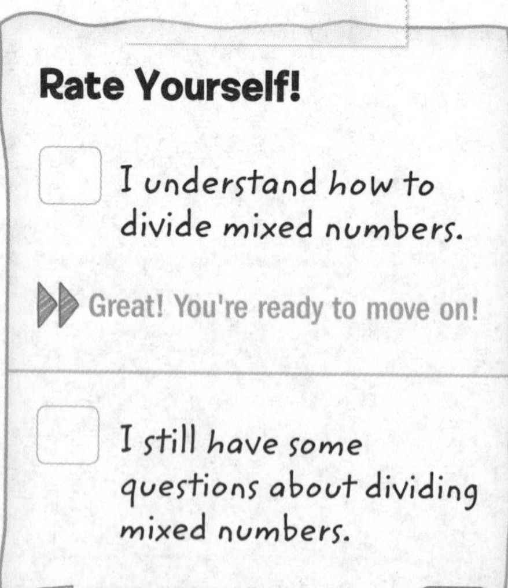

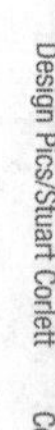

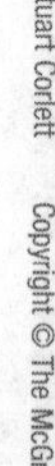

Name ______________________ My Homework ______________________

Independent Practice

Go online for Step-by-Step Solutions

Divide. Write in simplest form. Check by multiplying. (Examples 1–4)

1. $4\frac{1}{6} \div 10 =$ ______

Show your work.

2. $6\frac{1}{2} \div \frac{3}{4} =$ ______

3 $3\frac{3}{4} \div 5\frac{5}{8} =$ ______

4. The length of a kitchen wall is $24\frac{2}{3}$ feet long. A border will be placed along the wall of the kitchen. If the border comes in strips that are each $1\frac{3}{4}$ feet long, how many strips of border are needed? (Example 5)

____

____

5 Jay is cutting a roll of biscuit dough into slices that are $\frac{3}{8}$ inch thick. If the roll is $10\frac{1}{2}$ inches long, how many slices can he cut? (Example 5)

____

____

6. CCSS **Be Precise** Refer to the graphic novel frame below for Exercises a–c.

a. What is the total weight of the birdseed they bought? ______

b. If each bag contains $1\frac{1}{2}$ pounds, how many bags can they make?

____

c. Will there be any birdseed left over? Explain. ______

____

7. **Identify Structure** Complete the steps in dividing mixed numbers.

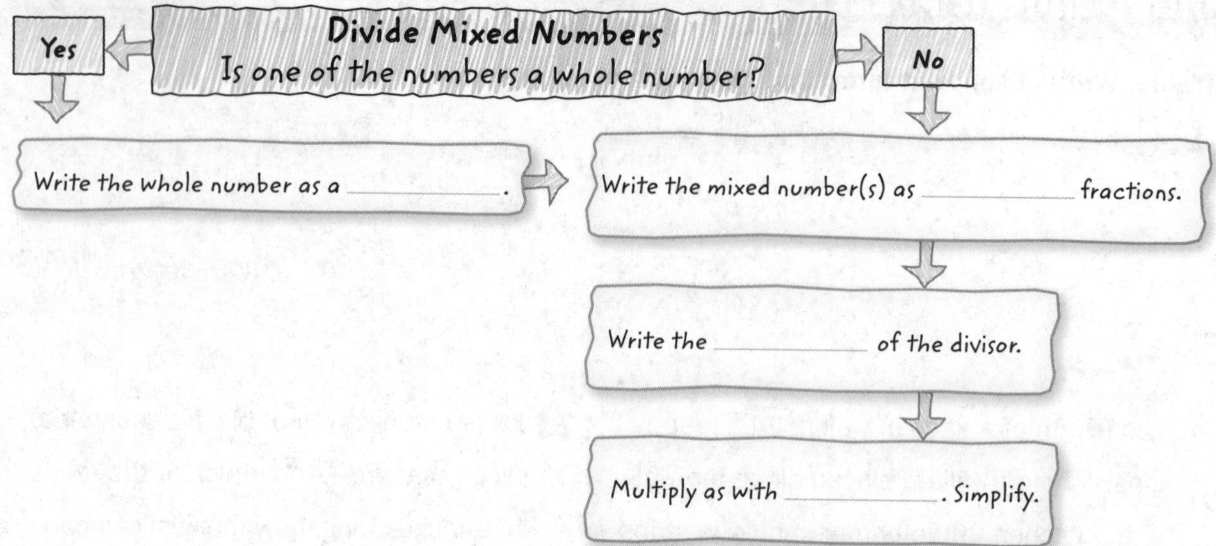

H.O.T. Problems Higher Order Thinking

8. **Which One Doesn't Belong?** Select the expression that has a quotient greater than 1. Explain your reasoning.

$4\frac{2}{3} \div 5\frac{1}{4}$	$3\frac{1}{8} \div 2\frac{2}{5}$	$1\frac{6}{7} \div 2\frac{1}{3}$	$5\frac{3}{4} \div 7\frac{3}{8}$

9. **Persevere with Problems** Without dividing, explain whether $5\frac{1}{6} \div 3\frac{5}{8}$ is greater than or less than $5\frac{1}{6} \div 2\frac{2}{5}$.

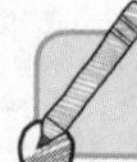

Standardized Test Practice

10. How many $\frac{3}{4}$-cup servings of cereal can be made from the box of cereal shown?

Ⓐ $8\frac{7}{16}$

Ⓑ $10\frac{1}{2}$

Ⓒ 12

Ⓓ 15

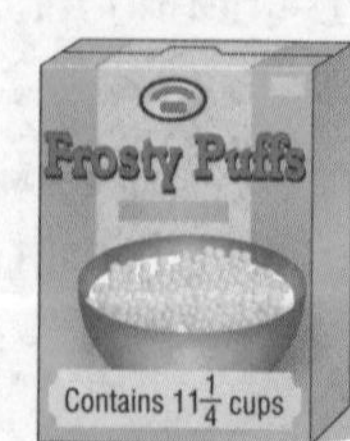

Extra Practice

Divide. Write in simplest form. Check by multiplying.

11. $5\frac{1}{2} \div 2 =$ $2\frac{3}{4}$

Homework Help

$5\frac{1}{2} \div 2 = \frac{11}{2} \div \frac{2}{1}$

$= \frac{11}{2} \times \frac{1}{2}$

$= \frac{11}{4}$ or $2\frac{3}{4}$

$\frac{11}{4} \times \frac{2}{1} = \frac{11}{2}$ or $5\frac{1}{2}$ ✓

12. $3 \div 4\frac{1}{2} =$ ______

13. $6 \div 2\frac{1}{4} =$ ______

14. $7\frac{4}{5} \div \frac{1}{5} =$ ______

15. $6\frac{1}{2} \div 3\frac{1}{4} =$ ______

16. $8\frac{3}{4} \div 2\frac{1}{6} =$ ______

17. $3\frac{3}{5} \div 1\frac{4}{5} =$ ______

18. $6\frac{3}{5} \div 2\frac{3}{4} =$ ______

19. $4\frac{2}{3} \div 2\frac{2}{9} =$ ______

20. **STEM** A human has 46 chromosomes. This is $5\frac{3}{4}$ times the number of chromosomes of a fruit fly. Write a division expression to find how many chromosomes a fruit fly has.

______

______

21. **CCSS Reason Abstractly** How many $\frac{3}{8}$-pound bags of trail mix can be made from $6\frac{3}{8}$ pounds of trail mix? Write a division expression.

______

22. Natasha is setting tiles along the baseboard in her bathroom. One side of the bathroom is $18\frac{3}{4}$ feet. Each tile is $1\frac{1}{2}$ feet long. How many tiles does she need for this section? ______

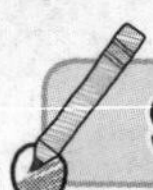

Standardized Test Practice

23. Lola used $1\frac{1}{2}$ cups of dried apricots to make $\frac{5}{6}$ of her trail mix. How many more cups of dried apricots does she need to finish making her trail mix?

Ⓐ 2 c
Ⓑ $1\frac{4}{5}$ c
Ⓒ $\frac{5}{9}$ c
Ⓓ $\frac{3}{10}$ c

24. How many $\frac{3}{4}$-ounce samples can be made from the bottle shown?

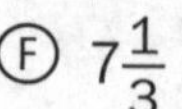
Ⓕ $7\frac{1}{3}$
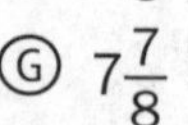
Ⓖ $7\frac{7}{8}$
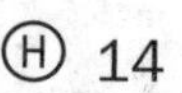
Ⓗ 14
Ⓘ 17

25. **Short Response** You have a bag that holds $25\frac{1}{2}$ pounds. How many $1\frac{1}{4}$-pound books can the bag hold? Explain your response.

____

____

CCSS Common Core Review

Multiply. Write in simplest form. 5.NF.4

26. $\frac{3}{4} \times 1 =$ ______

27. $\frac{3}{7} \times 2 =$ ______

28. $\frac{1}{2} \times \frac{1}{2} =$ ______

29. $\frac{1}{2} \times \frac{1}{4} =$ ______

30. $\frac{2}{5} \times \frac{1}{4} =$ ______

31. $\frac{3}{4} \times \frac{2}{3} =$ ______

32. Anna is planting corn on her farm. What is the area of cornfield? 5.NF.6

____

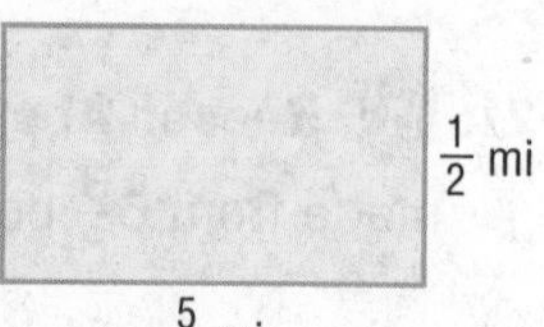

33. Owen has 24 pens and 18 notebooks. He wants to divide the pens and notebooks into equal groups. What is the greatest number of groups he can make using all the pens and notebooks? 4.OA.3

____

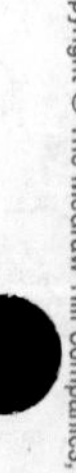

21ST CENTURY CAREER in Culinary Arts

Pastry Chef

Are you creative in the kitchen? Do you enjoy sharing your creations with others? If so, a career as a pastry chef might be perfect for you. Pastry chefs can have different responsibilities, such as creating desserts and dessert menus for restaurants; baking breads, pies, and cakes for neighborhood bakeries; or producing pastries and baked goods for grocery stores. Pastry chefs are not only artistic, but they are also precise and understand mathematics and the chemistry of the ingredients that they are using.

Explore college and careers at ccr.mcgraw-hill.com

Is This the Career for You?

Are you interested in a career as a pastry chef? Take some of the following courses in high school to get you started in the right direction.

- Algebra
- Chemistry
- Culinary Operations
- Food Science Technology

Find out how math relates to a career in Culinary Arts.

You be the Pastry Chef!

Use the information in the recipes below to solve each problem.

1. A chef is making only half of the frosting recipe. How much cream of tartar is needed? How much granulated sugar is needed? ______

2. The petits fours recipe is doubled. How much milk is needed for the cake? ______

3. If the frosting recipe is tripled, how much water is needed? Justify your procedure.

______

______

4. For a wedding, a pastry chef is increasing the cake recipe by $4\frac{1}{2}$ times. How much butter is needed? flour? ______

5. The recipe is increased to make $1\frac{3}{4}$ times the number of petits four than the original recipe. How much vanilla extract is needed for the cake and the frosting? Justify your procedure.

______

______

______

Petits Fours Frosting

- 3 cups granulated sugar
- $\frac{1}{4}$ teaspoon cream of tartar
- $1\frac{1}{2}$ cups water
- 1 cup powdered sugar, sifted
- $\frac{1}{2}$ teaspoon vanilla extract
- $\frac{1}{2}$ teaspoon almond extract

Petits Fours Cake

- $\frac{1}{4}$ cup butter, melted
- 1 cup shortening
- 1 cup granulated sugar
- 1 teaspoon vanilla extract
- $1\frac{1}{3}$ cups all-purpose flour
- 2 teaspoons baking powder
- $\frac{1}{2}$ teaspoon salt
- $\frac{2}{3}$ cup milk
- 3 egg whites

Career Project

It's time to update your career portfolio! Use the Internet or another source to research a career as a pastry chef. Write a paragraph that summarizes your findings.

______

______

______

______

______

List the strengths you have that would help you succeed in this career.

- ______
- ______
- ______
- ______
- ______

Chapter Review

Vocabulary Check

Fill in the blank with the correct vocabulary term. Then circle the word that completes the sentence in the word search.

1. A number that has a whole number part and a fraction part is a ________________.

2. The ________________ is the greatest of the common factors of two or more numbers.

3. The product of a number and its ________________ is one.

4. The number above the fraction bar is the ________________.

5. The number below the fraction bar is the ________________.

6. A ________________ is a number that represents part of a whole or part of a set.

7. A ________________ is a fraction with a denominator of 1.

8. A fraction with a numerator that is greater than or equal to the denominator is an ________________.

9. ________________ are numbers that are easy to divide mentally.

10. A fraction in which the GCF of the numerator and the denominator is 1 is written in ________________.

Key Concept Check

Use Your FOLDABLES®

Use your Foldable to help review the chapter.

Tape here

Tab 3	Multiply and Divide Fractions
Tab 2	
Tab 1	

Example	Example
fraction x mixed number	mixed number ÷ fraction

Got it?

The problems below may or may not contain an error. If the problem is correct, write a "✓" by the answer. If the problem is not correct, write an "X" over the answer and correct the problem.

1. $13 \times \frac{1}{3} = 4\frac{2}{3}$

The first one is done for you. $13 \times \frac{1}{3} = \frac{13}{3}$ or $4\frac{1}{3}$

2. $16 \times \frac{5}{6} = 19\frac{1}{5}$

3. $35 \times \frac{3}{7} = 15$

4. $\frac{5}{8} \div \frac{3}{4} = \frac{15}{32}$

5. $3\frac{2}{3} \div \frac{5}{6} = 4\frac{2}{5}$

6. $2\frac{2}{3} \div 1\frac{1}{4} = 2\frac{2}{3}$

Problem Solving

1. A game board measures $9\frac{1}{2}$ inches by $11\frac{3}{4}$ inches. Estimate the area of the game board. (Lesson 1) ______

2. In a two-week period it was sunny $\frac{3}{7}$ of the days. How many days were sunny? (Lesson 2) ______

3. Seven-eighths of the students in Mr. Klingel's class watched television last night. The table lists the fraction of those students that watched each type of show. What fraction of the entire class watched a reality show? (Lesson 3) ______

Type of Show	Fraction of Students
Reality	$\frac{1}{2}$
Sports	$\frac{1}{8}$
Comedy	$\frac{3}{8}$

4. Nathan deposited $\frac{7}{9}$ of his allowance into his savings account. He spent the remaining amount, or $2.50. How much did Nathan deposit into his savings account? (Lesson 3) ______

5. It is recommended that $\frac{3}{5}$ of the Calories a person consumes come from carbohydrates. If $\frac{1}{12}$ of those Calories should be from fiber, what fraction of the total number of Calories should come from fiber? (Lesson 3) ______

6. A pancake recipe calls for $2\frac{2}{3}$ cups of flour. If Vonetta wants to make $1\frac{1}{2}$ times the recipe, how much flour does she need? (Lesson 4) ______

7. CCSS **Be Precise** The largest telescope in the world is powerful enough to identify a penny that is 5 miles away. How many yards is this? (Lesson 5) ______

8. Each homeroom of Reedurban Middle School receives a $\frac{1}{12}$-acre plot of the land shown. How many homerooms receive a plot of land? (Lesson 7) ______

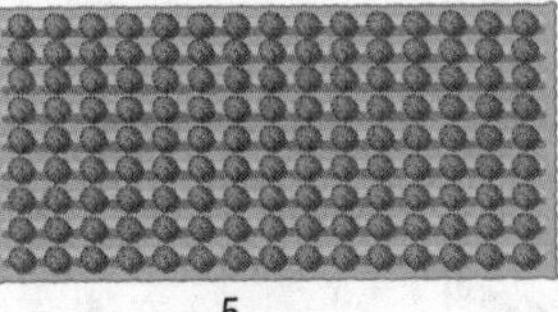
$\frac{5}{6}$ acre

9. To make $4\frac{1}{2}$ gallons of ice cream, it takes $6\frac{3}{10}$ gallons of milk. How many gallons of milk does it take to make one gallon of ice cream? (Lesson 8) ______

Reflect

Answering the Essential Question

Use what you learned about multiplying and dividing fractions to complete the graphic organizer.

Essential Question

WHAT does it mean to multiply and divide fractions?

Operation	Dividend and Divisor	Is the answer less than or greater than the dividend? Provide an example.
multiply	whole number by whole number	
multiply	fraction by fraction	
divide	whole number by whole number	
divide	fraction by fraction	

Answer the Essential Question. WHAT does it mean to multiply and divide fractions?

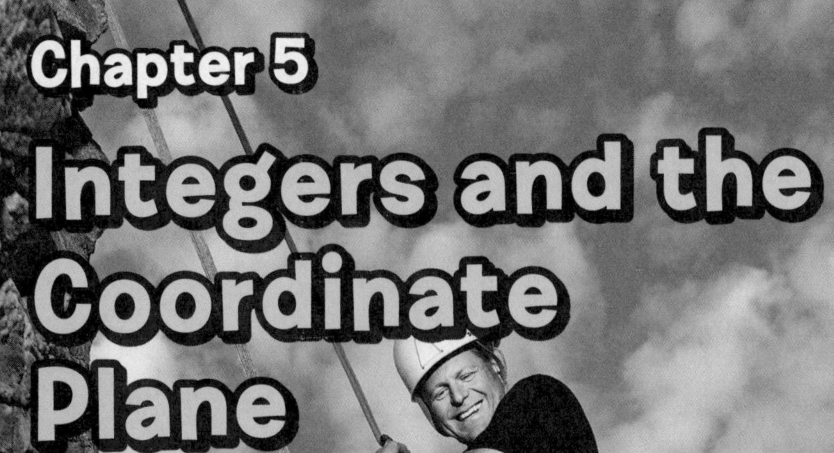

Chapter 5
Integers and the Coordinate Plane

Essential Question

HOW are integers and absolute value used in real-world situations?

Common Core State Standards

Content Standards
6.NS.5, 6.NS.6, 6.NS.6a, 6.NS.6b, 6.NS.6c, 6.NS.7, 6.NS.7a, 6.NS.7b, 6.NS.7c, 6.NS.7d, 6.NS.8

Mathematical Practices
1, 2, 3, 4, 5, 7, 8

Math in the Real World

Rappelling Two friends rappel 35 feet down into a canyon. Their starting position is represented by 0 on the number line. Their ending position can be represented by −35.

Graph −35 on the number line below.

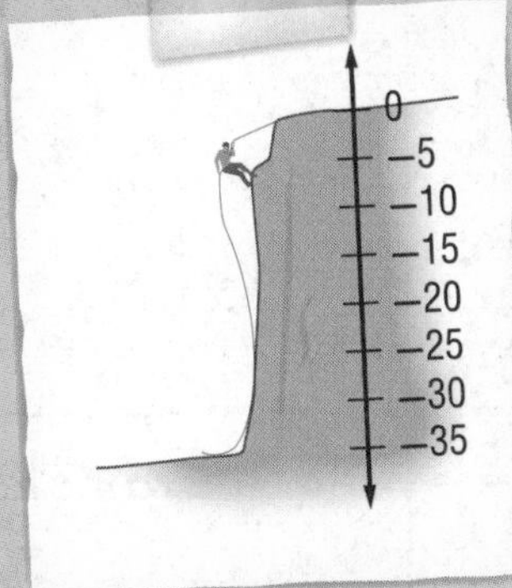

FOLDABLES® Study Organizer

1 Cut out the Foldable on page FL11 of this book.

Place your Foldable on page 418.

Use the Foldable throughout this chapter to help you learn about integers.

What Tools Do You Need?

Vocabulary

absolute value	positive integer
bar notation	quadrants
integer	rational number
negative integer	repeating decimal
opposites	terminating decimal

Review Vocabulary

Using a graphic organizer can help you to remember important vocabulary terms. Fill in the graphic organizer below for the word *decimal*.

Decimal	
Definition	
Math Example	Real World Example

When Will You Use This?

Your Turn! You will solve this problem in the chapter.

Are You Ready?

Try the Quick Check below.
Or, take the Online Readiness Quiz.

Quick Review

Common Core Review 4.NF.2, 5.NBT.3b

Example 1

Replace the ◯ with <, >, or = to make a true statement.

1.6 ◯ 1.3

1.3 1.6

1 1.2 1.4 1.6 1.8 2.0

Since 1.6 is to the right of 1.3, 1.6 > 1.3.

Example 2

Replace the ◯ with <, >, or = to make a true statement.

$\frac{2}{5}$ ◯ $\frac{7}{10}$

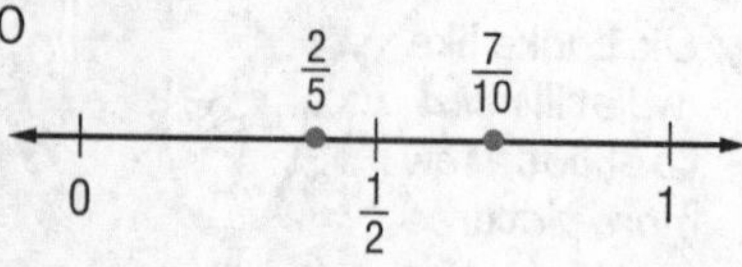

Since $\frac{2}{5}$ is less than $\frac{1}{2}$ and $\frac{7}{10}$ is greater than $\frac{1}{2}$, $\frac{2}{5} < \frac{7}{10}$.

Quick Check

Compare Decimals **Replace each ◯ with <, >, or = to make a true statement.**

Show your work.

1. 4.8 ◯ 4.80

2. 7.7 ◯ 7.5

3. 1.2 ◯ 2.1

Compare Fractions **Replace each ◯ with <, >, or = to make a true statement.**

4. $\frac{2}{11}$ ◯ $\frac{9}{10}$ =

5. $\frac{3}{5}$ ◯ $\frac{1}{4}$ =

6. $\frac{2}{3}$ ◯ $\frac{4}{6}$ =

7. Jahan bought $\frac{2}{3}$ pound of peanuts and $\frac{1}{4}$ pound of walnuts. Did Jahan buy more peanuts or more walnuts? ____________

Which problems did you answer correctly in the Quick Check? Shade those exercise numbers below.

① ② ③ ④ ⑤ ⑥ ⑦

Inquiry Lab

Integers

HOW can positive and negative values be represented?

CCSS Content Standards 6.NS.5, 6.NS.6, 6.NS.6c

Mathematical Practices 1, 3, 4

Ecosystem In coastal regions, some animals live above sea level and other animals live in the ocean. A sea star can be found at an ocean depth of two feet. How can you represent an ocean depth of two feet?

What do you know? ______________________________

____

What do you need to find? ______________________________

____

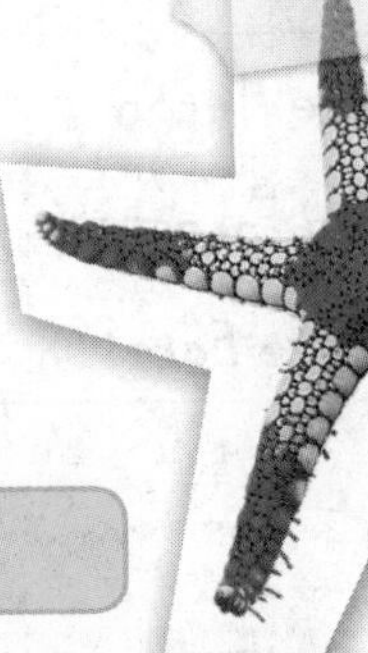

Investigation

Sea level can be represented with the number 0.

To represent a location above sea level, use a positive number. A positive number can be written with or without a positive sign, such as 5 or +5.

To represent a location below sea level, use a negative number. A negative number is written with a negative sign, such as −5.

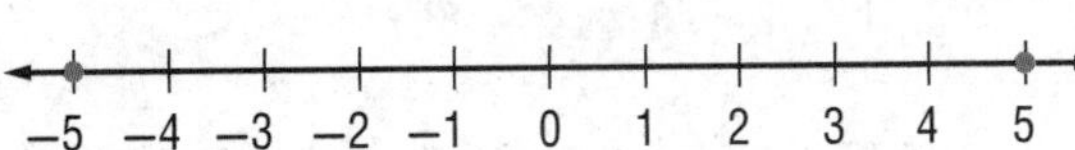

Write a number to represent an ocean depth of two feet.

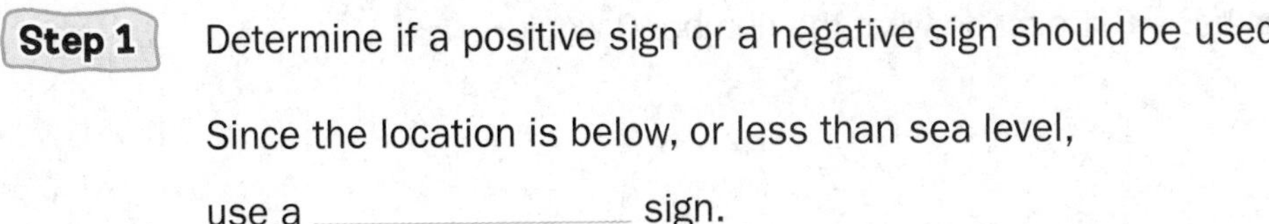

Step 1 Determine if a positive sign or a negative sign should be used.

Since the location is below, or less than sea level,

use a ______________ sign.

Step 2 Determine which number to use.

Use the number ☐ to represent two feet.

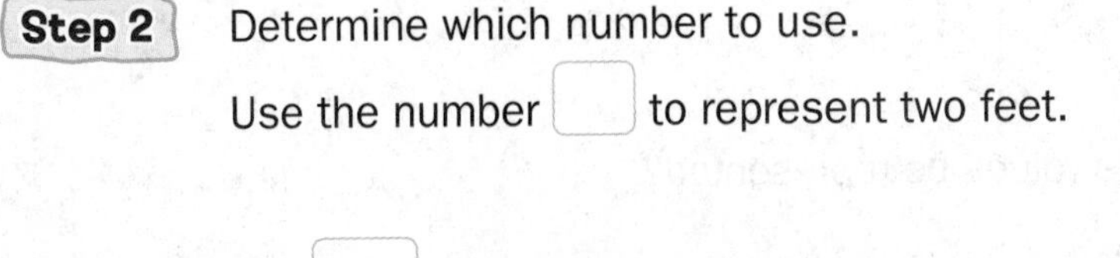

So, the number ☐ represents an ocean depth of two feet.

Work with a partner. Write the correct number to represent each location in relationship to sea level. The first one is done for you.

	Animal	Elevation (ft)	Above or Below Sea Level	Number
	Fiddler Crab	3	above sea level	+3
1.	Eagle's Nest	75	above sea level	
2.	Dolphin	10	below sea level	
3.	Spider Crab	375	below sea level	
4.	Blue Heron	4	above sea level	
5.	Kelp Forest	656	below sea level	
6.	White Egret	50	above sea level	

Reflect

7. CCSS **Reason Inductively** What negative number is the same distance from 0 as the number +4? Explain. Graph both numbers on the number line below.

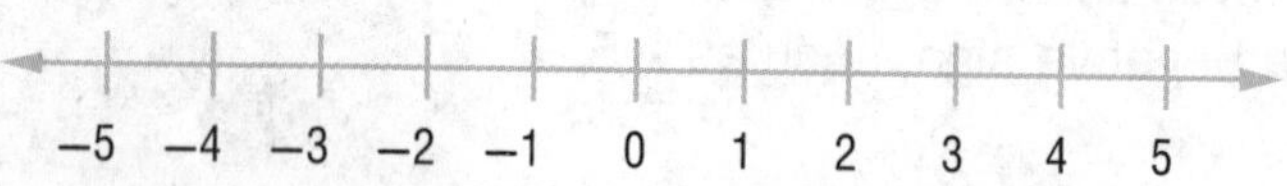

8. CCSS **Model with Mathematics** Write about a real world situation that can be described using the number –6. Describe what the number 0 would represent.

9. Inquiry HOW can positive and negative values be represented?

Lesson 1

Integers and Graphing

What You'll Learn

Scan the lesson. List two real-world scenarios in which you would use integers.

-
-

Essential Question

HOW are integers and absolute value used in real-world situations?

Vocabulary

integer
negative integer
positive integer

Common Core State Standards

Content Standards
6.NS.5, 6.NS.6, 6.NS.6a, 6.NS.6c

Mathematical Practices
1, 3, 4, 5, 7

Real-World Link

Money The bar graph shows the amount of money remaining in the clothing budgets of four students at the end of one month. A value of −$2 means that someone overspent the budget and owes his or her parents 2 dollars.

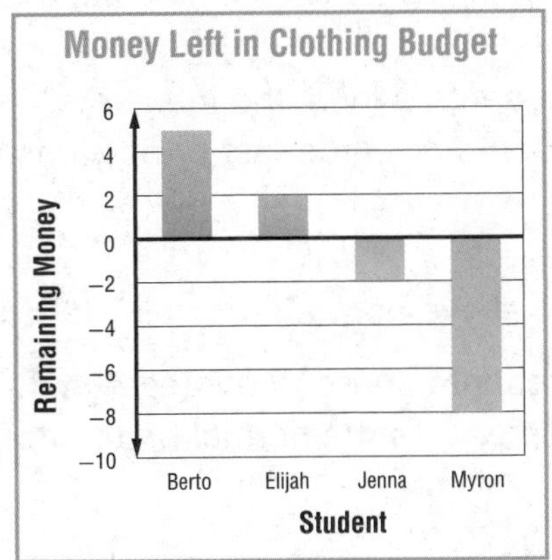

1. What number represents owing 8 dollars?
2. What number represents having 5 dollars left?
3. Who has the most money left? Who owes the most? Explain.

Work Zone

Use Integers to Represent Data

Positive whole numbers, their opposites, and zero are called **integers**. To represent data that are less than a 0, you can use **negative integers**. A negative integer is written with a – sign. Data that are greater than zero are represented by **positive integers**.

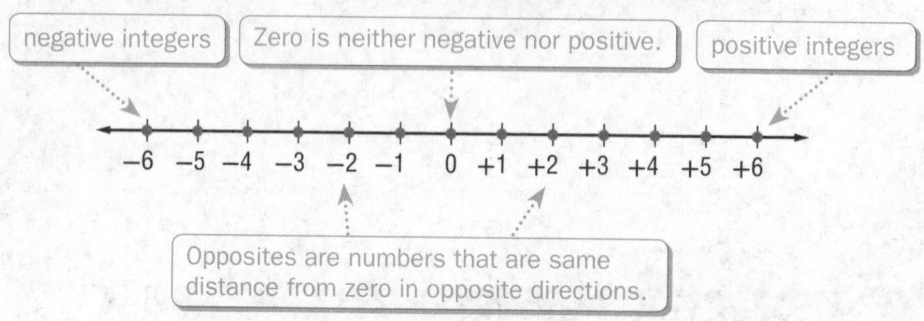

Examples

Write an integer for each situation. Explain the meaning of zero in each situation.

1. a 10-yard loss

Because it represents a loss, the integer is –10. In football, the integer 0 represents no yards lost or no yards gained.

2. 4 inches of rain above normal

Because it represents above, the integer is 4. In this situation, the integer 0 represents the normal amount of rain.

3. a $48 deposit into a savings account

Because it represents an increase, the integer is ______.

In this situation, the integer 0 represents ________________________________.

Zero

The number zero can have different meanings based on real-world context. Sometimes zero represents an amount that does not change. Zero can also be used to represent real-world ideas, such as sea level.

Got It? Do these problems to find out.

Write an integer for each situation. Explain the meaning of zero in each situation.

a. a gain of $2 a share

b. 10 degrees below zero

Graph Integers

Integers and sets of integers can be graphed on a horizontal or vertical number line. To graph a point on the number line, draw a point on the number line at its location. A set of integers is written using braces, such as {2, −9, 0}.

Examples

4. Graph −7 on a number line.

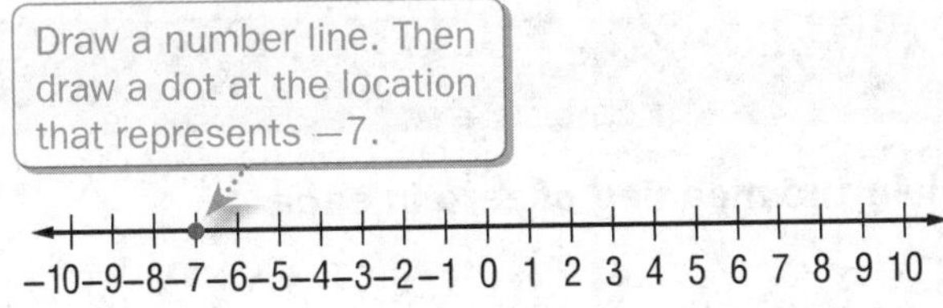

5. Graph the set of integers {−4, 2, −1} on a number line.

Draw a number line. Then draw a dot at the location of each integer.

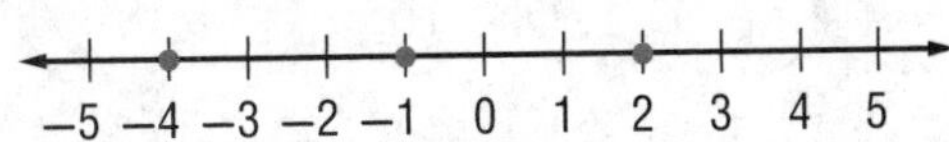

6. Graph the set of integers {0, 2, −3} on a number line.

Draw a number line. Then draw a dot at the location of each integer.

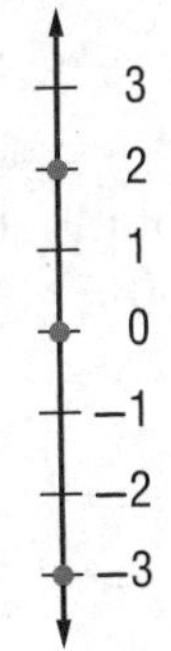

Got It? Do these problems to find out.

Graph each set of integers on a number line.

c. {−3, 0, −2, 4}

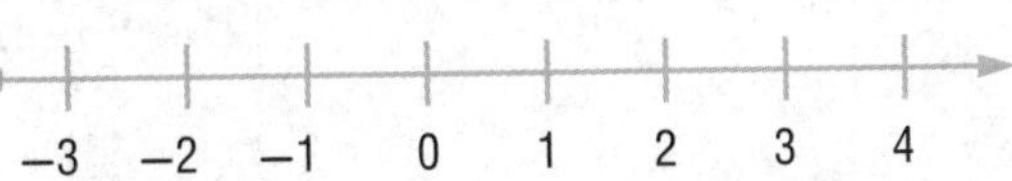

d. {8, −6, −9, 5}

Show your work.

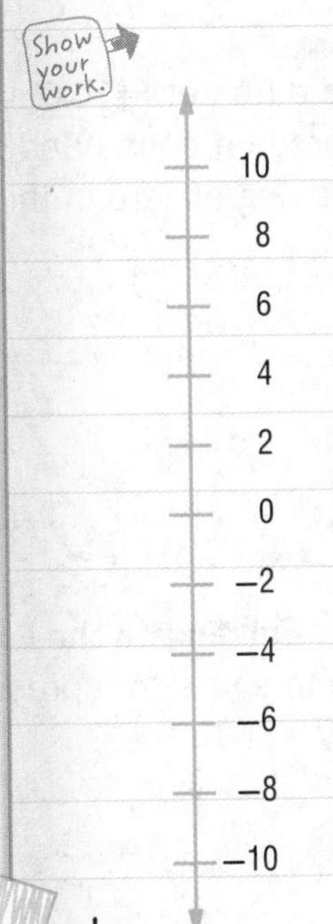

d. ________

Example

7. Alaina and her dad played golf on four different days. The data set {−1, +1, −3, +2} shows Alaina's scores in relation to par. Graph the scores. Explain the meaning of zero in this situation.

Draw a number line. Then draw a dot at the location of each golf score.

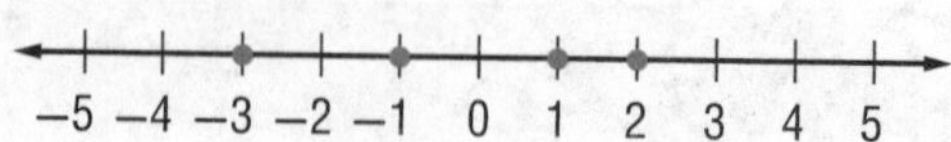

The integer 0 represents par.

Guided Practice

Write an integer for each situation. Explain the meaning of zero in each situation. (Examples 1–3)

1. 15-yard gain ______

2. loss of 2 hours ______

Graph each integer or set of integers on a number line. (Examples 4–6)

3. −2

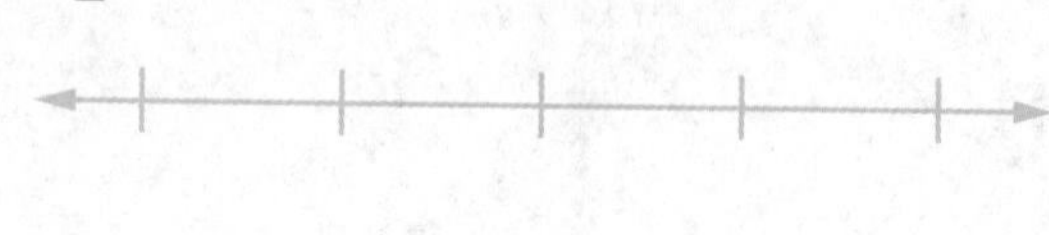

4. {−1, 1, 0}

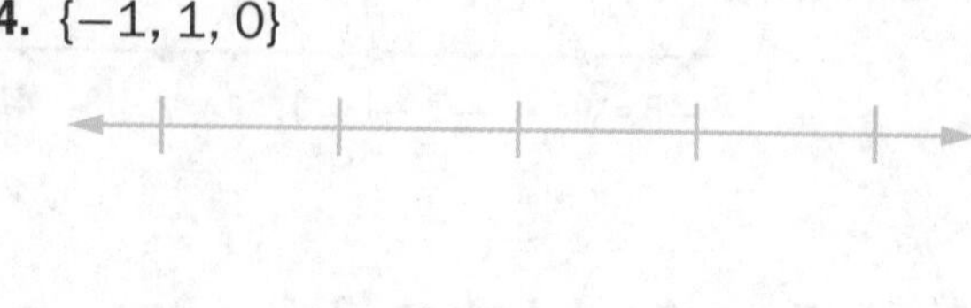

5. The data set {+5, 0, −15, +20} shows the number of points Delaney scored on each hand of a card game. Graph the scores. Explain the meaning of zero in this situation. (Example 7)

______

6. **Building on the Essential Question** How can you use integers to represent data?

______

Rate Yourself!

How confident are you about integers and graphing? Check the box that applies.

For more help, go online to access a Personal Tutor.

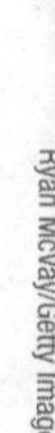

Name ______________________ My Homework ______________________

Independent Practice

Go online for Step-by-Step Solutions

Write an integer for each situation. Explain the meaning of zero in each situation. (Examples 1–3)

1. 3 miles below sea level ______________________

2. earning $45 ______________________

3. moving back 5 spaces on a game board ______________________

____

Graph each integer or set of integers on a number line. (Examples 4–6)

4. −5

5. {2, −3, 0, 1}

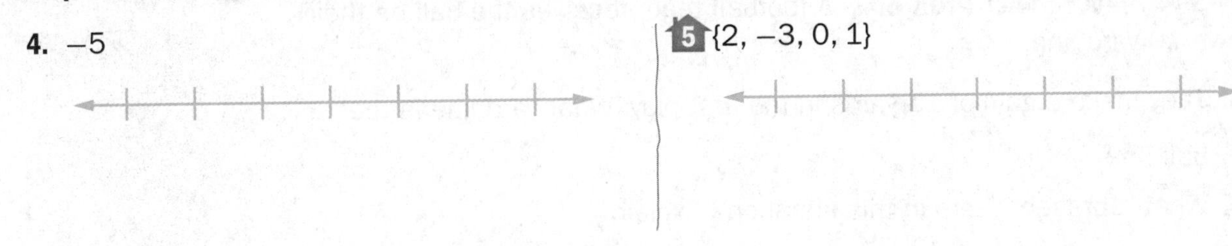

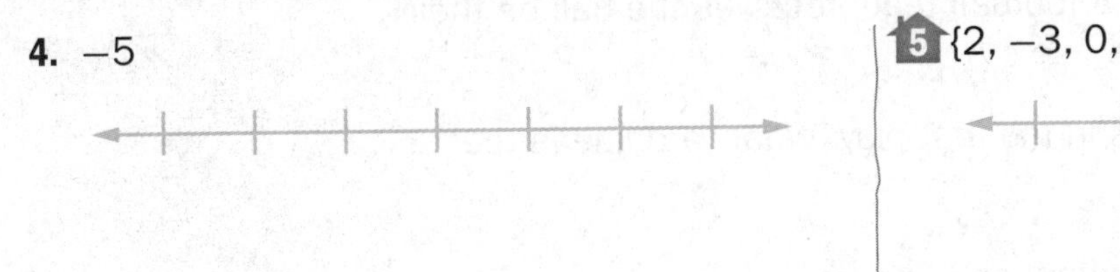

6. The data set {+4, −1, −2, 0} shows a change in number of state representatives for four states after the last census. Graph the change in number of representatives. Explain the meaning of zero in this situation. (Example 7)

____

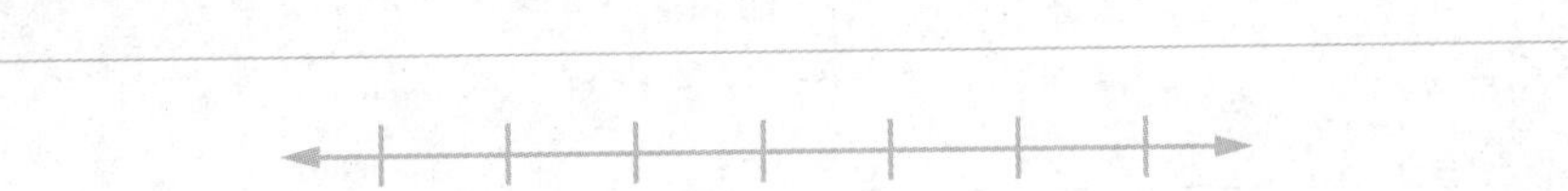

7. **CCSS Use Math Tools** The table shows the record low temperatures for several states. Graph the temperatures on a number line.

Record Low Temperature by State (°F)				
AL	AK	CT	NJ	VA
−27	−29	−32	−34	−30

8. **CCSS Use Math Tools** The table shows the number of points earned for each action in a video game. While playing the video game, Kevin fell in water, jumped over a rock, touched a cactus and climbed a mountain. Graph the number of points he earned for each action on the number line.

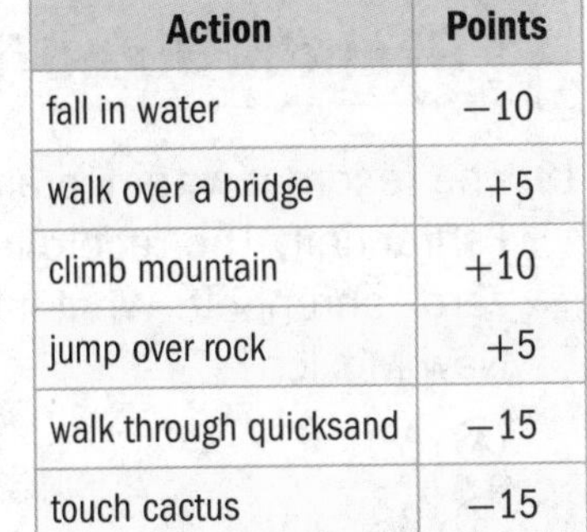

Action	Points
fall in water	−10
walk over a bridge	+5
climb mountain	+10
jump over rock	+5
walk through quicksand	−15
touch cactus	−15

9. **Model with Mathematics** Complete the graphic organizer by writing words or symbols used to represent positive and negative integers.

Positive Integer	Negative Integer
•	•
•	•
•	•
•	•

H.O.T. Problems Higher Order Thinking

10. **Persevere with Problems** A football team receives the ball on their own 10 yard line.

    a. They make a gain of 15 yards in the first play. What yard line is the ball on? ______

    b. What represents zero in this situation? Explain.

    ______

11. **Justify Conclusions** The temperature outside is 15°F. If the temperature drops 20°, will the outside temperature be represented by a positive or negative integer? Explain your reasoning.

    ______

12. **Identify Structure** Describe the characteristics of each set of numbers that make up the set of integers.

    ______

Standardized Test Practice

13. The record low temperature for New Mexico is 50 degrees below zero Fahrenheit. The record low temperature for Hawaii is 12 degrees above zero Fahrenheit. What integer represents the record low temperature for New Mexico?

    Ⓐ 50
    Ⓑ 38
    Ⓒ −38
    Ⓓ −50

Name ______________________ My Homework ______________________

Extra Practice

Write an integer for each situation. Explain the meaning of zero in each situation.

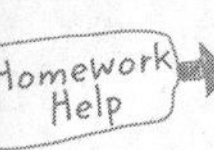

14. 13° below zero *−13; The integer 0 represents zero degrees.*

15. spending \$25 ______________________

16. 13-yard gain ______________________

Graph each integer or set of integers on a number line.

17. −8

18. {0, −3, 1, −1}

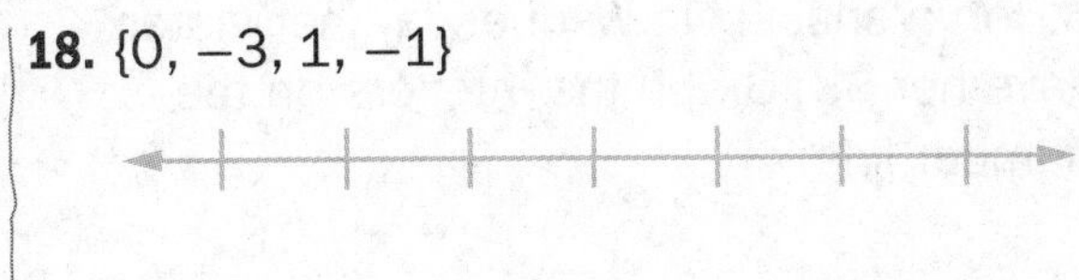

19. {−1, 1, −2}

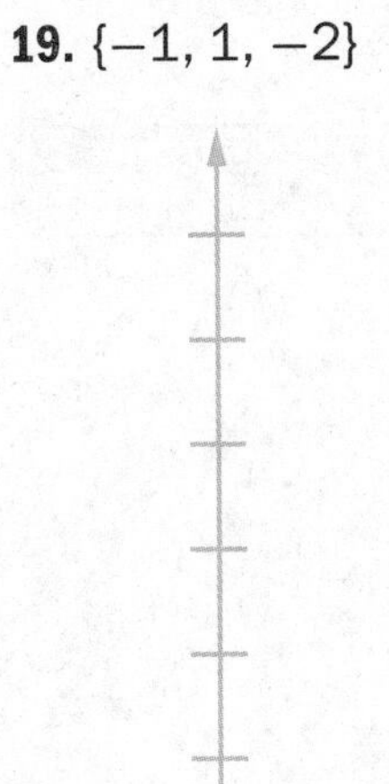

20. {3, −5, 4, −1}

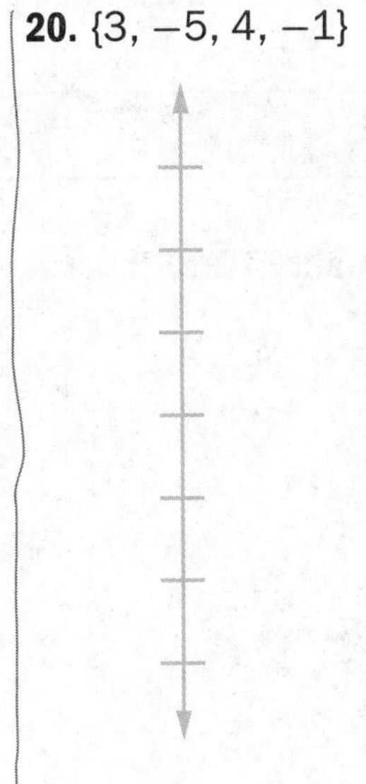

21. {4, −2, 2}

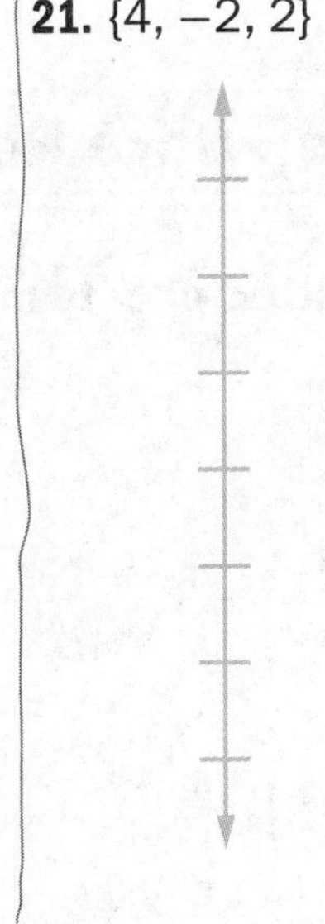

22. The data set {+3, −1, −2, +1} shows the moves a player made forward or backward in a board game. Graph the moves the player made. Explain the meaning of zero in this situation.

____

23. CCSS **Model with Mathematics** The table shows the overnight low temperatures for 5 days in Minneapolis. Graph the temperatures on a number line.

Overnight Low Temperatures (°F)				
1	−1	3	−6	0

Standardized Test Practice

24. The lowest elevation in Vermont is 95 feet above sea level. The lowest elevation in Louisiana is 8 feet below sea level. What integer represents the lowest elevation in Vermont?

Ⓐ 8　Ⓒ 95
Ⓑ −8　Ⓓ −95

25. Short Response On Monday Kennedy spent \$2 on lunch. On Tuesday she spent \$1 on a snack. On Wednesday, her sister gave her \$3. Graph the integers on the number line.

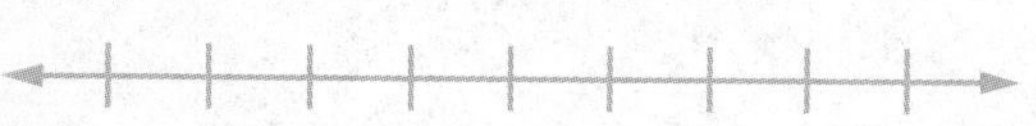

26. On Friday, a school spirit shop gave away a free T-shirt with each purchase over \$50. There were 47 purchases over \$50. Which integer represents the change in the number of free T-shirts the spirit shop had in stock at the end of the day on Friday?

Ⓕ −50　Ⓗ 47
Ⓖ −47　Ⓘ 50

27. Short Response Jackson owes his sister Monica \$15. Monica has a \$10 bill in her pocket. Explain the meaning of zero in this situation.

Common Core Review

Fill in each ◯ with < or > to make the inequality true. 4.NBT.2

28. 26 ◯ 22

29. 11 ◯ 13

30. 2.5 ◯ 3

31. 44 ◯ 4.4

32. 15 ◯ 6.8

33. 1.8 ◯ 1.9

34. Ally bought $\frac{1}{12}$ pound of cashews and $\frac{5}{6}$ pound of granola. Plot the fractions on the number line. Which quantity is greater? Explain. 4.NF.3d

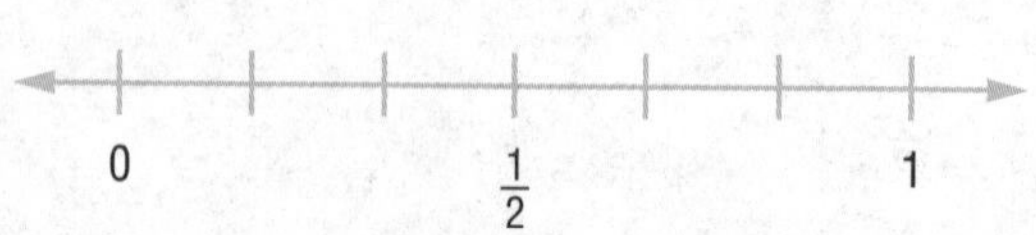

35. The number of raffle tickets the student council sold over three days is shown in the table. How many total raffle tickets did they sell? 4.NBT.4

Day	Tickets Sold
Wednesday	35
Thursday	23
Friday	46

Inquiry Lab
Absolute Value

HOW can a number line help you find two integers that are the same distance from zero?

Content Standards
6.NS.5, 6.NS.7, 6.NS.7c, 6.NS.7d

Mathematical Practices
1, 2, 3, 5

Hot Air Balloons Several hot air balloons were flying at the same height. The dashed line below represents their starting point. Which two balloons moved the same distance but in opposite directions?

Investigation

In the diagram below, +8 means Balloon A climbed 8 feet and −10 means Balloon B moved down 10 feet.

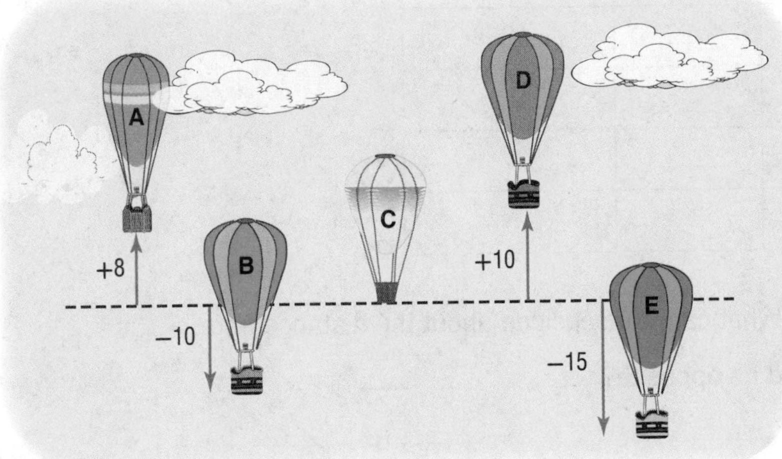

Use the diagram to compare the distance each balloon moved.

Step 1 Complete the chart to compare the distance each balloon moved from the dashed line.

Balloon	Integer	Direction	Distance Moved (ft)
C	0	none	0
D	+10		
E	−15		

Step 2 Determine which two balloons moved the same distance away from the dashed line.

So, Balloon ☐ and Balloon ☐ moved ☐ feet from the dashed line.

Collaborate

Use Math Tools **Use the number line to determine the distance between each integer and zero.**

−5 −4 −3 −2 −1 0 1 2 3 4 5

1. −2 ______

2. +3 ______

Work with a partner to complete the table. The first one is done for you.

	Integer	Distance Between Integer and Zero	Opposite Integer	Distance Between Opposite Integer and Zero
	3	3	−3	3
3.	4			
4.	7			
5.	−11			
6.	−13			
7.	19			
8.	−21			

9. **Reason Inductively** What can you conclude about the distance from zero for both an integer and its opposite? ______

Reflect

10. **Reason Abstractly** The movement of Balloon B in the Investigation was represented by the number −10. What does zero represent in the Investigation? ______

11. Inquiry HOW can a number line help you find two integers that are the same distance from zero?

Lesson 2

Absolute Value

What You'll Learn

Scan the lesson. Predict two things you will learn about absolute value.

-
-

Essential Question

HOW are integers and absolute value used in real-world situations?

Vocabulary

absolute value

opposites

Common Core State Standards

Content Standards
6.NS.6, 6.NS.6a, 6.NS.7, 6.NS.7c, 6.NS.7d

Mathematical Practices
1, 2, 3, 4

Vocabulary Start-Up

The distance between a number and 0 on the number line is called its **absolute value**.

1. Each mark on the number line indicates one yard. Draw a tree three yards west of the house. Draw a mailbox three yards east of the house.

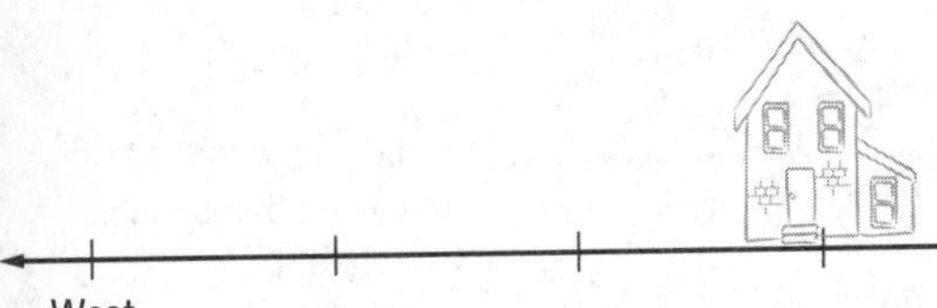

2. The distance between the house and the tree is ________ the distance between the house and the mailbox.

3. The tree and the mailbox are in ________ directions from the house.

4. How does the number line above help you to understand absolute value? ________

Real-World Link

5. **Errands** Jesse leaves home and walks 4 blocks west to the grocery store to buy milk then returns home. He then walks another 4 blocks east to the Post Office. Compare the distance and the direction of Jesse's house and the Post Office from the grocery store.

Work Zone

Find Opposites

Positive numbers, such as 2, are graphed to the right (or above) zero on a number line. Negative numbers, such as −2, are graphed to the left (or below) zero on a number line.

Opposites are numbers that are the same distance from zero in opposite directions. Since 0 is not negative nor positive, 0 is its own opposite. The opposite of the opposite of a number, is the number itself. For example, the opposite of the opposite of 3, −(−3), is 3.

−2 is 2 units to the left of zero.

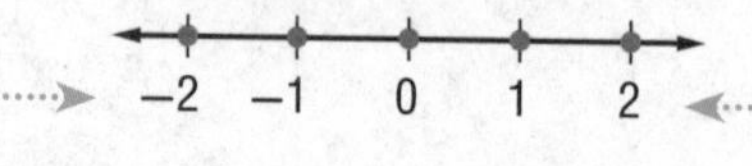

2 is 2 units to the right of zero.

Examples

1. Find the opposite of −5.

Method 1 **Use a number line.**

Draw a number line and graph −5.

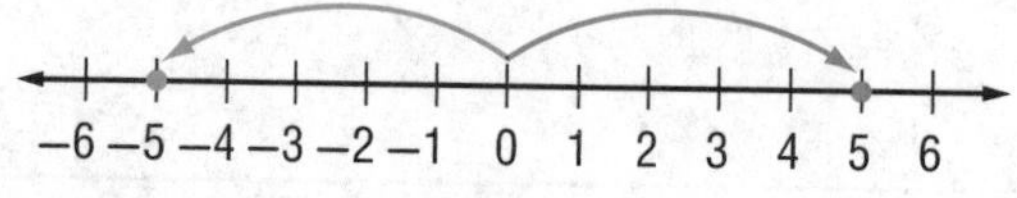

−5 is 5 units to the left of 0. The integer 5 is 5 units to the right of 0.

So, 5 is the opposite of −5.

Method 2 **Use symbols.**

The integer −5 uses the negative symbol.

The opposite of a negative symbol is a positive symbol.

So, the opposite of −5 is +5, or 5.

2. Find the opposite of the opposite of 4.

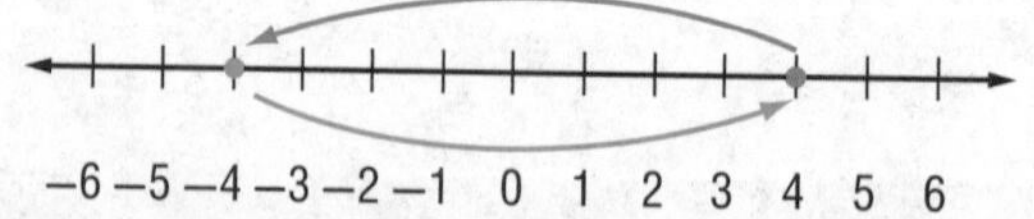

The opposite of 4 is −4.
The opposite of −4 is 4.

So, 4 is the opposite of the opposite of 4.

Show your work.

a. ____________

b. ____________

Got It? **Do these problems to find out.**

a. What is the opposite of 3?

b. What is the opposite of the opposite of −2?

Absolute Value

Key Concept

Words The absolute value of a number is the distance between the number and zero on a number line.

Model

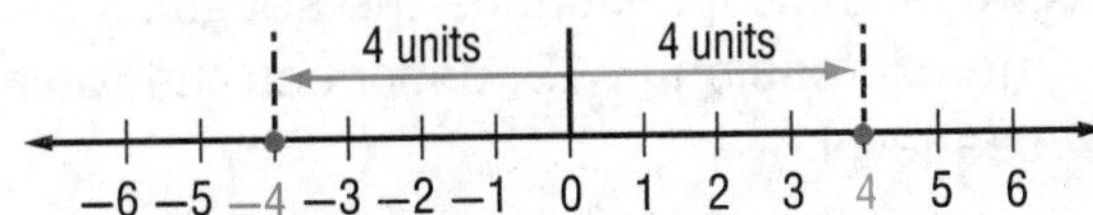

Symbols $|4| = 4$ The absolute value of 4 is 4.

$|-4| = 4$ The absolute value of −4 is 4.

Absolute Value
Since distance cannot be negative, the absolute value of a number is always positive or zero.

The integers −4 and 4 are each 4 units from 0, even though they are on opposite sides of 0. $|-4|$ is read *absolute value of negative four.*

Examples

3. Evaluate $|-7|$.

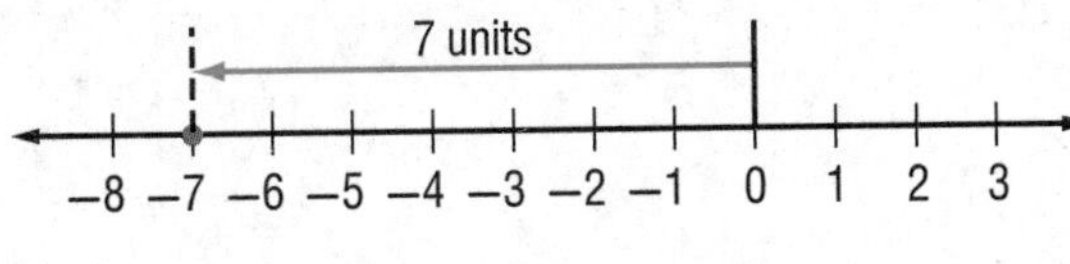

The graph of −7 is 7 units from 0 on the number line.

So, $|-7| = 7$.

4. Evaluate $|5| + |-6|$.

$|5| + |-6| = 5 + |-6|$ The absolute value of 5 is 5.

$= 5 + 6$ The absolute value of −6 is 6.

$= 11$ Simplify.

5. Evaluate $|-7| - |3|$.

$|-7| - |3| =$ ☐ − ☐ Find the absolute value of −7 and 3.

$=$ ☐ Simplify.

Got It? Do these problems to find out.

c. $|14|$ **d.** $|-9| + |3|$ **e.** $|-8| - |-2|$

Show your work.

c. ________

d. ________

e. ________

Example

6. A seagull is flying 25 feet above sea level. Nevaeh is diving 15 feet below sea level. What is the distance between Nevaeh and the seagull?

The expression $|25|$ describes the seagull's distance above sea level. The expression $|-15|$ describes the Nevaeh's distance below sea level.

To find the distance, add the absolute values.

$|25| + |-15| = 25 + |-15|$ The absolute value of 25 is 25.

$= 25 + 15$ The absolute value of -15 is 15.

$= 40$ Add.

So, the total distance is 40 feet.

Guided Practice

1. What is the opposite of 0? (Example 1)

2. What is the opposite of the opposite of 6? (Example 2)

Evaluate each expression. (Examples 3–5)

3. $|-5| =$ ______

4. $|20| - |-3| =$ ______

5. $|-16| + |-12| =$ ______

6. A game show contestant lost 15 points. He answered another question incorrectly and lost another 15 points. How many total points has he lost? (Example 6)

7. **Building on the Essential Question** How can absolute value help you to understand the size of a quantity? Give an example.

Rate Yourself!

How well do you understand opposites and absolute value? Circle the image that applies.

Clear

Somewhat Clear

Not So Clear

For more help, go online to access a Personal Tutor.

Name ________________ My Homework ________________

Independent Practice

Go online for Step-by-Step Solutions 

Find the opposite of each integer. (Example 1)

1. 6 ______

2. −3 ______

3. 0 ______

Find the opposite of the opposite of each integer. (Example 2)

4. 12 ______

5. −9 ______

6. −17 ______

Evaluate each expression. (Examples 3–5)

7. $|-14| =$ ______

8. $|31| - |-1| =$ ______

9 $|-15| + |-6| =$ ______

10. Jayson spent $18 on a shirt. Then he spent $24 on a pair of pants. What is the total amount he spent? (Example 6) ______

11. Lilly saw a jelly fish at 6 feet below sea level. She saw a bright blue fish at 10 feet below sea level. What is the distance between the blue fish and the jelly fish? (Example 6) ______

12. **STEM** The table shows the melting points of various elements. Is the absolute value of the melting point of neon greater than or less than the absolute value of the melting point of hydrogen? ______

Element	Melting Point (°C)
Hydrogen	−259
Neon	−248
Oxygen	−218

13 **STEM** The surface of Jupiter is made of colorful clouds created by various chemicals in the atmosphere. The temperature at the top of the clouds is −230°F. The temperature below the clouds is 70°F. Which temperature has the lower absolute value? ______

 **Reason Abstractly** **Evaluate each expression.**

14. $-|3| =$ ______

15. $|5 + 9| =$ ______

16. $|17 - 8| =$ ______

H.O.T. Problems Higher Order Thinking

17. **Find the Error** Mei is evaluating an expression using absolute value. Find her mistake and correct it.

18. **Which One Doesn't Belong?** Identify the phrase that *cannot* be described by the same absolute value as the other three. Explain your reasoning.

a loss of 8 pounds	8 miles above sea level	giving away \$8	18° below normal

Persevere with Problems **Determine whether each statement is *always, sometimes,* or *never* true. Explain.**

19. The absolute value of a positive integer is a negative integer.

20. If a and b are integers and $a > b$, then $|a| > |b|$.

21. **Reason Abstractly** Explain why the absolute value of a number is never negative.

22. **Reason Abstractly** Explain why an account balance less than −40 dollars represents a debt greater than 40 dollars.

Standardized Test Practice

23. If $x = -2$ and $y = 2$, then which of the following statements is false?

Ⓐ $|x| > 1$

Ⓑ $|x| = |y|$

Ⓒ $|y| < 1$

Ⓓ $|x| = y$

Name ______ My Homework ______

Extra Practice

Find the opposite of each integer.

24. −2 2

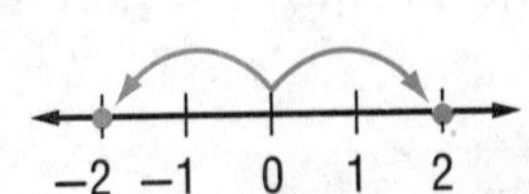

25. 15 ______

26. 42 ______

Find the opposite of the opposite of each integer.

27. 9 ______

28. 0 ______

29. −8 ______

Evaluate each expression.

30. $|18| =$ ______

31. $|0| =$ ______

32. $|25|$ ______

33. $|2| + |-13| =$ ______

34. $|-20| - |17| =$ ______

35. $|-16| - |5| =$ ______

36. The balance of Bryce's account is $16. Jada's account is $5 overdrawn. What is the difference between their account balances?

______

37. A football team lost 3 yards on their first play and 6 yards on their second play. How many total yards did they lose?

______

38. The table shows the lowest elevations for several states. Is the absolute value of the lowest elevation of California greater than or less than the absolute value of the lowest elevation of Illinois?

State	Lowest Elevation (ft)
Oklahoma	289
Illinois	279
Kentucky	257
California	−282

______

CCSS **Reason Abstractly** **Evaluate each expression.**

39. $-|-10| =$ ______

40. $|13 - 6| =$ ______

Standardized Test Practice

41. If $x = -1$ and $y = -2$, then which of the following statements is true?

Ⓐ $|x| > 1$
Ⓑ $|x| < |y|$
Ⓒ $|y| < 1$
Ⓓ $|y| < x$

42. Refer to the number line below. Which point represents the number with the greatest absolute value?

M (−5), F (−2), C (1), T (4)

−5 −4 −3 −2 −1 0 1 2 3 4 5

Ⓕ point M
Ⓖ point F
Ⓗ point C
Ⓘ point T

43. A video game has different point values associated with different actions. The table shows some of the actions. Which action has associated points with the greatest absolute value?

Action	Points
collect gem	+5
fall in water	−10
build bridge	+12
climb tree	−15

Ⓐ collect gem
Ⓑ fall in water
Ⓒ build bridge
Ⓓ climb tree

44. Which expression has the greatest value?

Ⓕ $|-25|$
Ⓖ $|-16|$
Ⓗ $|18|$
Ⓘ $|22|$

45. Short Response The table shows the freezing point of different liquids. What liquid's freezing point has the greatest absolute value?

Liquid	Freezing Point (°F)
Water	32
Acetic Acid	62
Linseed Oil	−4
Acetone	−94

CCSS Common Core Review

Fill in each ○ with >, <, or = to make a true statement. 4.NBT.2

46. 69.23 ○ 69.25

47. 171.10 ○ 171.09

48. 47.74 ○ 47.740

49. Part of a sauce recipe is shown. If all the ingredients are mixed together, how much sauce will be made? 5.NF.1

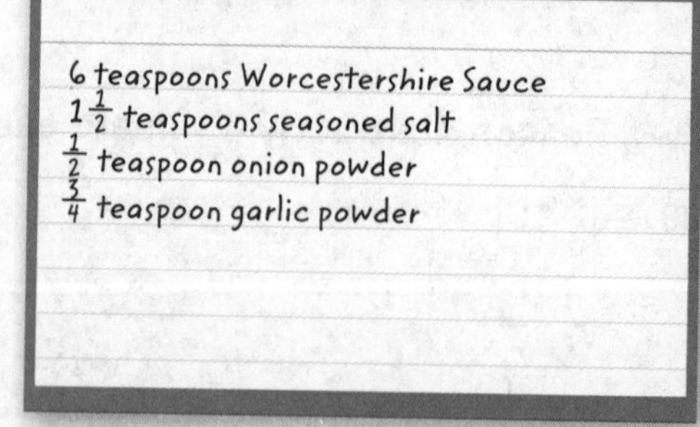

50. Caroline's soccer practice starts at quarter after 4 P.M. and ends at 5 P.M. How many minutes does her soccer practice last? 4.MD.2

Lesson 3

Compare and Order Integers

What You'll Learn

Scan the lesson. List two real-world scenarios in which you would compare integers.

- ____________________
- ____________________

Essential Question

HOW are integers and absolute value used in real-world situations?

Common Core State Standards

Content Standards
6.NS.7, 6.NS.7a, 6.NS.7b, 6.NS.7d

Mathematical Practices
1, 2, 3, 4, 5

Real-World Link

Watch

Winter Fairbanks is located in interior Alaska. The average temperature for several months is shown.

1. The average temperature for December is –6.5°F and the average temperature for March is 11°F. Label December and March on the thermometer.

2. Which months have a greater average temperature than February? ____________________

3a. Which months have a lower average temperature than November? ____________________

3b. Complete the inequality to compare the temperatures of November and February.

3 > ☐

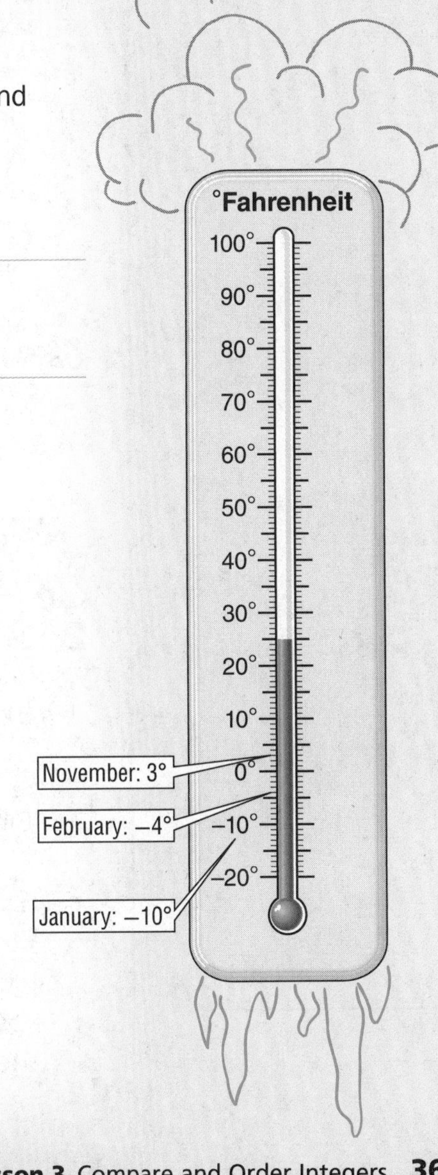

Work Zone

Compare Integers

To compare integers, you can compare the signs as well as the magnitude, or size, of the numbers. Greater numbers are graphed farther to the right.

Compare the signs.

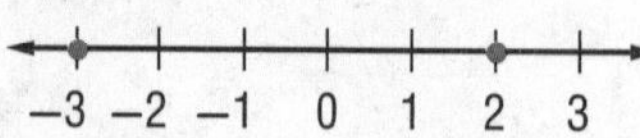

Positive numbers are greater than negative numbers. So, 2 > −3.

Compare the position on the number line.

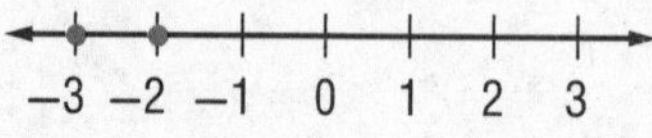

Since −2 is farther to the right, −2 > −3.

Example

Fill in the ◯ with <, >, or = to make a true sentence.

1. **12 ◯ −4**

Graph 12 and −4 on a number line. Then compare.

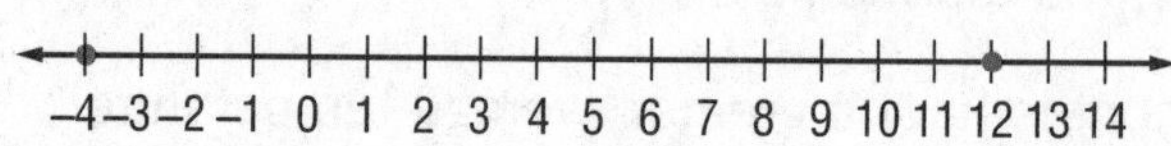

Since 12 is to the right of −4, 12 > −4.

Got It? **Do these problems to find out.**

a. −3 ◯ −5 **b.** −5 ◯ 0 **c.** 6 ◯ −1

Absolute Value

Although −5 is the least value in Example 2, it represents the greater point deficit.

|−5| > |−4|

Real World Example

2. **Justin has a score of −4 on the Trueville Trivia game. Desiree's score is −5. Write an inequality to compare the scores. Explain the meaning of the inequality.**

−4 > −5 −4 is farther to the right on a number line than −5.

Since −4 > −5, Justin has a higher score than Desiree.

Got It? **Do this problem to find out.**

d. The temperature on Tuesday was 2°F. The temperature on Wednesday was −2°F. Write an inequality to compare the temperatures. Explain the meaning of the inequality.

d. ____________

Order Integers

You can use a number line to order a set of integers. Integers can be ordered from least to greatest or from greatest to least.

Example

3. Order the set {–9, 6, –3, 0} from least to greatest.

Method 1 **Use a number line.**

Graph the numbers on a number line.

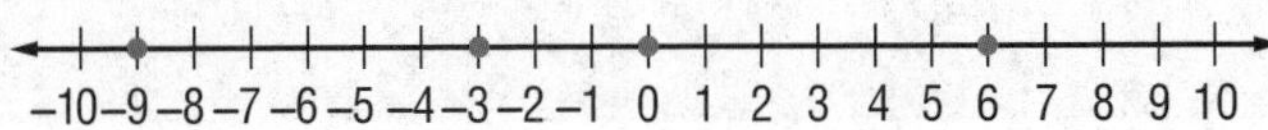

The order from left to right is –9, –3, 0, and 6.

Method 2 **Compare signs and values.**

Compare negative numbers. Then compare positive numbers.

The negative integers are –9 and –3. $-9 < -3$

The integer 0 is neither positive nor negative.

The positive integer is 6.

So, the order from least to greatest is –9, –3, 0, and 6.

Absolute Value
Since absolute value is always positive, it is not used to compare and order integers.

Got It? **Do these problems to find out.**

e. Order the set {–4, 3, 11, –25} from greatest to least.

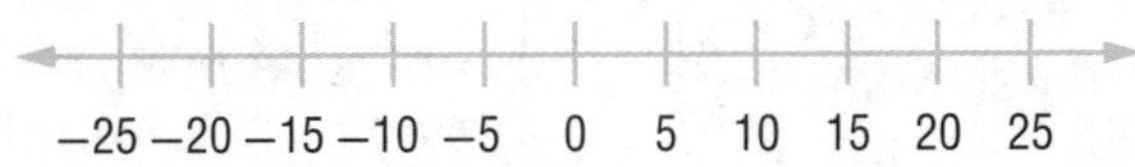

f. Order the set {–18, 30, 12, –6, 3} from least to greatest.

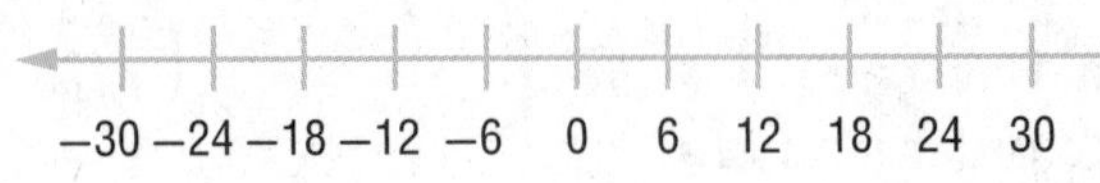

e. ____________

f. ____________

Example

4. STEM The table shows the lowest elevation for several continents. Order the elevations from least to greatest.

First, graph each integer. Then, write the integers as they appear on the number line from left to right.

Continent	Lowest Elevation (m)
Africa	−156
Asia	−418
Australia	−12
Europe	−28
North America	−86
South America	−105

−500 −450 −400 −350 −300 −250 −200 −150 −100 −50 0

The elevations from least to greatest are −418, −156, −105, −86, −28, and −12.

Guided Practice

Fill in each ◯ with <, >, or = to make a true statement. (Example 1)

1. 17 ◯ 31
2. −6 ◯ −10
3. −83 ◯ −38

4. Andrew and his father are scuba diving at −38 feet and Tackle Box Canyon has an elevation of −83 feet. Write an inequality to compare the depths. Explain the meaning of the inequality. (Example 2)

5. STEM The daily low temperatures in Kate's hometown last week were 2°C, −9°C, −18°C, −6°C, 3°C, 0°C, and −7°C. Order the temperatures from greatest to least. (Examples 3 and 4)

6. **Building on the Essential Question** How can symbols and absolute value help you to order sets of integers?

Rate Yourself!

How confident are you about comparing and ordering integers? Shade the ring on the target.

For more help, go online to access a Personal Tutor.

FOLDABLES *Time to update your Foldable!*

Name ______________________ My Homework ______________________

Independent Practice

Go online for Step-by-Step Solutions 

Fill in each ◯ with <, >, or = to make a true statement. (Example 1)

1. −2 ◯ −4

2. 1 ◯ −3

3. 5 ◯ 0

4. Amy is building a house. The basement floor is at −15 feet. The roof of the house is above the ground 25 feet. Write an inequality to compare the heights. Explain the meaning of the inequality. (Example 2)

5. The low temperature in Anchorage, Alaska, one day was −9°F. On the same day, the low temperature in Flagstaff, Arizona, was 26°F. Write an inequality to compare the temperatures. Explain the meaning of the inequality. (Example 2)

Order each set of integers from least to greatest. (Example 3)

6. {15, 17, 21, 6, 3}

7. {−55, 143, 18, −79, 44, 101}

8. The table indicates Xavier's cell phone use over the last four months. Positive values indicate the number of minutes he went over his allotted time, and negative values indicate the number of minutes he was under. Arrange the months from least to most minutes used. (Example 4)

Month	Time (min)
February	−156
March	12
April	0
May	−45

9. CCSS **Use Math Tools** Refer to the table and the following information. The apparent magnitude of an object measures how bright the object appears to the human eye. A negative magnitude identifies a brighter object than a positive magnitude.

Object	Approximate Apparent Magnitude
100-Watt Bulb	−19
Alpha Centauri	4
Andromeda Galaxy	0
Full Moon	−13
Sun	−27
Venus	−5

a. Which object appears the brightest to the human eye?

b. Order the objects from the brightest to the faintest.

c. Find the least apparent magnitude of this data set.

10. **Justify Conclusions** Refer to the graphic novel frame below for exercises a–c.

a. If about 32,834.5 kilobytes of memory is still available, how many more pictures can they take? ______

b. Write an inequality to compare the number of pictures taken during school to the number of pictures taken after school. ______

c. Explain the meaning of the inequality. ______

H.O.T. Problems Higher Order Thinking

11. **Model with Mathematics** Write a real-world situation to explain the inequality $-\$15 < \$7$. ______

12. **Reason Abstractly** Explain why -11 is less than -7, but $|-11|$ is greater than $|-7|$. ______

13. **Persevere with Problems** Order the fractions $-\frac{1}{2}, \frac{5}{2}, -\frac{12}{4}, \frac{1}{6}$, and $\frac{7}{8}$ from least to greatest. ______

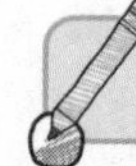

Standardized Test Practice

14. Order the set {−5, 3, 2, −7} from greatest to least.

Ⓐ −7, −5, 3, 2
Ⓑ −7, −5, 2, 3
Ⓒ 2, 3, −5, −7
Ⓓ 3, 2, −5, −7

Name ______________________ My Homework ______________________

Extra Practice

Fill in each ◯ with <, >, or = to make a true statement.

15. −2 ⓐ 4

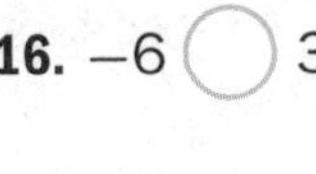

Since −2 is to the left of 4, −2 < 4.

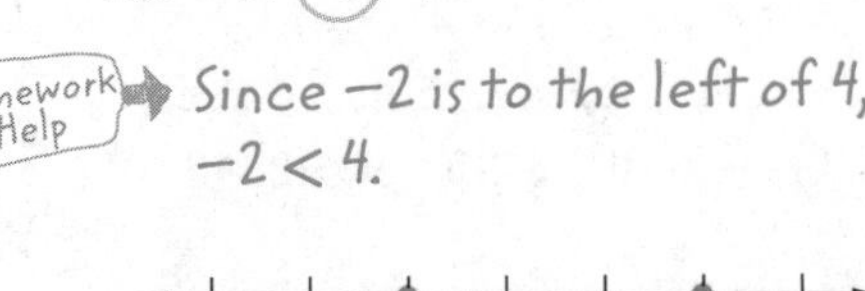

16. −6 ◯ 3

17. −3 ◯ 2

18. The elevation of Driskill Mountain, Louisiana, is 163 meters above sea level. Death Valley has an elevation of −86 meters. Write an inequality to compare the elevations. Explain the meaning of the inequality.

19. Yvonne owes her sister $25. Michael's checking account balance is −$20. Write an inequality to compare the amounts. Explain the meaning of the inequality.

Order each set of integers from least to greatest.

20. {14, 1, 6, 23, 7, 5}

21. {−221, 63, 54, −89, −71, −10}

22. Gary, Sindhu, and Beth are all waiting for their trains to arrive. Gary's train leaves at 5 minutes before noon, Sindhu's leaves at 25 minutes after noon, and Beth's leaves 5 minutes before Sindhu's train. Order the three by who will leave first.

23. 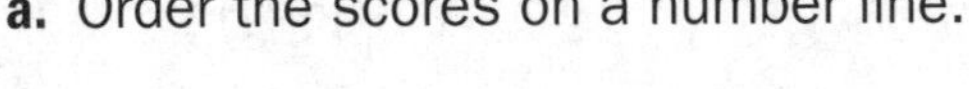

Use Math Tools Use the bar graph and the information below. The bar graph gives the scores of four golfers (A, B, C, and D). The numbers indicate scores above and below par.

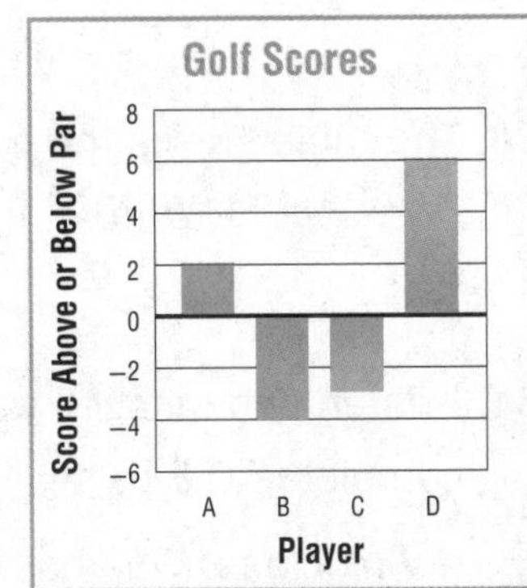

a. Order the scores on a number line.

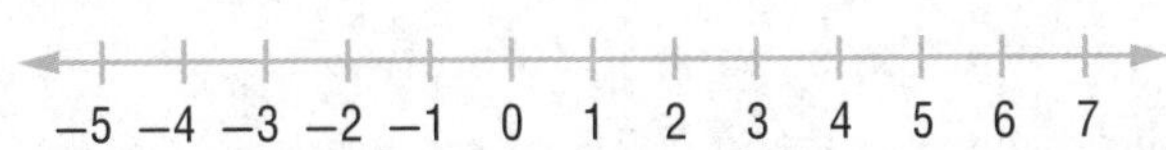

b. Which player had the worst score? Explain your answer.

Standardized Test Practice

24. The table shows the temperatures for a four-day period.

Temperature (°F)	
Monday	−7
Tuesday	8
Wednesday	−2
Thursday	−1

Which list shows the temperatures from least to greatest?

Ⓐ 8, −2, −1, −7

Ⓑ 8, −1, −2, −7

Ⓒ −7, −2, −1, 8

Ⓓ −7, −1, −2, 8

25. Verónica (V) was 12 minutes early to class, Deshawn (D) was right on time, and Kendis (K) was 3 minutes late. Which time line represents the students' arrival to class?

Ⓕ 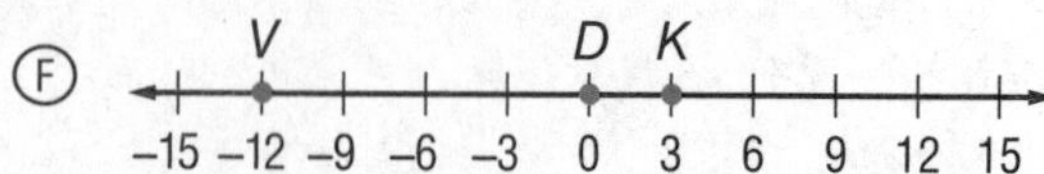

Ⓖ 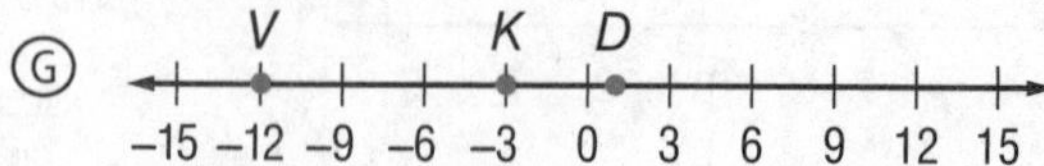

Ⓗ 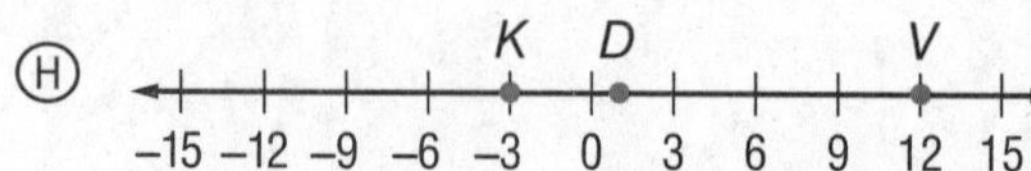

Ⓘ 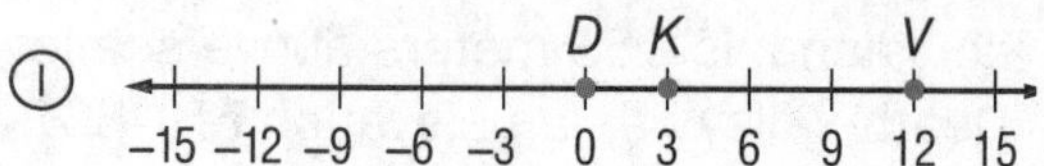

26. **Short Response** The table shows the scores for a game of miniature golf. The integer 0 represents par. Arrange the players from least shots taken to most shots taken.

Player	Score
Cristian	−6
Bailey	2
Liam	−3
Marisol	5

Common Core Review

Write each fraction as a decimal. 5.NF.5b

27. $\frac{3}{4} =$ ______

28. $\frac{1}{5} =$ ______

29. $\frac{3}{20} =$ ______

30. The table shows the heights of Sonya's siblings. Who is taller, Frieda or Julio? Compare their heights using the symbol >. 4.NBT.2

Member	Height (ft)
Frieda	$5\frac{1}{4}$
Julio	$5\frac{5}{6}$

31. Kristen and Mitchell were given the same math assignment. Kristen completes 0.8 of her work in class. Mitchell completes 0.75 of his work during class. Who has more homework remaining after class? 5.NBT.3

Problem-Solving Investigation

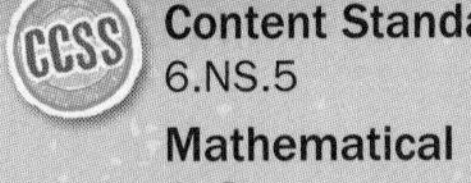

Content Standards
6.NS.5

Mathematical Practices
1, 3

Case #1 Hit the Slopes!

Marissa and her family are on a ski trip at Mount Washington in New Hampshire. They returned from the slopes at 6 P.M. By 9 P.M., the temperature had fallen 18° to the day's low temperature of −8°F.

What was the temperature at 6 P.M.?

Understand What are the facts?

- By 9 P.M., the temperature had fallen 18°.
- The day's low temperature was −8°F.

Plan What is your strategy to solve this problem?

Work backward from the low temperature at 9 P.M. Use a thermometer diagram to find the temperature at 6 P.M.

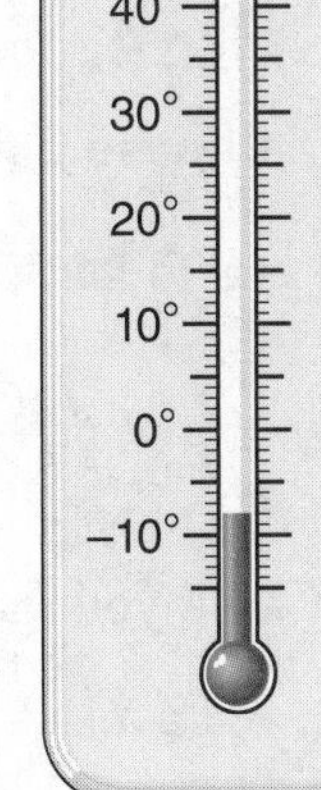

Solve How can you apply the strategy?

Start at ☐ °F. Shade the thermometer ☐ degrees to find the temperature at 6 P.M.

So, the temperature at 6 P.M. was ______.

Check Does the answer make sense?

−8°F is 8 degrees away from 0°F. 18 − 8 = 10

So, add to ☐ to 0°F. Since 0 + ☐ = ☐, the answer is reasonable.

Analyze the Strategy

Justify Conclusions The high temperature was 36°F. How far away from −8°F is 36°F? Explain. ______________________________

Case #2 Get Ready, Get Set, Go

The table shows the amount of time it takes Henry to do different activities before going to soccer practice.

If he needs to be at practice at 8:15 A.M., what time should he wake up in the morning to get to the soccer field?

Activity	Time (minutes)
Travel to field from home	15
Eating breakfast	35
Changing into uniform	10
Checking E-mail messages	20

Understand

Read the problem. What are you being asked to find?

I need to find ______________________________.

Underline key words and values. What information do you need to know?

The table shows the time it takes Henry to do each activity. It takes him ☐ minutes to get to the field, ☐ minutes to eat, ☐ minutes to change for practice, and ☐ minutes to check his email.

Plan

Choose a problem-solving strategy.

I will use the ______________________________ *strategy.*

Solve

Use your problem-solving strategy to solve the problem.

8:15 A.M. − 20 min = ________ A.M. 7:45 A.M. − 35 min = ________ A.M.

7:55 A.M. − 10 min = ________ A.M. 7:10 A.M. − 15 min = ________ A.M.

So, Henry should wake up at ________.

Check

Use information from the problem to check your answer.

Begin at ________ A.M. and add the minutes from the table.

☐ + ☐ + ☐ + ☐ = ☐ minutes

________ A.M. plus ☐ minutes is 8:15 A.M.

Collaborate Work with a small group to solve the following cases. Show your work on a separate piece of paper.

Case #3 Sea Level

Mr. Ignacio went diving along the coral reef in Oahu. He descended 12 meters below sea level. The difference between this point on the coral reef and the highest point on the island, Mount Ka'ala, is 1,232 meters.

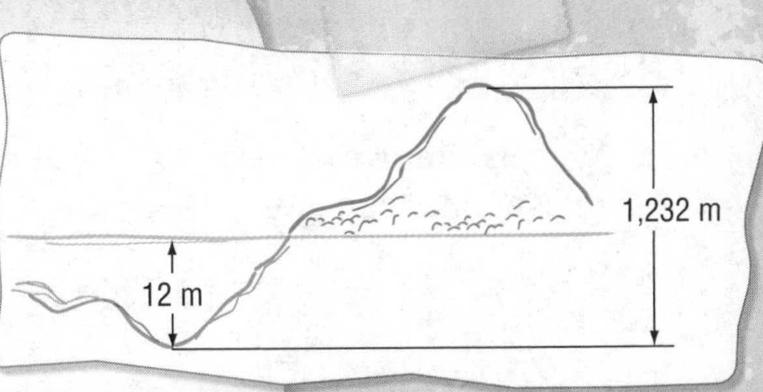

How far above sea level is Mount Ka'ala?

Case #4 Cameras

Adamo saved 13 pictures on his digital camera, and deleted 32 pictures.

If there are now 108 pictures, how many pictures had he saved originally?

Case #5 Mystery Number

A number is multiplied by 4, and then 6 is added to the product. The result is 18.

What is the number?

Case #6 Ladders

You are standing on the middle rung of a ladder.

If you first climb up 3 rungs, then down 5 rungs, and then up 10 rungs to get onto the top rung, how many rungs are on the ladder?

Mid-Chapter Check

Vocabulary Check

1. **Be Precise** Define *negative integer*. Give an example of a negative integer and then give its opposite. (Lesson 1)

2. Fill in the blank in the sentence below with the correct term. (Lesson 2)

   The ______________ of the numbers −4 and 4 is 4.

Skills Check and Problem Solving

Model with Mathematics Graph each set of integers on a number line. (Lesson 1)

3. {−4, −6, 0, 3}

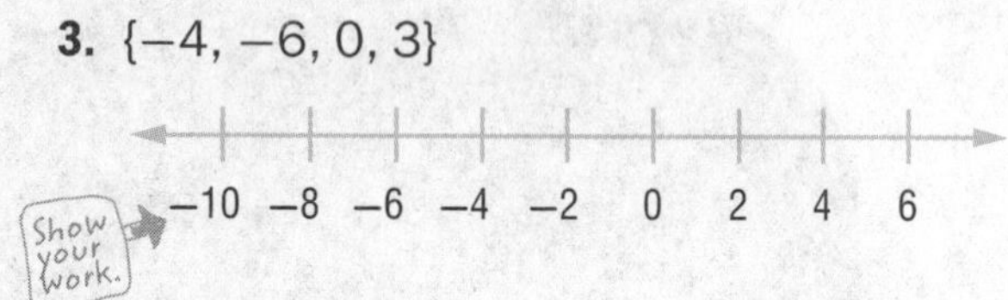

4. {2, −3, 1, −1}

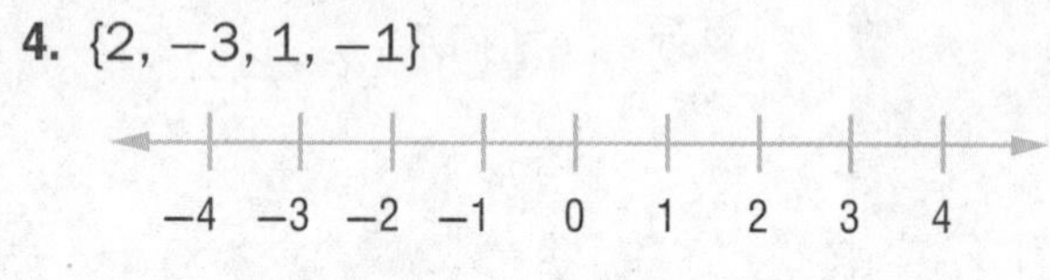

Evaluate each expression. (Lesson 2)

5. $|-12| =$ ______

6. $|-4| + |-10| =$ ______

7. $|9| + |-2| =$ ______

8. $|13| - |-5| =$ ______

9. $|-16| - |-2| =$ ______

10. $|-15| + |-7| =$ ______

11. Hailey, Priya, and Shetal are auditioning for the same role. Hailey auditions at 10 minutes before four, Priya auditions 30 minutes before Hailey, and Shetal auditions at 5 minutes before four. Order the three by who will audition first. (Lesson 3) ______________

12. **Standardized Test Practice** The table shows the overnight low temperatures for a four-day period. (Lesson 3)

Temperature (°F)	
Thursday	−8
Friday	7
Saturday	18
Sunday	−11

Which list shows the temperatures from least to greatest?

Ⓐ −8, −11, 7, 18

Ⓑ 7, −8, −11, 18

Ⓒ −11, −8, 7, 18

Ⓓ −11, 7, −8, 18

Inquiry Lab
Number Lines

HOW can you use a number line to model and compare positive and negative rational numbers?

CCSS **Content Standards**
6.NS.6, 6.NS.6c, 6.NS.7

Mathematical Practices
1, 3, 4

Beach Marcus and Silvio are at the beach. Marcus builds a sandcastle 0.6 meter high. Silvio digs a hole in the sand 0.8 meter deep.

Investigation 1

Just as you can graph integers on a number line, you can graph positive and negative fractions and decimals. Recall that positive numbers are to the right of zero on the number line and negative numbers are to the left of zero.

Step 1 Complete the number line from −1 to 1, with increments of 0.2.

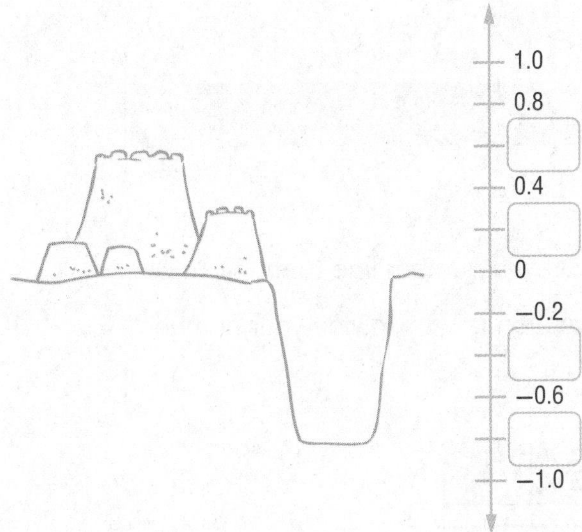

Step 2 The sandcastle is above sea level. Its height is *greater than zero* on the number line, so draw a dot at ______ to represent the sandcastle.

Step 3 The hole is below sea level. Its depth is *less than zero* on the number line. So draw a dot at

______ to represent the hole.

Collaborate

Model with Mathematics **Work with a partner. Graph each number on a number line.**

1. −2.4

2. 0.1

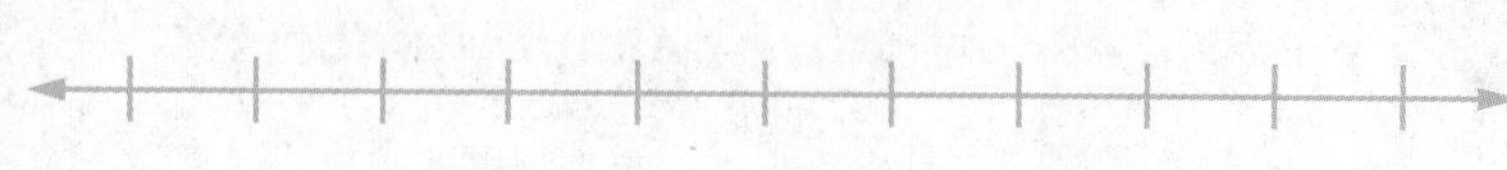

3. −4.5

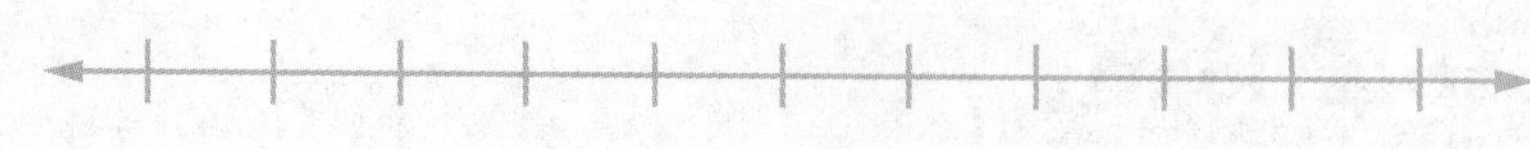

4. −6.8

Investigation 2

Graph $-\frac{3}{4}$ on a number line.

Step 1 Model $-\frac{3}{4}$ using fraction tiles. Draw a number line from −1 to 0. Since the denominator of the fraction is ☐, divide your number line into ☐ equal parts.

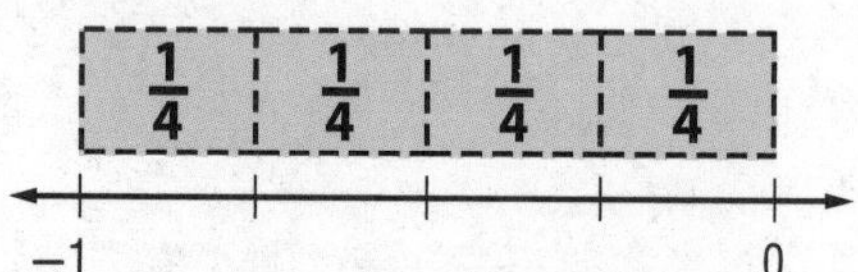

Step 2 Each mark on the number line represents $\frac{\square}{\square}$. Label the number line with $-\frac{3}{4}$, $-\frac{2}{4}$, and $-\frac{1}{4}$.

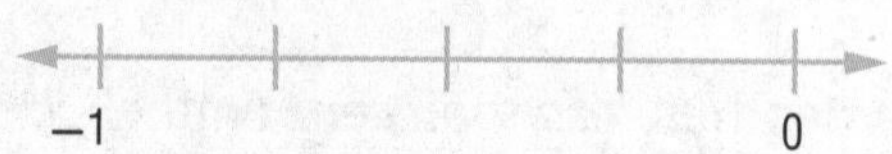

Step 3 Draw a dot to graph $-\frac{3}{4}$ on the number line above.

Model with Mathematics Work with a partner. Graph each number on a number line.

5. $-\frac{4}{5}$

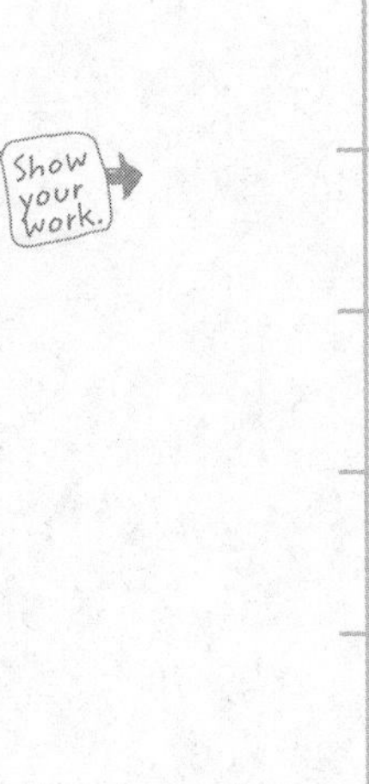

6. -5.75

7. $\frac{7}{10}$

8. $-\frac{3}{8}$

9. 8.75

10. $-\frac{3}{10}$

11. $-\frac{5}{12}$

Analyze

Work with a partner to complete the table. The first one is done for you.

	Number	Positive or Negative	Greater Than or Less Than Zero	Left or Right of 0 on the Number Line
	-3.5	negative	$<$	left
12.	$+\frac{4}{5}$			
13.	$-\frac{1}{3}$			
14.	$+0.3$			

15. CCSS **Reason Inductively** Which number is greater, 0.3 or -0.7? Explain.

16. CCSS **Reason Inductively** Jacyln thinks that $-\frac{1}{2}$ is greater than $\frac{1}{4}$ because it is farther from zero on the number line. Is her thinking correct? Explain.

Reflect

17. CCSS **Model with Mathematics** Write a real-world problem that involves a positive and a negative value. Then graph the values used in the problem on a number line. Compare the values.

18. 

HOW can you use a number line to model and compare positive and negative rational numbers?

Lesson 4

Terminating and Repeating Decimals

What You'll Learn

Scan the lesson. Predict two things you will learn about terminating and repeating decimals.

- ______________________________________
  ____
- ______________________________________
  ____

Essential Question

HOW are integers and absolute value used in real-world situations?

Vocabulary

terminating decimal
repeating decimal
bar notation
rational number

CCSS Common Core State Standards

Content Standards
Preparation for 6.NS.6c and 6.NS.7a

Mathematical Practices
1, 2, 3, 4, 7, 8

Vocabulary Start-Up

Any number that can be written as a fraction is called a **rational number**. Every rational number can be written as either a **terminating decimal** or a **repeating decimal**.

Draw lines from each word to its matching statement.

terminating decimal	the decimal form of a rational number; 0.33333...
repeating decimal	the decimal form of a rational number which has a repeating digit of zero; 06.25

Real-World Link

Party Favors Jude is buying fruit snacks for party favors. He asks the cashier for a half pound of fruit snacks.

1. Express one half as a fraction.

   ____

2. Write the decimal that represents half a pound.

   ____

3. Suppose Jude wanted to buy one third of a pound. What decimal would the scale show?

   ____

Key Concept

Rational Numbers

Watch

Work Zone

Words Rational numbers can be written as fractions.

Algebra $\frac{a}{b}$, where a and b are integers and $b \neq 0$.

Model

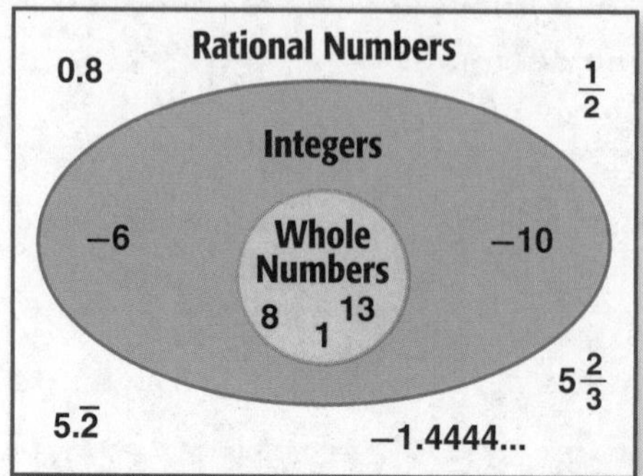

Fractions, terminating and repeating decimals, percents, and integers are all rational numbers. Every rational number can be expressed as a decimal by dividing the numerator by the denominator.

Rational Number	Repeating Decimal	Terminating Decimal
$\frac{3}{10}$	0.300...	0.3
$\frac{4}{5}$	0.800...	0.8
$\frac{5}{6}$	0.833...	does not terminate

To indicate the number pattern that repeats indefinitely, use bar notation. **Bar notation** is a bar placed over the digits that repeat.

$0.545454\ldots = 0.\overline{54}$ $0.583333\ldots = 0.58\overline{3}$

Example

Tutor

1. Write $\frac{5}{12}$ as a decimal.

$$\begin{array}{r} 0.4166 \\ 12\overline{)5.000} \\ -\underline{48} \\ 20 \\ -\underline{12} \\ 80 \\ -\underline{72} \\ 80 \\ -\underline{72} \\ 8 \end{array}$$

Divide 5 by 12.

The remainder will never be zero.

So, $\frac{5}{12} = 0.4166\ldots$ or $0.41\overline{6}$.

Got It? Do these problems to find out.

Write each fraction as a decimal. Use bar notation if necessary.

a. $\frac{1}{6}$

b. $\frac{8}{9}$

c. $\frac{2}{11}$

a. ______

b. ______

c. ______

Write a Negative Fraction as a Decimal

When writing negative fractions as decimals, the process is the same. Divide as with positive fractions. Write the negative sign in front of the decimal.

Examples

Tutor

2. Write $-\frac{2}{9}$ as a decimal.

$$\begin{array}{r} 0.222 \\ 9\overline{)2.000} \\ -\underline{18} \\ 20 \\ -\underline{18} \\ 20 \\ -\underline{18} \\ 2 \end{array}$$

Divide 2 by 9.

The remainder will never be zero.

So, $-\frac{2}{9} = -0.222...$ or $-0.\overline{2}$.

Repeating Decimals
When dividing, it is sometimes helpful to divide until the repeated pattern is shown at least three times.

3. Write $-2\frac{2}{3}$ as a decimal.

$-2\frac{2}{3}$ can be rewritten as $-\frac{8}{3}$.

The mixed number $-2\frac{2}{3}$ can be written as $-2.\overline{6}$.

$$\begin{array}{r} 2.6... \\ 3\overline{)8.0} \\ -\underline{6} \\ 20 \\ -\underline{18} \\ 2 \end{array}$$

Got It? Do these problems to find out.

Write each fraction as a decimal. Use bar notation if necessary.

d. $-\frac{1}{4}$

e. $-\frac{5}{6}$

f. $-2\frac{1}{6}$

d. ______

e. ______

f. ______

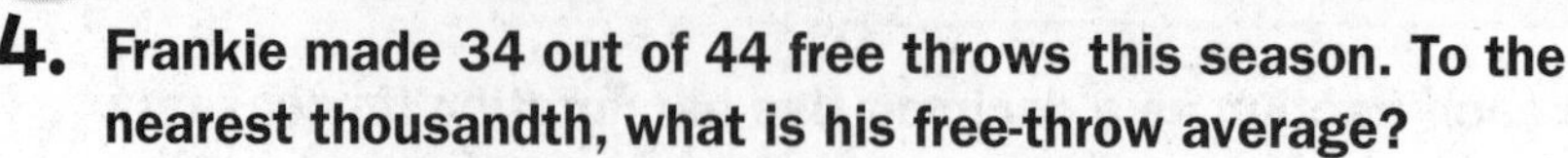

Example

4. Frankie made 34 out of 44 free throws this season. To the nearest thousandth, what is his free-throw average?

Using a calculator, divide 34 by 44.

34 [÷] 44 [ENTER] 0.77272727

To the nearest thousandth, his free-throw average is 0.773.

Got It? Do this problem to find out.

g. Of nine students surveyed, four said they prefer exercising in the morning rather than in the evening. Express this fraction as a decimal. Use bar notation if necessary.

Show your work.

g. ______

Guided Practice

Write each fraction as a decimal. Use bar notation if necessary. (Examples 1–3)

1. $\frac{7}{9} =$ ______

2. $-\frac{1}{33} =$ ______

3. $-2\frac{5}{6} =$ ______

4. $\frac{10}{15} =$ ______

5. $-\frac{4}{5} =$ ______

6. $1\frac{5}{9} =$ ______

7. Dana bought $\frac{2}{3}$ yard of fabric to make a new purse. Write the amount of fabric she used as a decimal. (Example 4)

______

8. **Building on the Essential Question** How are repeating decimals used in real-world situations?

______

Rate Yourself!

Are you ready to move on? Shade the section that applies.

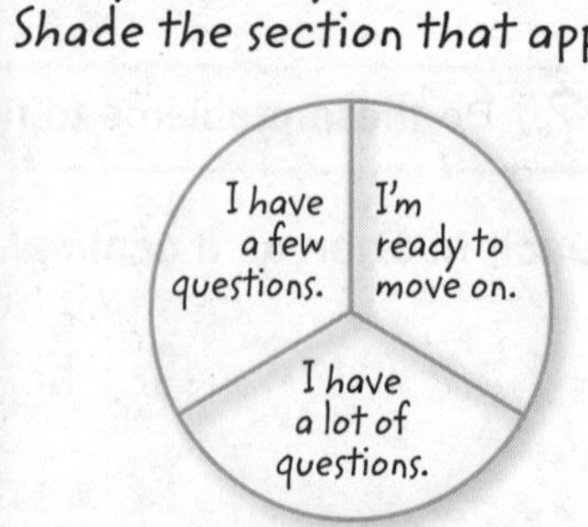

For more help, go online to access a Personal Tutor.

Name ______________________ My Homework ______________________

Independent Practice

Go online for Step-by-Step Solutions 

Write each fraction as a decimal. Use bar notation if necessary. (Examples 1–3)

1. $\frac{7}{15}$ = ________

2. $\frac{8}{18}$ = ________

3. $-\frac{8}{12}$ = ________

4. $-\frac{6}{7}$ = ________

5. $3\frac{15}{44}$ = ________

6. $-2\frac{5}{22}$ = ________

7. Sarafina had 34 out of 99 hits when she was at bat during the softball season. What was her batting average? (Example 4)

____

8. Shiv and his friends ate $3\frac{1}{6}$ pizzas. Write this amount as a decimal. (Example 4)

____

Write each decimal as a fraction or mixed number in simplest form.

9. −0.9 = ________

10. −0.85 = ________

11. −3.8 = ________

Evaluate each expression.

12. $|-2.3|$ = ________

13. $\left|\frac{4}{13}\right|$ = ________

14. $\left|-8\frac{7}{11}\right|$ = ________

15. **STEM** There are over 2,700 species of snakes in the world. Over 600 species are venomous. Write the fraction of species that are *not* venomous as a decimal. ________

16. **CCSS Justify Conclusions** The ratio of the circumference of a circle to its diameter is represented by the number π. The number π is a decimal that does not repeat. The fraction $\frac{22}{7}$ is sometimes used as an estimate of π. Is $\frac{22}{7}$ a repeating decimal? Explain.

____

17. **Reason Abstractly** Refer to the graphic novel frame below for Exercises a–b.

a. How many total photos were taken? ______

b. What fraction of the photos were taken after school? Write this fraction as a decimal. Round to the nearest thousandth. ______

H.O.T. Problems Higher Order Thinking

18. **Identify Structure** Write a fraction and an equivalent terminating decimal between 0.2 and 0.6. ______

19. **Persevere with Problems** Predict whether or not the decimal equivalent to $\frac{17}{36}$ is terminating. Explain your reasoning. Check your prediction with a calculator. ______

20. **Which One Doesn't Belong?** Identify the decimal equivalent that does *not* have the same characteristic as the other three. Explain.

$\frac{1}{12}$	$\frac{2}{12}$	$\frac{3}{12}$	$\frac{4}{12}$

______

Standardized Test Practice

21. Which decimal represents the shaded portion of the figure below?

Ⓐ 0.4
Ⓒ 0.5
Ⓑ $0.\overline{4}$
Ⓓ $0.\overline{5}$

Extra Practice

Write each fraction as a decimal. Use bar notation if necessary.

22. $\frac{32}{75} = 0.42\overline{6}$

Homework Help

$$\begin{array}{r} 0.42\overline{6} \\ 75\overline{)32.000} \\ -300 \\ \hline 200 \\ -150 \\ \hline 500 \\ -450 \\ \hline 50 \end{array}$$

23. $\frac{3}{11} =$ ________

24. $-\frac{5}{8} =$ ________

25. $-\frac{7}{10} =$ ________

26. $2\frac{5}{7} =$ ________

27. $-1\frac{80}{99} =$ ________

28. Cris answered 61 out of 66 questions correctly on a test. What is his test average to the nearest thousandth? ________

Write each decimal as a fraction or mixed number in simplest form.

29. $-0.15 =$ ________

30. $-7.75 =$ ________

31. $-12.54 =$ ________

32. CCSS **Identify Repeated Reasoning** The table shows the decimal equivalent to fractions with a denominator of 7.

Fraction	Decimal	Fraction	Decimal
$\frac{1}{7}$	$0.\overline{142857}$	$\frac{4}{7}$	$0.\overline{571428}$
$\frac{2}{7}$	$0.\overline{285714}$	$\frac{5}{7}$	$0.\overline{714285}$
$\frac{3}{7}$	$0.\overline{428571}$	$\frac{6}{7}$	$0.\overline{857142}$

a. What do you notice about the pattern of the six repeated numbers?

____

b. Using the decimals, add the first half of each pattern to the numbers in the last half. For example, $\frac{1}{7} = 0.\overline{142857}$, so add 142 + 857.

What pattern do you notice? ________

c. Using a calculator, try the same experiment with $\frac{5}{13}$. Is the result the same? Justify your reasoning. ________

Standardized Test Practice

33. Which decimal represents the shaded portion of the figure below?

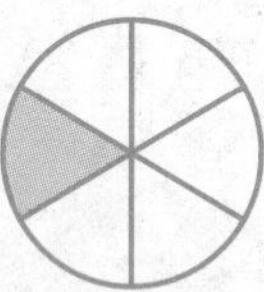

Ⓐ 0.16 Ⓒ 0.17

Ⓑ $0.1\overline{6}$ Ⓓ 1.6

34. Which of the following is *not* equivalent to $0.\overline{3}$?

Ⓕ $\frac{1}{3}$

Ⓖ $\frac{3}{9}$

Ⓗ $\frac{3}{10}$

Ⓘ $\frac{11}{33}$

35. **Short Response** Write −1.25 as a fraction.

Common Core Review

Fill in each ◯ with < or > to make a true statement. 4.NBT.2

36. 4,556 ◯ 4,565

37. 8,698 ◯ 8,689

38. 47,872 ◯ 47,871

39. 26,525 ◯ 26,522

40. 1,123,004 ◯ 1,123,040

41. 5,776,050 ◯ 5,775,005

42. The table shows the number of miles Katie walked for two weeks. Compare the distances using the < symbol. 5.NBT.3b

Week	Number of Miles
1	5.78
2	5.691

43. The table shows the amount of different colored paints in a bin in art class. Compare the amount of blue and orange paint using the > symbol. 5.NBT.3

Color	Number of Ounces
Blue	47.362
Green	47.637
Orange	47.394
Yellow	47.583

Lesson 5

Compare and Order Rational Numbers

What You'll Learn

Scan the lesson. List two headings you would use to make an outline of the lesson.

- ______________________________

- ______________________________

Essential Question

HOW are integers and absolute value used in real-world situations?

Common Core State Standards

Content Standards
6.NS.6, 6.NS.6c, 6.NS.7, 6.NS.7a, 6.NS.7b

Mathematical Practices
1, 2, 3, 4, 5, 7

Real-World Link

Insects The lengths of several common types of insects are shown in the table.

Insect	Length (in.)
Green June Beetle	$\frac{3}{4}$
Cricket	$\frac{1}{1}$
Fire ant	$\frac{1}{3}$
Firefly	$\frac{3}{4}$
Housefly	$\frac{1}{4}$
Japanese beetle	$\frac{1}{2}$
Mosquito	$\frac{5}{8}$

1. Which of the insects is the longest?

____

2. Shade each fraction strip to represent the lengths of a fire ant and a housefly. Which is longer, the fire ant or housefly?

____

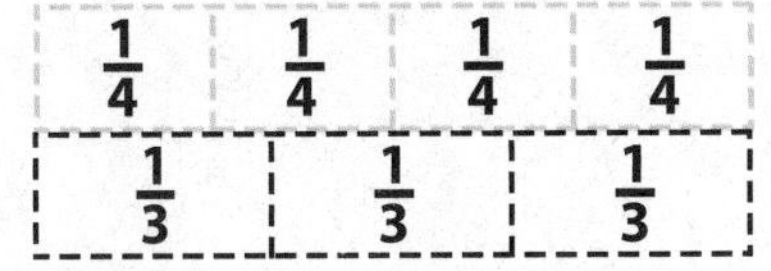

3. How many of the insects are longer than 0.5 inch?

____

4. Order the lengths of a housefly, a Green June beetle, and a fire ant from the shortest to longest.

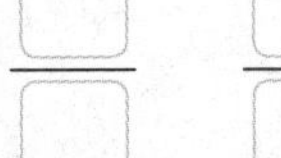 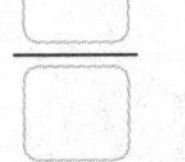 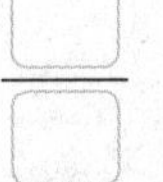

Work Zone

Compare Decimals and Fractions

Positive and negative rational numbers can be represented on a number line. You can use a number line to help you compare and order rational numbers.

Examples

Fill in each ◯ with <, >, or = to make a true statement.

1. **−1.2 ◯ 0.8**

Graph the decimals on a number line.

Since −1.2 is to the left of 0.8, −1.2 < 0.8.

2. **−1.40 ◯ −1.25**

Graph the decimals on a number line.

Since −1.40 is below −1.25, −1.40 < −1.25.

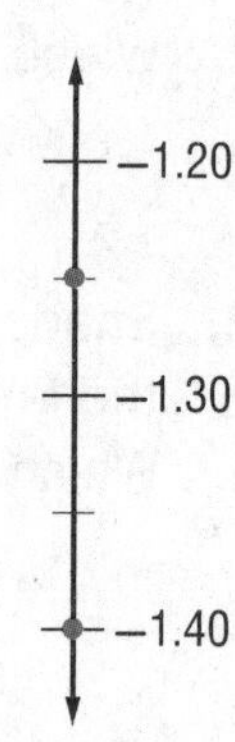

3. $-\frac{3}{8}$ ◯ $-\frac{5}{16}$

Rename the fractions using the least common denominator.

$-\frac{3}{8} = -\frac{3 \times 2}{8 \times 2} = -\frac{6}{16}$ $\quad$ $-\frac{5}{16} = -\frac{5 \times 1}{16 \times 1} = -\frac{5}{16}$

Since $-6 < -5$, $-\frac{6}{16} < -\frac{5}{16}$ and $-\frac{3}{8} < -\frac{5}{16}$.

Check

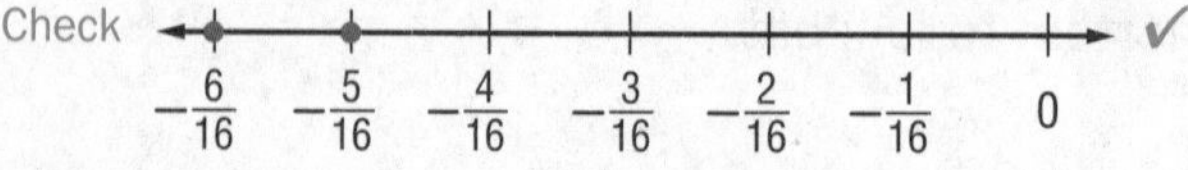

Least Common Multiple (LCM)

Find the LCM of 8 and 16.

8: 8, 16, 24

16: 16, 32, 48

The LCM is 16.

Got It? **Do these problems to find out.**

a. 3.1 ◯ −3.7

b. −4.5 ◯ −4.49

c. $\frac{9}{16}$ ◯ $\frac{12}{16}$

d. $-\frac{7}{10}$ ◯ $-\frac{4}{5}$

Compare and Order Rational Numbers

To compare and order rational numbers, first write them in the same form.

Examples

Fill in each ◯ with <, >, or = to make a true statement.

4. $-0.51 \bigcirc -\frac{8}{15}$

Rename $-\frac{8}{15}$ as a decimal. Then graph both decimals on a number line.

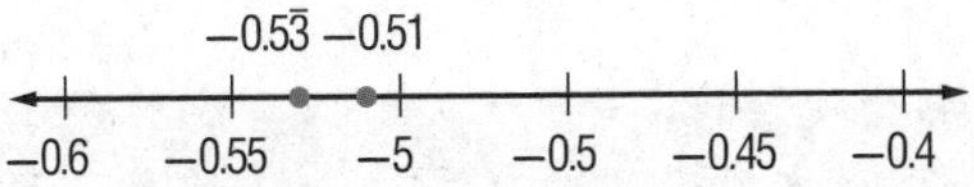

$-\frac{8}{15} = -0.5\overline{3}$

Number line: $-0.5\overline{3}$ and -0.51 graphed; ticks at −0.6, −0.55, −5, −0.5, −0.45, −0.4

Since −0.51 is to the right of $-0.5\overline{3}$ on the number line, $-0.51 > -\frac{8}{15}$.

5. Order the set $\left\{-2.46, -2\frac{22}{25}, -2\frac{1}{10}\right\}$ from least to greatest.

Write $-2\frac{22}{25}$ and $-2\frac{1}{10}$ as decimals to the hundredths place.

$-2\frac{22}{25} = -2.88$ $\quad -2\frac{1}{10} = -2.1$

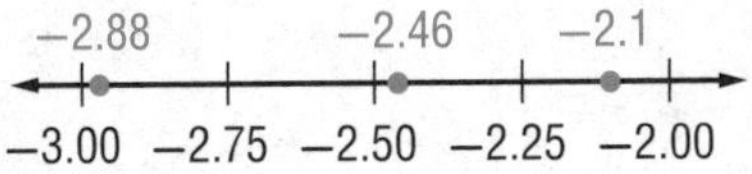

Number line: −2.88, −2.46, −2.1 graphed; ticks at −3.00, −2.75, −2.50, −2.25, −2.00

Graph the decimals on the number line.

From least to greatest, the order is $-2\frac{22}{25}$, −2.46, and $-2\frac{1}{10}$.

and Reflect

How could you represent that −8.3 feet is deeper than −5.7 feet? Explain.

Got It? Do these problems to find out.

Fill in each ◯ with <, >, or = to make a true statement.

e. $-3\frac{5}{8} \bigcirc -3.625$

f. $\frac{3}{7} \bigcirc 0.413$

g. Order the set $\left\{-7\frac{13}{20}, -7.78, -7\frac{17}{100}\right\}$ from greatest to least.

g. ____________

Example

6. Mr. Plum's science class is growing plants under different conditions. The table shows the difference from the average for some students' plants. Order the differences from least to greatest.

Student	Difference (in.)
Ricky	$3\frac{1}{4}$
Debbie	-2.2
Suni	1.7
Leonora	$-1\frac{7}{10}$

Express each number as a decimal.

Ricky's plant: $3\frac{1}{4} = 3.25$

Debbie's plant: -2.2

Suni's plant: 1.7

Leonora's plant: $-1\frac{7}{10} = -1.7$

From least to greatest, the differences are -2.2, $-1\frac{7}{10}$, 1.7, and $3\frac{1}{4}$.

Guided Practice

Fill in each ◯ with <, >, or = to make a true statement. (Examples 1–4)

1. 9.7 ◯ -10.3
2. $\frac{5}{8}$ ◯ $-\frac{3}{8}$
3. -6.7 ◯ $-6\frac{7}{10}$
4. $-\frac{5}{6}$ ◯ -0.94

Show your work.

Order the following sets of numbers from least to greatest. (Example 5)

5. $\left\{-3\frac{1}{3}, 3.3, -3\frac{3}{4}, 3.5\right\}$ ____________

6. $\left\{2.\overline{1}, -2.1, 2\frac{1}{11}, -2\right\}$ ____________

7. **Financial Literacy** Steve recorded these amounts in his checkbook: $-\$6.50$, $\$7.00$, $-\$6.75$, and $\$7.25$. Order these amounts from least to greatest. (Example 6)

____

8. **Building on the Essential Question** How can a number line help in ordering rational numbers?

____

____

____

Rate Yourself!

Are you ready to move on? Shade the section that applies.

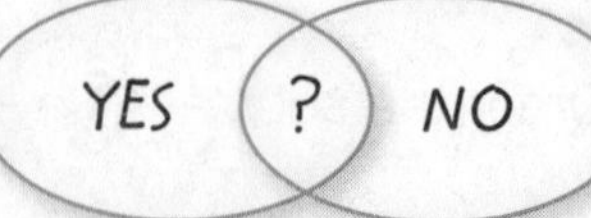

For more help, go online to access a Personal Tutor.

FOLDABLES *Time to update your Foldable!*

Name ____________________ My Homework ____________

Independent Practice

Go online for Step-by-Step Solutions

Fill in each ◯ with <, >, or = to make a true statement. (Examples 1–4)

1. $\frac{5}{4}$ ◯ $-\frac{1}{4}$

2. $-6\frac{1}{3}$ ◯ -6.375

3. $-\frac{3}{5}$ ◯ -0.6

4. $-9\frac{2}{7}$ ◯ -9.3

Order each set of numbers from least to greatest. (Example 5)

5. $\left\{2.8, -2\frac{3}{4}, 3\frac{1}{8}, -2.\overline{2}\right\}$ ____________

6. $\left\{\frac{2}{3}, -0.6, 0.65, \frac{4}{5}\right\}$ ____________

7. **Financial Literacy** The change in four stocks during a day are: $-4\frac{1}{2}$, 5.6, $-2\frac{3}{8}$, and 1.35.

   Order the changes from least to greatest. (Example 6)

   ____

8.  **Multiple Representations** Consider the inequality $-3.5 < -1.5$.

   a. **Words** Write a real-world problem that could be represented by the inequality.

   ____

   ____

   b. **Number Line** Graph -3.5 and -1.5 on the number line.

   c. **Symbols** Use the symbol > to compare -3.5 and -1.5.

   ____

9. For a STEM competition, Julienne constructed a model rocket. The rocket can reach an average height of 545 feet. Find the differences between the average height and the actual heights reached. Then write them as positive and negative rational numbers. Order the differences from least to greatest.

   ____

Trials	Actual Height (ft)
1	534.2
2	556.4
3	554.0
4	535.3

10. **Identify Structure** Fill in the diagram with appropriate numbers.

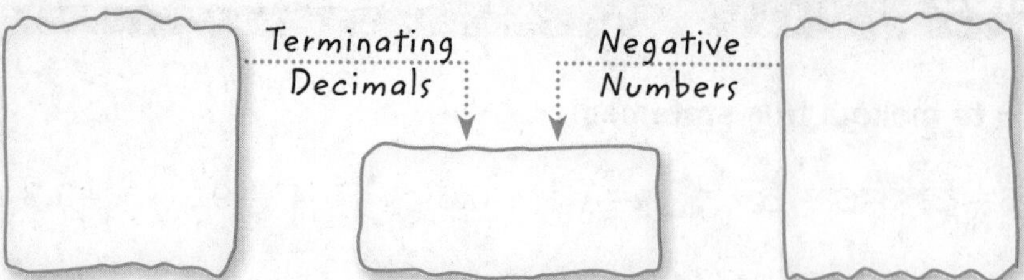

H.O.T. Problems Higher Order Thinking

11. **Reason Inductively** Determine whether the following statement is *always*, *sometimes*, or *never* true. Give examples to justify your answer. *If x and y are both greater than zero and $x > y$, then $-x < -y$.*

12. **Justify Conclusions** Determine whether the fractions $-\frac{4}{5}$, $-\frac{4}{6}$, $-\frac{4}{7}$, and $-\frac{4}{8}$ are arranged in order from least to greatest. Explain.

13. **Reason Abstractly** Explain why -0.33 is greater than $-0.\overline{33}$.

14. **Persevere with Problems** Compare the set $\left\{-0.\overline{7}, -0.\overline{67}, -\frac{7}{9}, -\frac{2}{3}\right\}$. Explain your answer.

Standardized Test Practice

15. Which of the following numbers is less than $-\frac{2}{3}$?

Ⓐ 0.6

Ⓑ $\frac{1}{3}$

Ⓒ $0.\overline{6}$

Ⓓ $-\frac{5}{6}$

Name ________________ My Homework ________________

Extra Practice

Fill in each ◯ with <, >, or = to make a true statement.

16. -18.6 (<) -18.06

−18.6 −18.4 −18.2 −18

17. -4.08 ◯ -4.7

18. $-\frac{3}{7}$ ◯ $-\frac{2}{5}$

19. -3.375 ◯ $-3\frac{4}{10}$

20. $-5\frac{1}{5}$ ◯ -5.2

21. $-8\frac{2}{5}$ ◯ -8.3

Order the following sets of numbers from least to greatest.

22. $\left\{\frac{1}{8}, -0.02\overline{5}, 0.2, -\frac{1}{7}\right\}$ ________________

23. $\left\{1.25, 1\frac{3}{4}, 1.2\overline{5}, 1\frac{1}{5}\right\}$ ________________

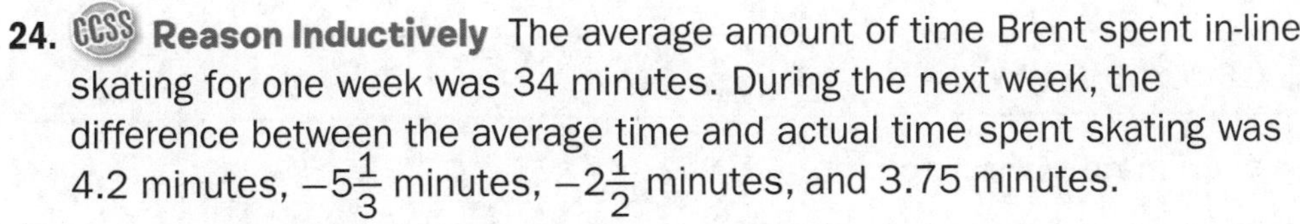

24. CCSS **Reason Inductively** The average amount of time Brent spent in-line skating for one week was 34 minutes. During the next week, the difference between the average time and actual time spent skating was 4.2 minutes, $-5\frac{1}{3}$ minutes, $-2\frac{1}{2}$ minutes, and 3.75 minutes.

Order these differences from least to greatest. ________________

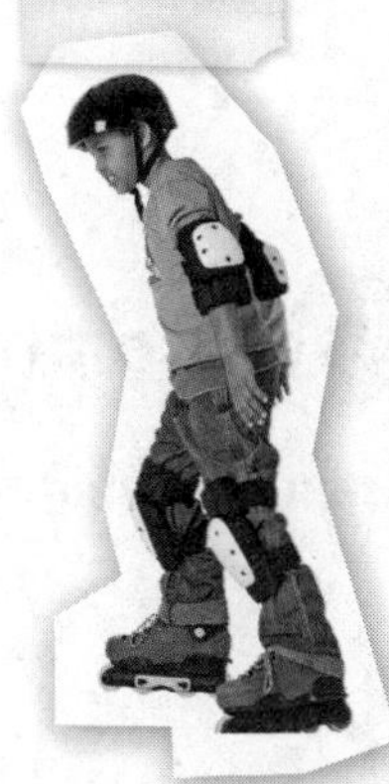

Fill in each ◯ with <, >, or = to make a true statement.

25. $-4\frac{4}{5}$ ◯ $-4.\overline{7}$

26. $-3.2\overline{5}$ ◯ $-3.\overline{2}$

27. $-5.\overline{31}$ ◯ $-5.\overline{313}$

28. The table shows the profit or loss of the after-school snack stand.

a. Write each profit as a positive number and each loss as a negative number. ________________

b. Order the numbers from least to greatest.

____

Day	Profit or Loss	($)
1	Profit	7.50
2	Loss	3.50
3	Loss	6.00
4	Profit	4.50

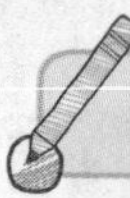

Standardized Test Practice

29. Which point shows the approximate location of $-\frac{1}{3}$?

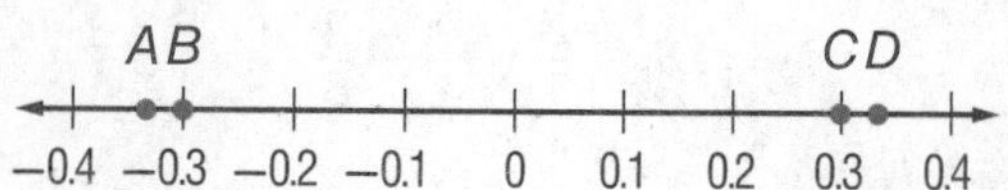

Ⓐ point *A*
Ⓑ point *B*
Ⓒ point *C*
Ⓓ point *D*

30. Which number has the least value?

Ⓕ $-2\frac{3}{10}$
Ⓖ −2.47
Ⓗ −3.62
Ⓘ $-3\frac{17}{20}$

31. Refer to the number line. Which inequality is true?

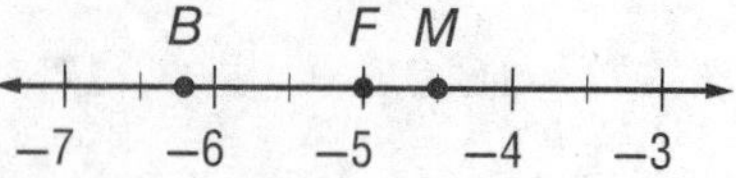

Ⓐ point $B >$ point M
Ⓑ point $F <$ point B
Ⓒ point $M <$ point F
Ⓓ point $F <$ point M

32. Which of the following numbers is the greatest?

Ⓕ −0.73
Ⓖ 0.32
Ⓗ −0.21
Ⓘ 0.19

33. Short Response Student Council's goal was to raise $50 each week for 4 weeks to have enough money for the school dance. The table shows the difference between the goal and the actual amount raised. Order these amounts from least to greatest.

Week	1	2	3	4
Difference ($)	5.50	−6.25	7.80	−2.45

CCSS Common Core Review

Graph the points on the coordinate plane. 5.G.1

34. *H*(1, 6)

35. *M*(7, 0)

36. *I*(5, 8)

37. *N*(4, 9)

38. *J*(6, 3)

39. *O*(7, 5)

40. *L*(3, 1)

41. *P*(2, 2)

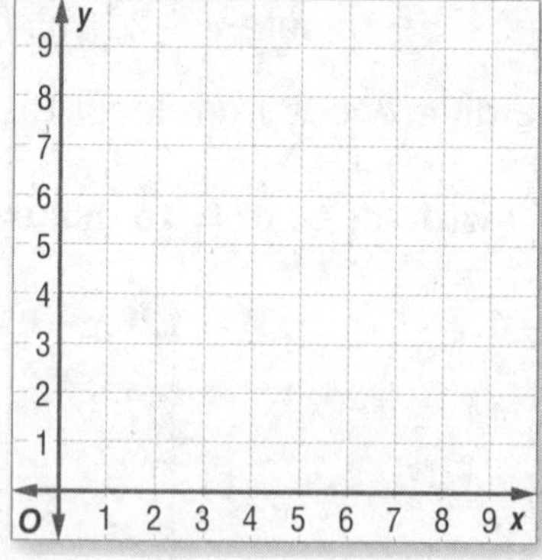

42. Graph the point on the number line that represents $\frac{3}{10}$ and label it *A*. 4.NF.6

Lesson 6

The Coordinate Plane

What You'll Learn

Scan the lesson. List two real-word scenarios where you would use the coordinate plane.

-
-

Essential Question

HOW are integers and absolute value used in real-world situations?

Vocabulary

quadrants

Common Core State Standards

Content Standards
6.NS.6, 6.NS.6b, 6.NS.6c

Mathematical Practices
1, 3, 4, 5, 7

Real-World Link

Watch

Maps The map shows the layout of a small town. The locations of buildings are described in respect to the town hall. Each unit on the grid represents one block.

1. Describe the location of the barber shop in relation to the town hall. ______

2. What building is located 7 blocks east and 5 blocks north of the town hall? ______

3. Violeta is at the library. Describe how many blocks and in what direction she should travel to get to the supermarket. ______

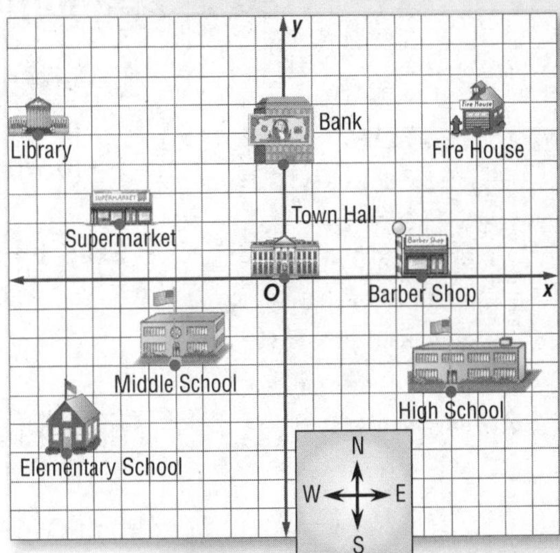

4. Town Hall and the bank are both located on the same vertical number line. The number 0 represents the location of Town Hall on the number line. What number represents the location of the bank? ☐

Work Zone

Identify Points and Ordered Pairs

A coordinate plane is formed when the *x*-axis and *y*-axis intersect at their zero points. The axes separate the coordinate plane into four regions called **quadrants**.

You can use the location on the plane or use the *x*-coordinates and *y*-coordinates to identify the quadrant in which a point is located.

Quadrant	*x*-coordinate	*y*-coordinate	Example
I	positive	positive	(2, 5)
II	negative	positive	(−2, 5)
III	negative	negative	(−2, −5)
IV	positive	negative	(2, −5)

Examples

1. Identify the ordered pair that names point *C*. Then identify the quadrant in which it is located.

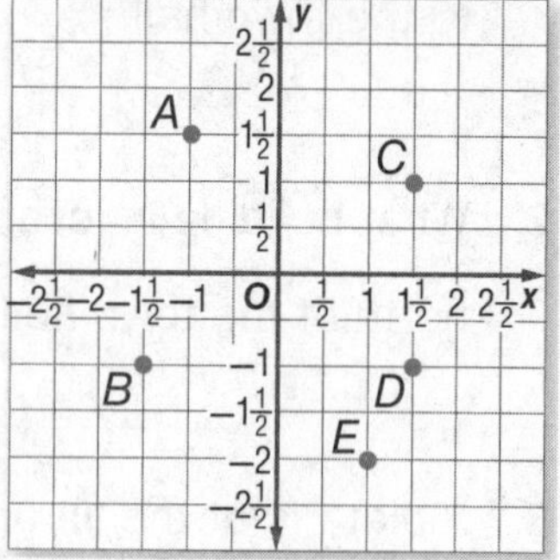

Step 1 Start at the origin. Move right on the *x*-axis. The *x*-coordinate of point *C* is $1\frac{1}{2}$.

Step 2 Move up the *y*-axis. The *y*-coordinate is 1.

Point *C* is located at $\left(1\frac{1}{2}, 1\right)$. Both coordinates are positive. So, point *C* is in Quadrant I.

2. Identify the point located at $\left(-1\frac{1}{2}, -1\right)$. Then identify the quadrant in which it is located.

Step 1 Start at the origin. Move left on the *x*-axis. The *x*-coordinate is $-1\frac{1}{2}$.

Step 2 Move down the *y*-axis. The *y*-coordinate is −1.

Point *B* is located at $\left(-1\frac{1}{2}, -1\right)$. Both coordinates are negative. So, point *B* is in Quadrant III.

Ordered Pairs
A point located on the *x*-axis will have a *y*-coordinate of 0. A point located on the *y*-axis will have an *x*-coordinate of 0. Points located on an axis are not in any quadrant.

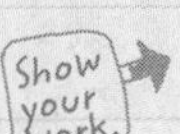

Got It? Do these problems to find out.

a. Identify the ordered pair that names point *A*. Then identify the quadrant in which it is located.

b. Identify the point located at (1, −2). Then identify the quadrant in which it is located.

a. __________

b. __________

Reflections on the Coordinate Plane

You can use what you know about number lines and opposites to compare locations on the coordinate plane. Consider the number line and coordinate plane below.

The number line shows that −4 and 4 are opposites.

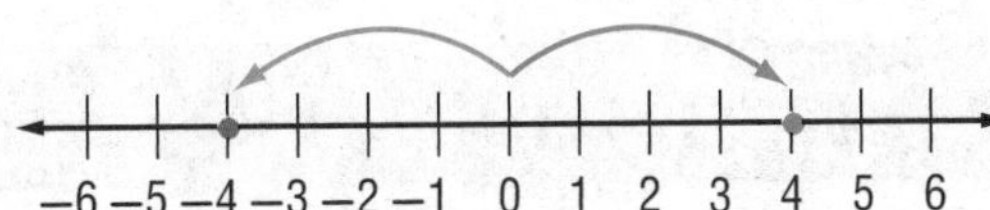

The coordinate plane shows that the points (−4, 0) and (4, 0) are the same distance from the *y*-axis in opposite directions. So, they are *reflected* across the *y*-axis. Notice that the *y*-coordinates did not change and that the *x*-coordinates are opposites.

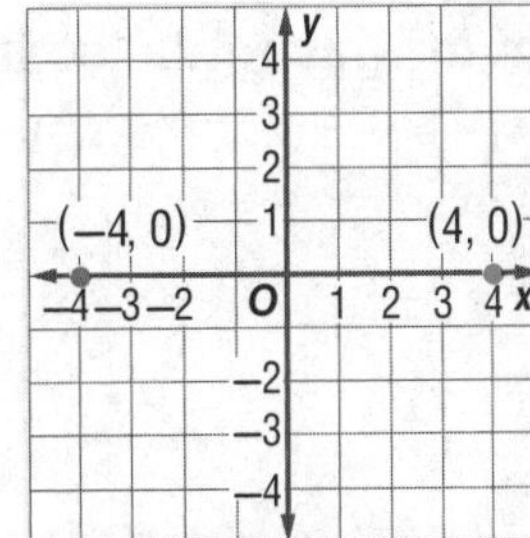

Example

3. Name the ordered pair that is a reflection of (−3, 2) across the *x*-axis.

To reflect across the *x*-axis, keep the same *x*-coordinate, −3, and take the opposite of the *y*-coordinate. The opposite of +2 is −2.

So, (−3, 2) reflected across the *x*-axis is located at (−3, −2).

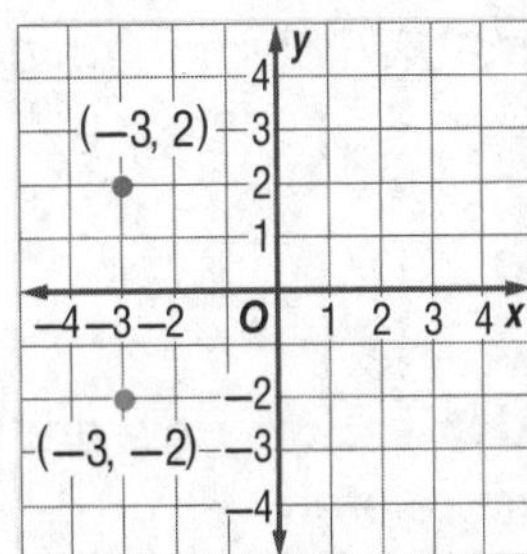

Got It? Do these problems to find out.

Name the ordered pair that is a reflection of each point across the *x*-axis.

c. (1, −4) **d.** (−2, 5) **e.** (−3, −1)

c. __________

d. __________

e. __________

Real World Example

4. Kendall is building a square fence. She places fence posts at the locations indicated on the grid. What is the location of the post that reflects (−4, 4) across the y-axis?

To reflect across the *y*-axis, keep the same *y*-coordinate, 4.

The opposite of the *x*-coordinate, −4, is 4.

So, (−4, 4) reflected across the *y*-axis is (4, 4).

Got It? **Do this problem to find out.**

Show your work.

f. ______

f. Kendall also placed a fence post at (−4, −4). What is the location of the post that reflects (−4, −4) across the *y*-axis?

Guided Practice

Identify the ordered pair that names each point or the name of each point. Then identify the quadrant in which it is located. (Examples 1 and 2)

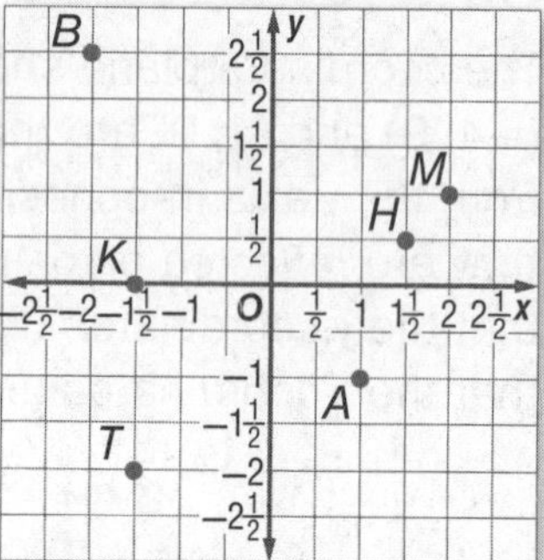

1. *T* ______

2. $\left(-1\frac{1}{2}, 0\right)$ ______

3. $\left(-2, 2\frac{1}{2}\right)$ ______

4. Refer to the diagram of a school. (Examples 3 and 4)

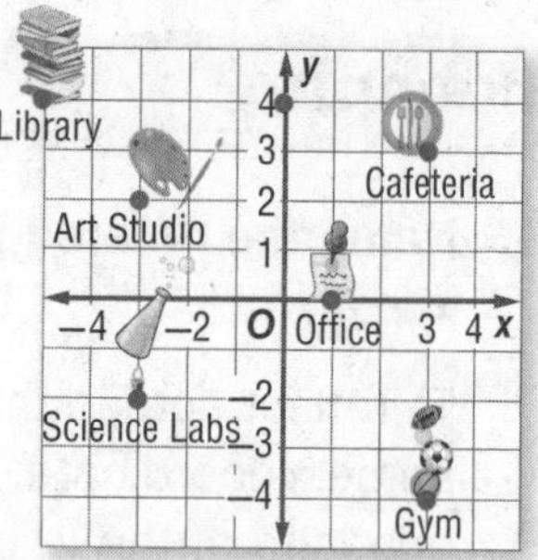

a. What is located at the reflection of (−3, −4) across the *y*-axis. What are the coordinates of this location?

______

b. What is located at the reflection of the science labs across the *x*-axis? What are the coordinates of this location? ______

5. **Building on the Essential Question** How are number lines and the coordinate plane related?

______

Rate Yourself!

Are you ready to move on? Shade the section that applies.

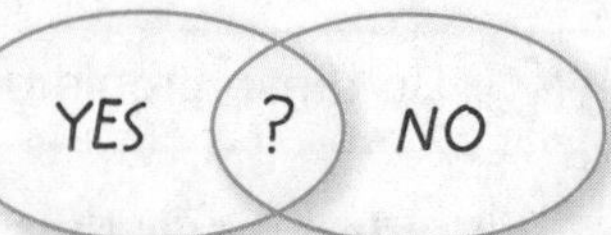

For more help, go online to access a Personal Tutor.

Independent Practice

Go online for Step-by-Step Solutions

Identify the ordered pair that names each point. Then identify the quadrant in which it is located. (Example 1)

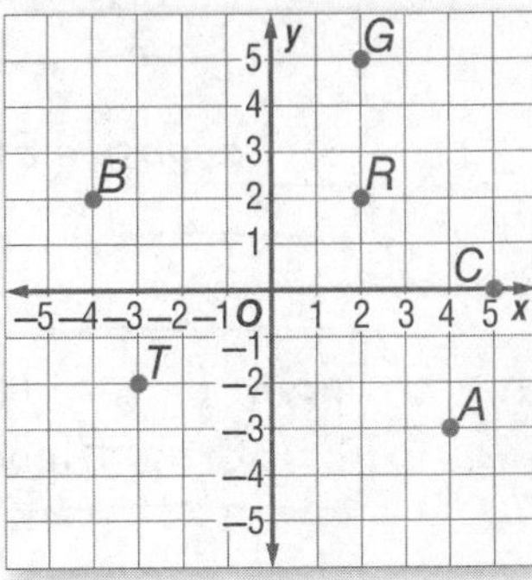

Show your work.

1. *R* ______

2. *G* ______

3. *B* ______

4. *T* ______

5. *C* ______

6. *A* ______

Identify the name of each point. Then identify the quadrant in which it is located. (Example 2)

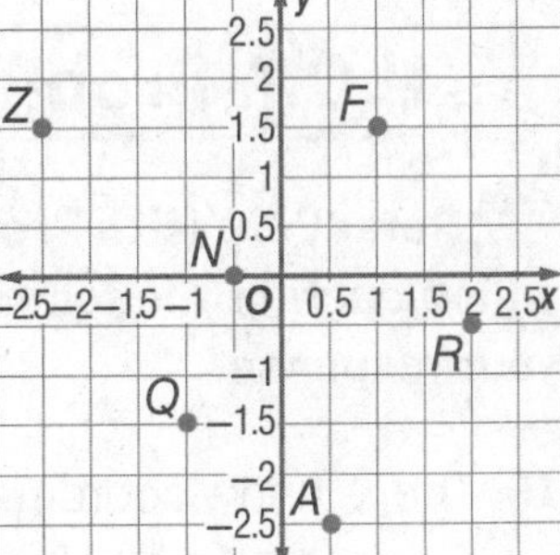

7. (−2.5, 1.5) ______

8. (1, 1.5) ______

9. (0.5, −2.5) ______

10. (2, −0.5) ______

11. (−0.5, 0) ______

12. (−1, −1.5) ______

13. CCSS **Use Math Tools** Refer to the map of Wonderland Park. (Examples 3 and 4)

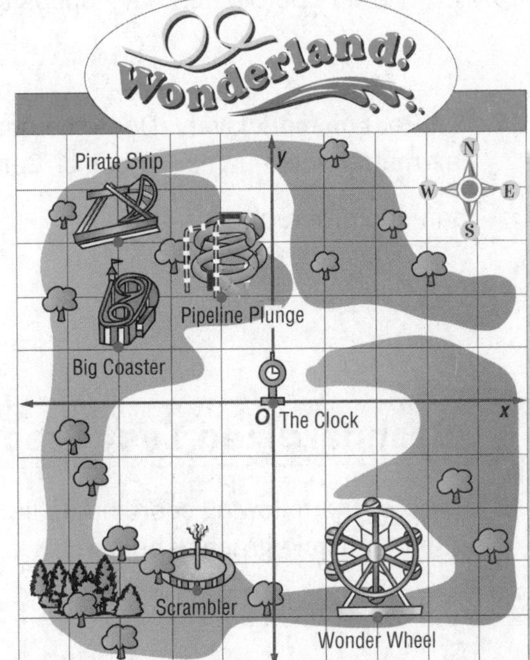

a. What is located closest to the origin?

______

b. Liza is standing at (2, 4). What is located at the reflection of (2, 4) across the *x*-axis? What are the coordinates of this location?

______

c. What is located at the reflection of (3, 1) across the *y*-axis? What are the coordinates of this location? ______

d. The Pipeline Plunge is reflected across the *x*-axis. What are the coordinates of its new location? ______

14. Identify Structure Fill in the graphic organizer below. Consider the point (−3, 2).

Action		Result
opposite of −3	→	
point (−3, 2) reflected across the y-axis	→	
point (−3, 2) reflected across the x-axis	→	

H.O.T. Problems Higher Order Thinking

Persevere with Problems Without graphing, identify the quadrant(s) for which each of the following statements is true for any point (*x*, *y*). Justify your response.

15. The *x*- and *y*-coordinates have the same sign.

16. The *x*- and *y*-coordinates have opposite signs.

17. Reason Inductively Does the order of the numbers in an ordered pair matter when naming a point? Can that point be represented by more than one ordered pair?

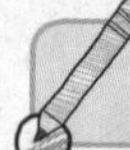

Standardized Test Practice

18. Which of the following coordinates lie within the circle graphed below?

Ⓐ (−1, 1.5) Ⓒ (−0.5, 1)

Ⓑ (−1.5, −2) Ⓓ (−1.5, 2)

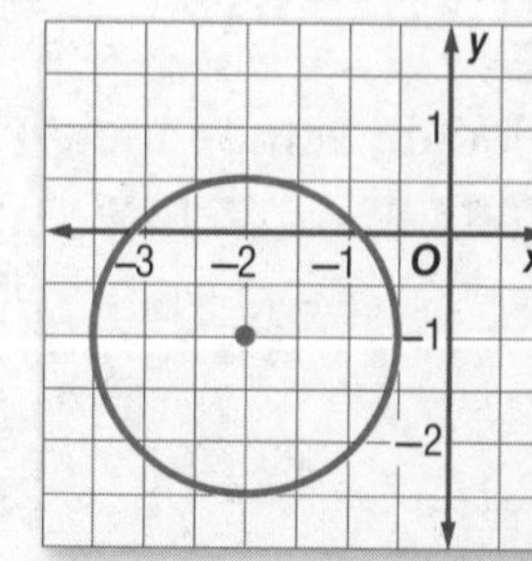

Name _______________ My Homework _______________

Extra Practice

Identify the ordered pair that names each point. Then identify the quadrant in which it is located.

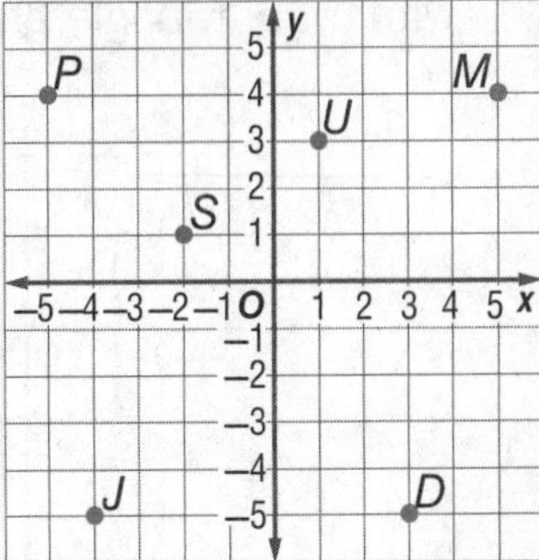

19. U

(1, 3); I

Both numbers are positive so it is in the first quadrant.

20. D

21. S

22. P

23. J

24. M

Identify the name of each point. Then identify the quadrant in which it is located.

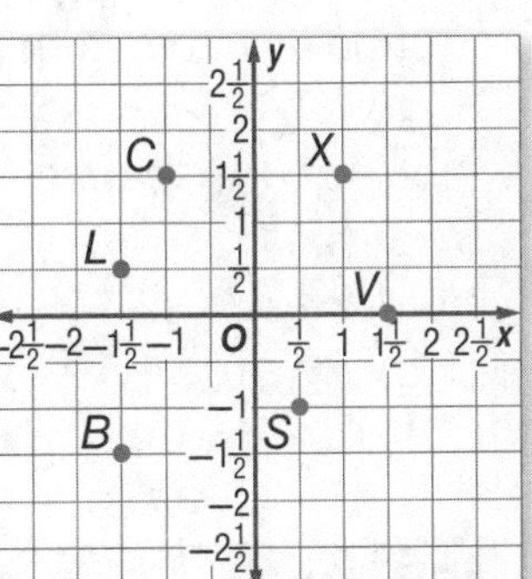

25. $\left(-1\frac{1}{2}, \frac{1}{2}\right)$

26. $\left(1, 1\frac{1}{2}\right)$

27. $\left(\frac{1}{2}, -1\right)$

28. $\left(1\frac{1}{2}, 0\right)$

29. $\left(-1\frac{1}{2}, -1\frac{1}{2}\right)$

30. $\left(-1, 1\frac{1}{2}\right)$

31.  **Model with Mathematics** Luke is making a model of a park. He has the basketball court drawn on his model.

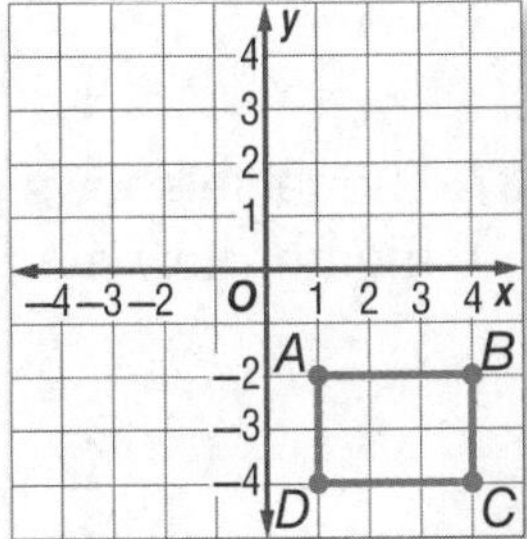

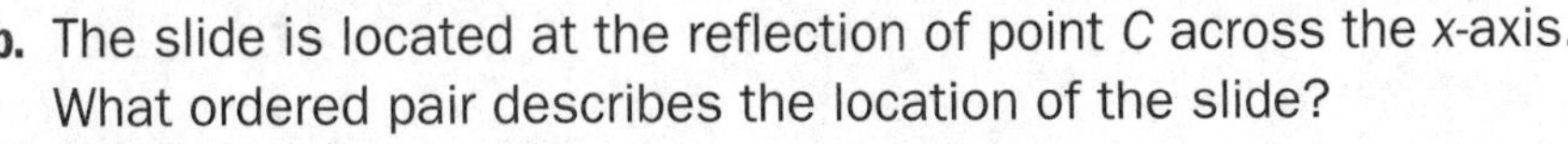

a. The swing set is located at the reflection of point B across the x-axis. What ordered pair describes the location of the swing set?

b. The slide is located at the reflection of point C across the x-axis. What ordered pair describes the location of the slide?

c. A water fountain is located at the reflection of point D across the y-axis. What ordered pair describes the location of the water fountain?

Standardized Test Practice

32. Which of the following coordinates lie within the triangle graphed below?

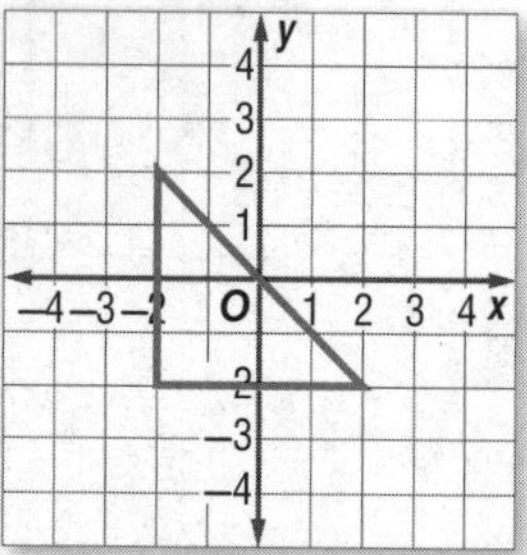

Ⓐ (1, 0)

Ⓑ (−2, −3)

Ⓒ (−1, 2)

Ⓓ (−1, −1)

33. Identify the point for the ordered pair (−3, 5).

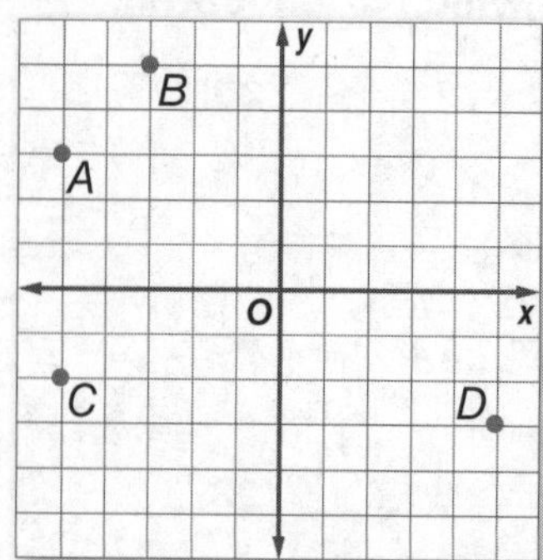

Ⓕ Point *A*

Ⓖ Point *B*

Ⓗ Point *C*

Ⓘ Point *D*

34. Which ordered pair represents the reflection of point *J* across the *y*-axis?

Ⓐ (−4, −2)

Ⓑ (4, 4)

Ⓒ (4, 2)

Ⓓ (2, 2)

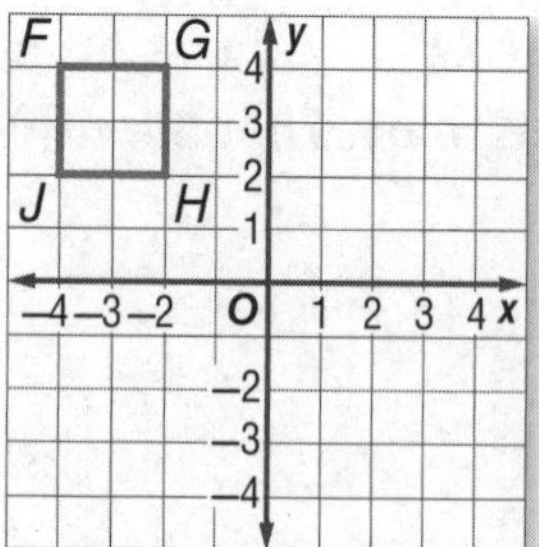

Common Core Review

Represent the set of numbers as decimals on the number line. 4.NF.6

35. $\left\{3\frac{1}{10}, 2\frac{7}{10}, 2\frac{9}{10}\right\}$

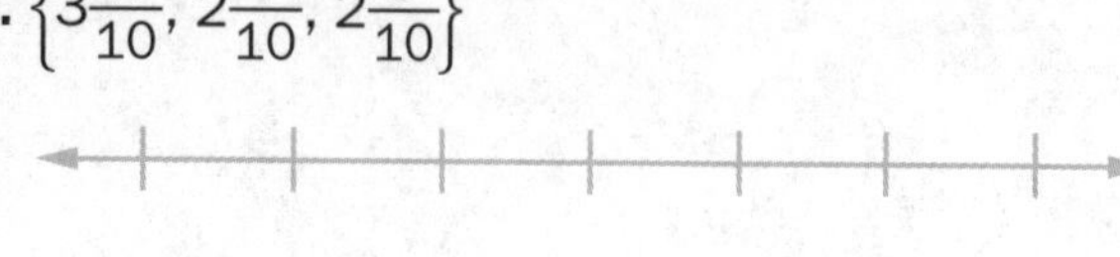

36. $\left\{5\frac{3}{10}, 5\frac{1}{10}, 5\right\}$

37. The table shows how many magazines three co-workers sold in one month. How many magazines did they sell in total? 4.NBT.4

Name	Number of Magazines
Julie	12
Dion	0
Calvin	7

38. Draw a line of symmetry on the figure shown. 4.G.3

Lesson 7

Graph on the Coordinate Plane

What You'll Learn

Scan the lesson. Predict two things you will learn about graphing on the coordinate plane.

- ______________________________
- ______________________________

Essential Question

HOW are integers and absolute value used in real-world situations?

Common Core State Standards

Content Standards
6.NS.6, 6.NS.6b, 6.NS.6c, 6.NS.8

Mathematical Practices
1, 2, 3, 4, 7

Real-World Link

Scavenger Hunt Maria hid the clues to a scavenger hunt for her hiking club. Use the map to show where she hid the clues. Identify the location of each clue.

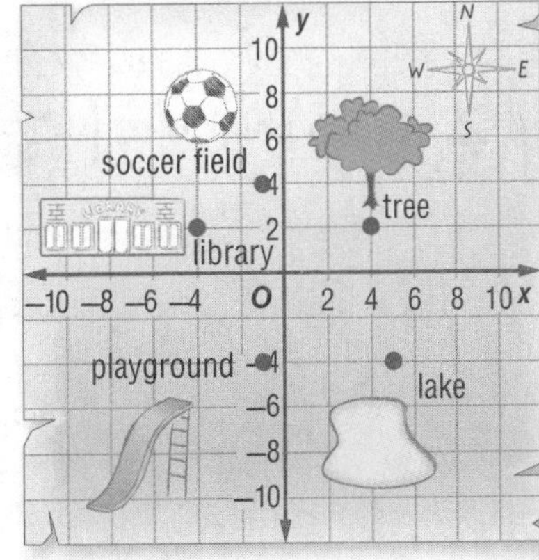

1. The first clue is hidden near a tree. What ordered pair describes its location?

2. Maria hid the next clue at a location reflected across the *y*-axis. Where is it hidden?

3. She walks 3 blocks east and 2 blocks north to place the next clue. Where is it hidden?

4. The next clue is at a location reflected across the *x*-axis. Where is it hidden?

5. Maria hid the next clue under a rock by the lake. How many blocks east did she walk to the lake?

6. The final clue tells the hikers to walk 5 blocks north and three blocks east to find the prize. What ordered pair describes the location of the prize?

Work Zone

Graph Ordered Pairs

To graph an ordered pair, draw a dot at the point that corresponds to the coordinates.

Examples

Tutor

1. Graph point M at $(-3, 5)$.

Start at the origin. The x-coordinate is –3. So, move 3 units to the left.

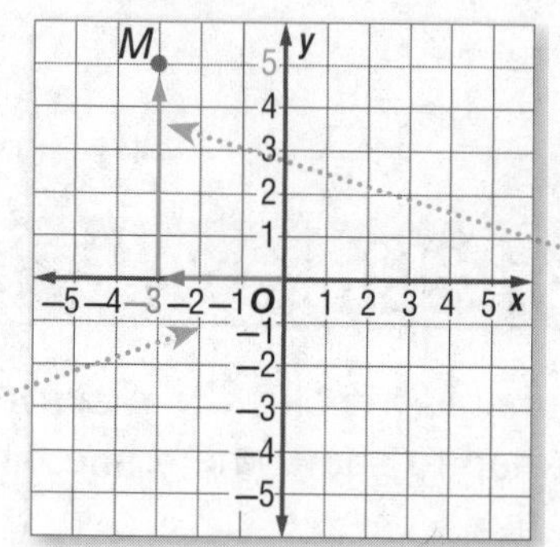

Next, since the y-coordinate is 5, move 5 units up. Draw a dot.

2. Graph point N at $\left(-2\frac{1}{2}, -3\frac{1}{2}\right)$.

The x-coordinate $-2\frac{1}{2}$ is between –2 and –3.

Start at the origin and move $2\frac{1}{2}$ units left.

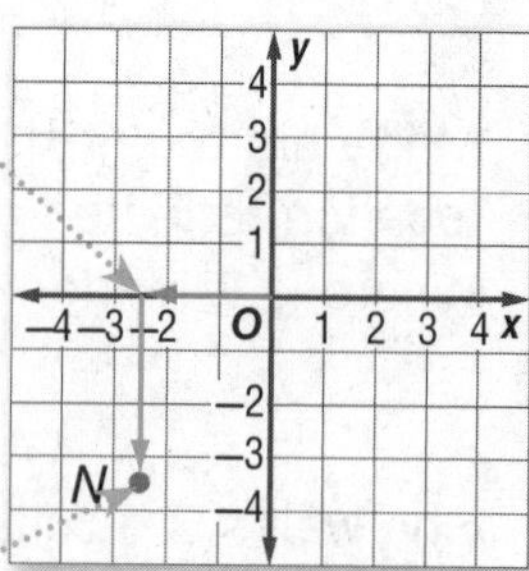

The y-coordinate $-3\frac{1}{2}$ is between –3 and –4.

Next, move $3\frac{1}{2}$ units down. Draw a dot.

Got It? Do these problems to find out.

Graph and label each point on the coordinate plane below.

a. $P(-2, 4)$

b. $Q(0, -4)$

c. $R\left(-\frac{1}{2}, -2\frac{1}{2}\right)$

d. $S(4.5, 1)$

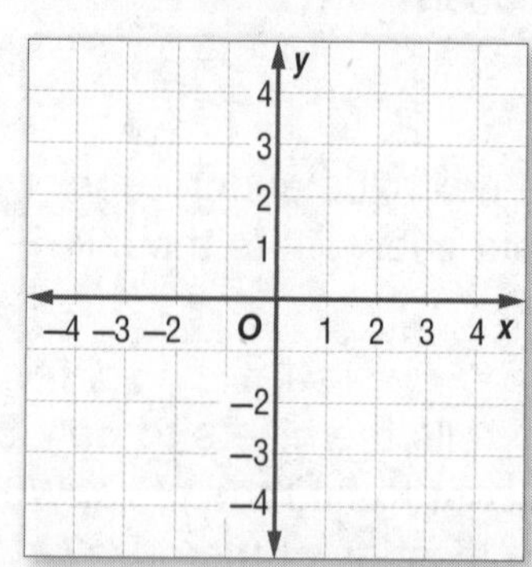

Graph Reflections on the Coordinate Plane

You can graph points that are reflected across the *x*- and *y*-axes. Remember that points reflected across the *x*-axis will have the same *x*-coordinates and their *y*-coordinates will be opposites. Points reflected across the *y*-axis will have the same *y*-coordinates and their *x*-coordinates will be opposites.

Symbols

Use the notation A' to label the reflection of a point A.

Examples

3. Graph $A(2, -4)$. Then graph its reflection across the *x*-axis.

Graph point *A*.

To reflect across the *x*-axis, keep the same *x*-coordinate, 2, and take the opposite of the *y*-coordinate.

The opposite of -4 is 4.

So, point *A* reflected across the *x*-axis is located at point $A'(2, 4)$. Graph point A'.

A' (2, 4)

A (2, −4)

4. Graph $B(-1.5, 3)$. Then graph its reflection across the *y*-axis.

Graph point *B*.

To reflect across the *y*-axis, keep the same *y*-coordinate and take the opposite of the *x*-coordinate.

The opposite of -1.5 is 1.5.

So, point *B* reflected across the *y*-axis is point $B'(1.5, 3)$.

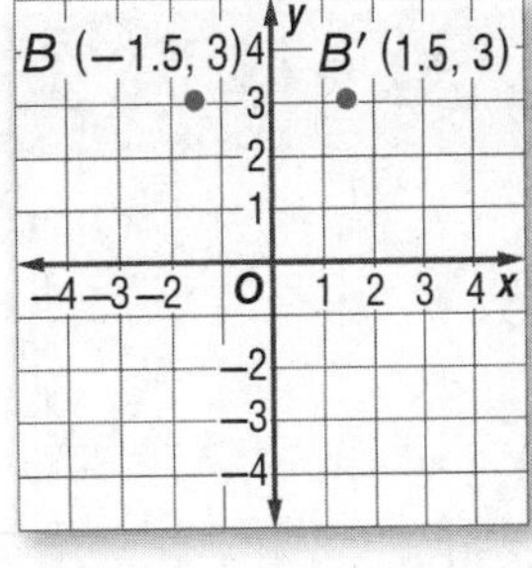

Got It? Do these problems to find out.

e. Graph $C(-1, 5)$. Then graph its reflection across the *x*-axis.

f. Graph $D\left(2, 3\frac{1}{2}\right)$. Then graph its reflection across the *y*-axis.

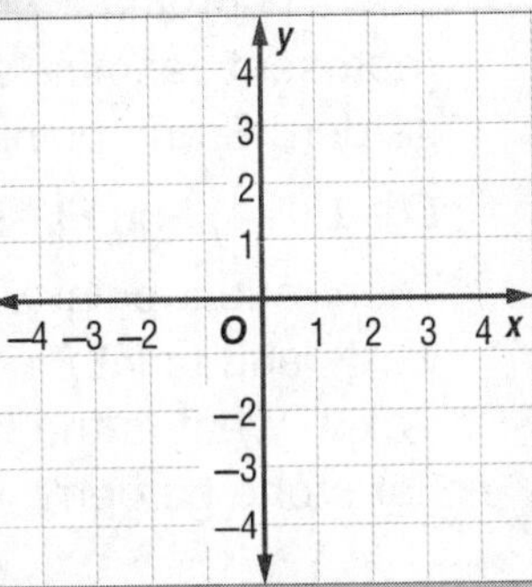

Example

5. Mr. Martin is using a coordinate plane to design a logo. He graphs points at (2, 4) and (2, −2). He reflects (2, −2) across the *y*-axis. Then he reflects the new point across the *x*-axis. What figure is Mr. Martin using for his logo?

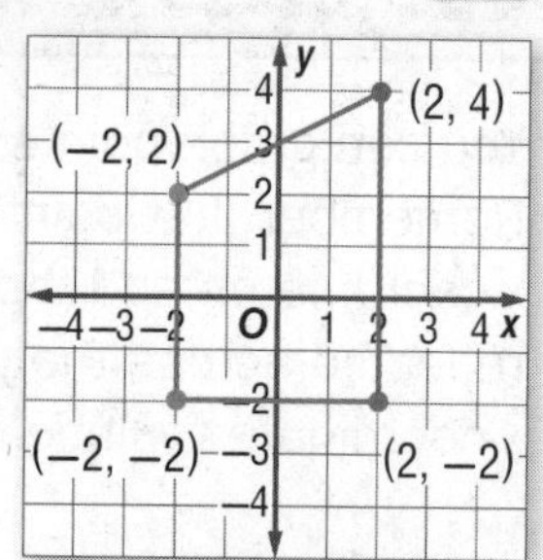

Graph (2, 4) and (2, −2). (2, −2) reflected across the *y*-axis is (−2, −2).

Graph (−2, −2). (−2, −2) reflected across the *x*-axis is (−2, 2).

Graph (−2, 2).

So, the figure is a trapezoid.

Got It? Do this problem to find out.

g. Ms. Shaull is drawing a map of the school. Her room is at (−3, 4) and the gym is at (3, 4). The library is a reflection of (3, 4) across the *x*-axis. This point is reflected across the *y*-axis to graph the office. What figure is graphed on the map?

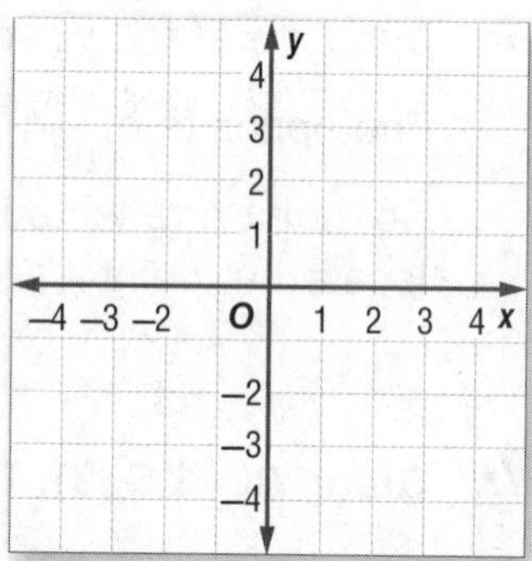

g. ____________

Guided Practice

1. Use a coordinate plane to represent Jasmine's stone garden. Graph points $E(-1, -4)$ and $F\left(-3\frac{1}{2}, 4\right)$. Then reflect point *E* across the *y*-axis and point *F* across the *x*-axis. What is the shape of her stone garden? (Examples 1–5)

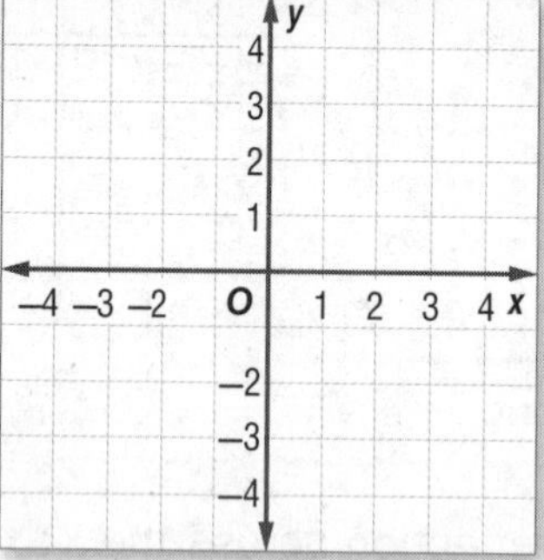

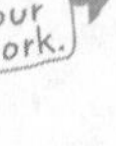

____

2. **Building on the Essential Question** How can the coordinate plane be used to represent geometric figures?

____

____

____

Rate Yourself!

How confident are you about graphing on the coordinate plane? Check the box that applies.

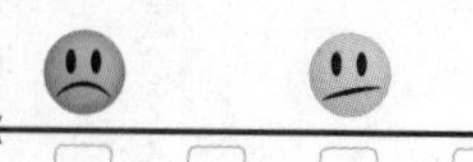

For more help, go online to access a Personal Tutor.

Name ______________________ My Homework ______________________

Independent Practice

Go online for Step-by-Step Solutions  eHelp

Graph and label each point on the coordinate plane to the right.
(Examples 1 and 2)

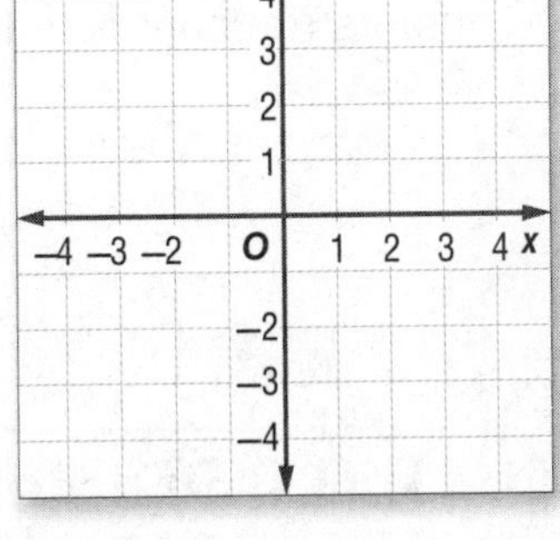

1. $T(0, 0)$

2. $D(2, 1)$

3. $K(-3.25, 3)$

4. $N\left(0, -1\frac{1}{2}\right)$

5 $F(-4.5, 0)$

6. $A\left(-3\frac{1}{2}, -3\right)$

7. $L(2.5, -3.5)$

8. $S\left(4, 2\frac{1}{2}\right)$

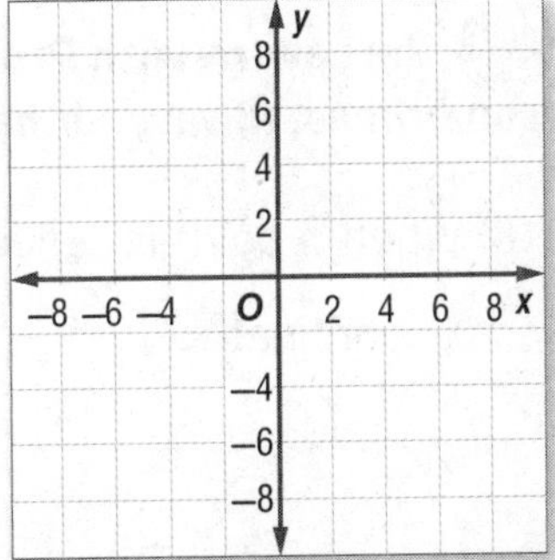

9 Graph $U(3.5, -3)$ on the coordinate plane to the right. Then graph its reflection across the x-axis. (Example 3)

10. Graph $B(-7, 6)$ on the coordinate plane on the right. Then graph its reflection across the x-axis. (Example 3)

11. Graph $R(-2, 5)$ on the coordinate plane to the right. Then graph its reflection across the y-axis. (Example 4)

12. Amelia is drawing a map of the park. She graphs the entrance at $(2, -3)$. She reflects $(2, -3)$ across the y-axis. Then Amelia reflects the new point across the x-axis. What figure is graphed on the map? (Example 5)

____

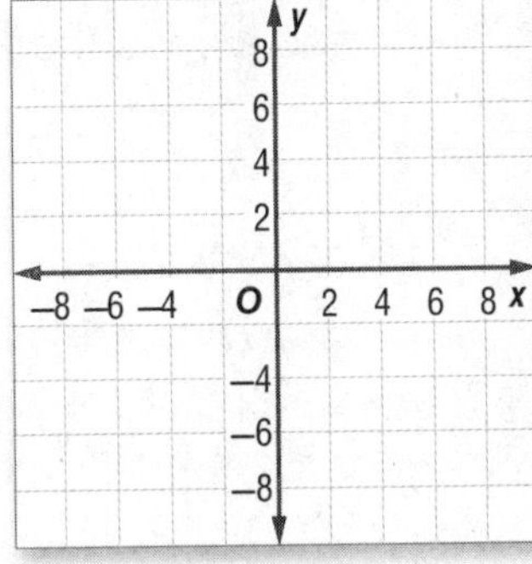

13. A point is reflected across the y-axis. The new point is located at $(-4.25, -1.75)$. Write the ordered pair that represents the original point. ______________________

14. CCSS **Model with Mathematics** A point is reflected across the x-axis. The new point is $(-7.5, 6)$. What is the distance between the two points?

____

15 On a coordinate plane, draw triangle *ABC* with vertices *A*(−1, −1), *B*(3, −1), and *C*(−1, 2). Find the area of the triangle in square units.

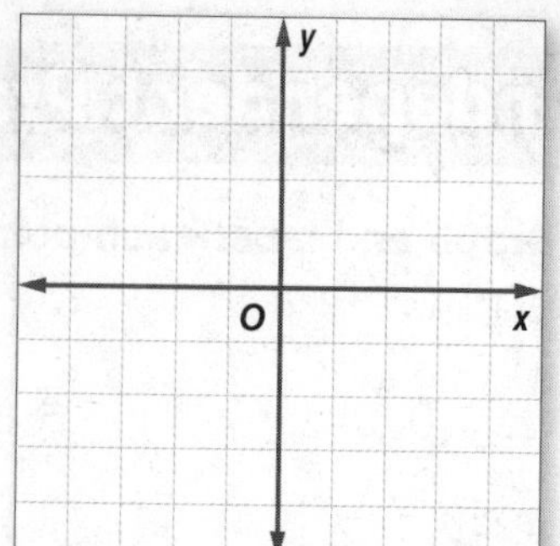

16. The points (4, 3) and (−4, 0) are graphed on a coordinate plane. The point (4, 3) is reflected across the *x*- and *y*-axes. If all four points are connected, what figure is graphed?

H.O.T. Problems Higher Order Thinking

17. CCSS **Identify Structure** Three vertices of a quadrilateral are (−1 −1), (1, 2), and (5, −1). What are the coordinates of two vertices that will form two different parallelograms?

CCSS **Persevere with Problems Determine whether each statement is *sometimes*, *always*, or *never* true. Give an example or a counterexample.**

18. When a point is reflected across the *y*-axis, the new point has a negative *x*-coordinate.

19. The point (*x*, *y*) is reflected across the *x*-axis. Then the new point is reflected across the *y*-axis. The location of the point after both reflections is (−*x*, −*y*).

Standardized Test Practice

20. What are the coordinates of *Y*′ after *Y*(−3.5, 5) is reflected across the *x*-axis?

Ⓐ (3.5, −5)

Ⓑ (−3.5, −5)

Ⓒ (5, −3.5)

Ⓓ (3.5, 5)

Extra Practice

Graph and label each point on the coordinate plane to the right.

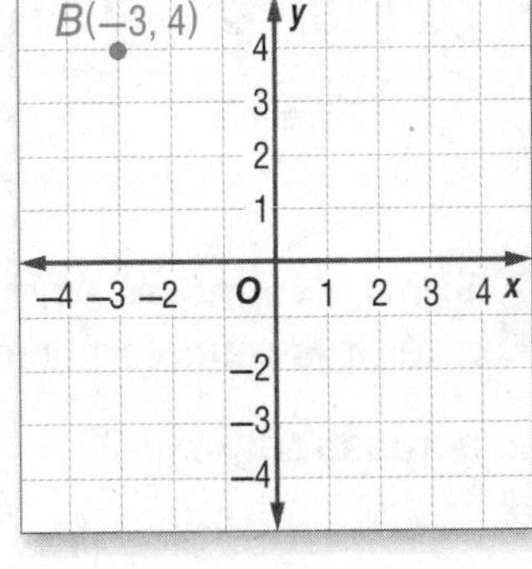

21. $B(-3, 4)$ *The x-coordinate is −3. The y-coordinate is 4.*

22. $D(-1.5, 2.5)$

23. $A\left(4\frac{3}{4}, -1\frac{1}{4}\right)$

24. $J\left(2\frac{1}{2}, -2\frac{1}{2}\right)$

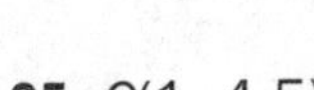

25. $C(1, 4.5)$

26. $F(-4, -3.5)$

27. $G\left(3\frac{1}{2}, 3\right)$

28. $H\left(-3, -1\frac{1}{2}\right)$

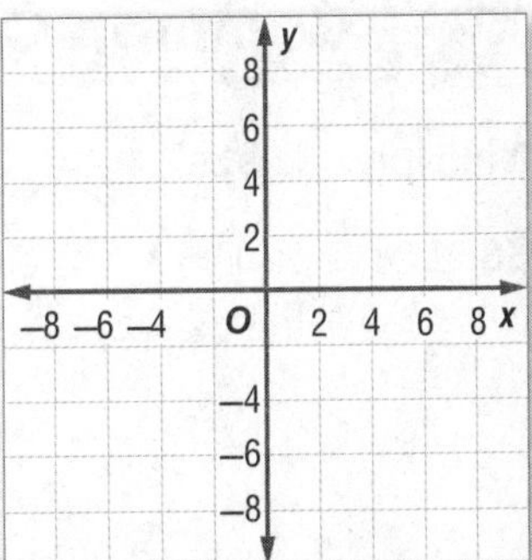

29. Graph $N(1, -3)$ on the coordinate plane to the right. Then graph its reflection across the y-axis.

30. Graph $H(7, 8)$ on the coordinate plane on the right. Then graph its reflection across the x-axis.

31. Graph $F(-6, 5.5)$ on the coordinate plane to the right. Then graph its reflection across the x-axis.

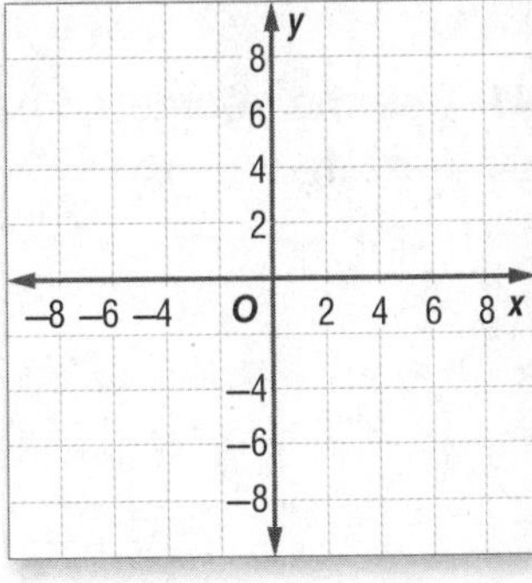

32. Marcus is drawing a plan for his vegetable garden. He graphs one corner at $(-7.5, 2)$ and one corner at $(7.5, 2)$. He reflects $(-7.5, 2)$ across the x-axis. Then Marcus reflects the new point across the y-axis. What shape is the vegetable garden?

____

33. A point is reflected across the x-axis. The new point is located at $(4.75, -2.25)$. Write the ordered pair that represents the original point.

____

34. CCSS **Model with Mathematics** A point is reflected across the x-axis. The new point is $(5, -3.5)$. What is the distance between the two points?

____

Standardized Test Practice

35. What are the coordinates of $B(-0.5, 2)$ after it is reflected across the y-axis?

Ⓐ (0.5, −2) Ⓒ (2, −0.5)

Ⓑ (−0.5, −2) Ⓓ (0.5, 2)

36. What figure is made when the points (−1, 2), (2, 2), (2, −1), and (−1, −1) are connected?

Ⓕ triangle Ⓗ trapezoid

Ⓖ rectangle Ⓘ square

37. Short Response What are the coordinates of point H after it is reflected across the x-axis, and then reflected across the y-axis? ______

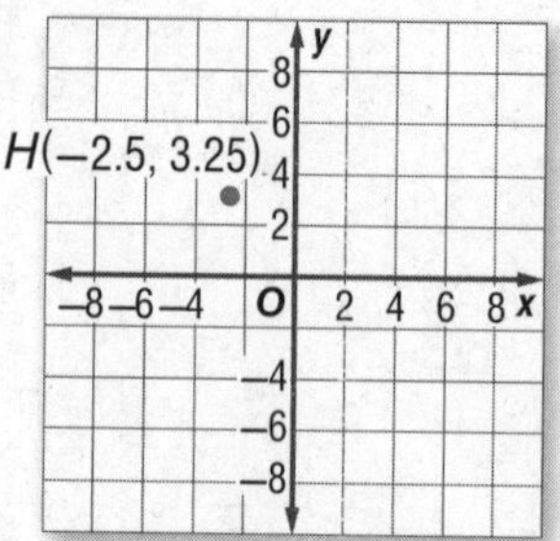

Common Core Review

Multiply. 4.NBT.5

38. $1 \times 1 \times 1 =$ ______

39. $3 \times 3 \times 3 =$ ______

40. $6 \times 6 \times 6 =$ ______

41. Use the geometric pattern below to find the number of squares in the next figure. 4.OA.5

______

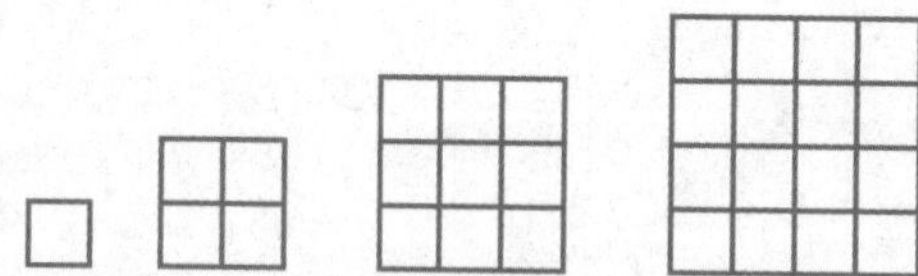

42. Alexa saved a total of $210. Each week she saved the same amount of money. She has been saving for 7 weeks. How much money did Alexa save each week? 5.NBT.6

______

Inquiry Lab

Find Distance on the Coordinate Plane

Inquiry WHAT is the relationship between coordinates and distance?

CCSS Content Standards 6.NS.8

Mathematical Practices 1, 3, 4

Maps Taylor's house and school are each shown on the map. What is the distance between the two points?

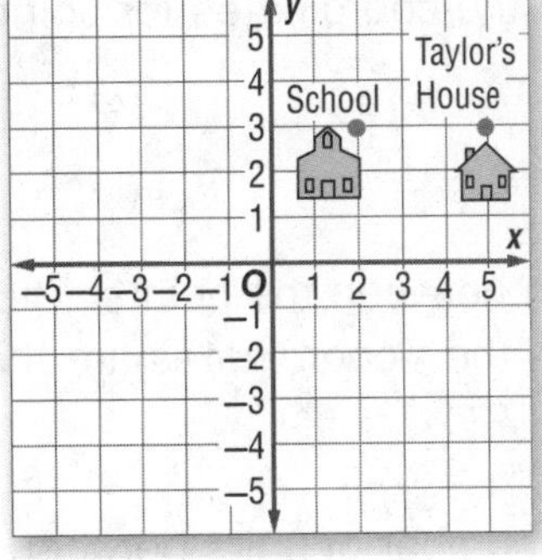

What do you know? ______________________

What do you need to find? ______________________

Investigation 1

Find the distance between Taylor's house and the school.

Step 1 Find the coordinates of Taylor's house.

______

Step 2 Find the coordinates of the school.

______

Step 3 Draw a line between the points. The line is horizontal, so the *y*-coordinates are the same.

Step 4 To find the distance, count the number of units between the *x*-coordinates.

Location	*x*-coordinate
house	
school	

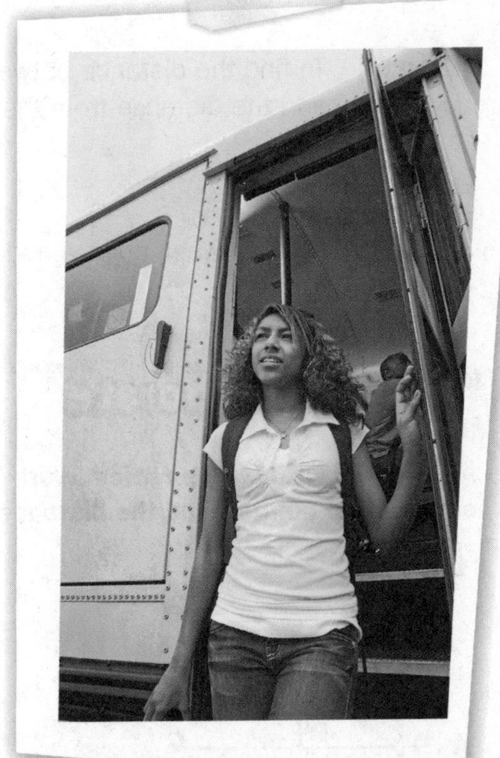

So, there are ☐ units between Taylor's house and the school.

Investigation 2

Find the distance between point *A* and point *B* on the coordinate plane.

Step 1 Determine the coordinates for point *A*.

______

Step 2 Determine the coordinates for point *B*.

______

Step 3 Draw a line between the points. The line is vertical, so the *x*-coordinates are the same.

Step 4 Count the number of units between each *y*-coordinate and the *x*-axis.

Point	*y*-coordinate	Distance from *x*-axis
A		
B		

Step 5 To find the distance between the two points, add the distance from the *x*-axis to each point.

So, the distance between point *A* and point *B* is ☐ units.

Collaborate

CCSS Model with Mathematics Work with a partner. Draw a line between each pair of points. Find the distance between each pair of points.

1. ______

Show your work.

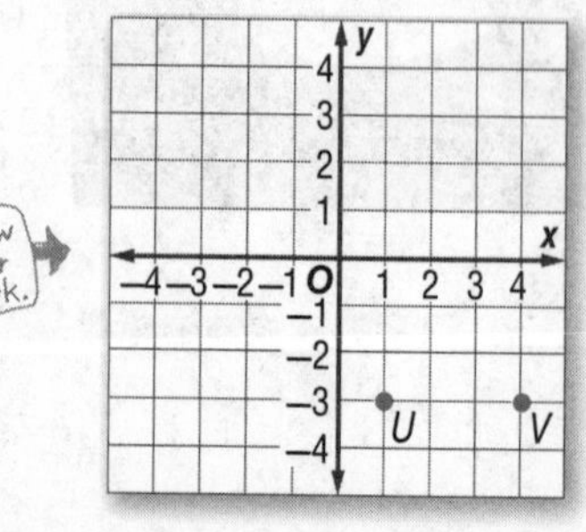

2. ______

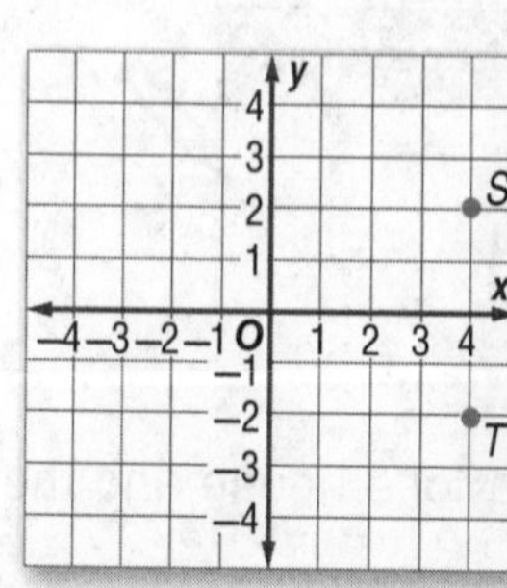

CCSS Model with Mathematics Work with a partner. Plot each pair of points on the coordinate plane. Find the distance between each pair of points.

3. $C(-3, -6)$, $D(-3, -1)$ ________

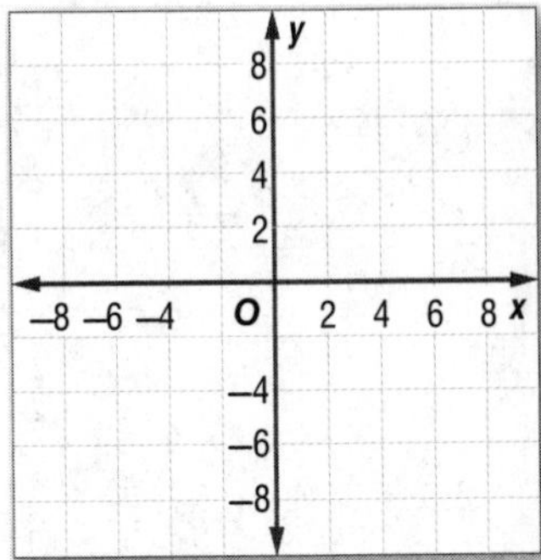

4. $E(-6, -2)$, $F(1, -2)$ ________

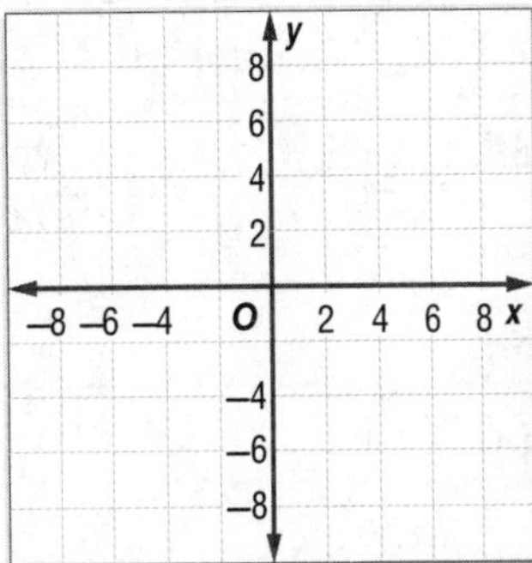

5. $G(1, -4)$, $H(4, -4)$ ________

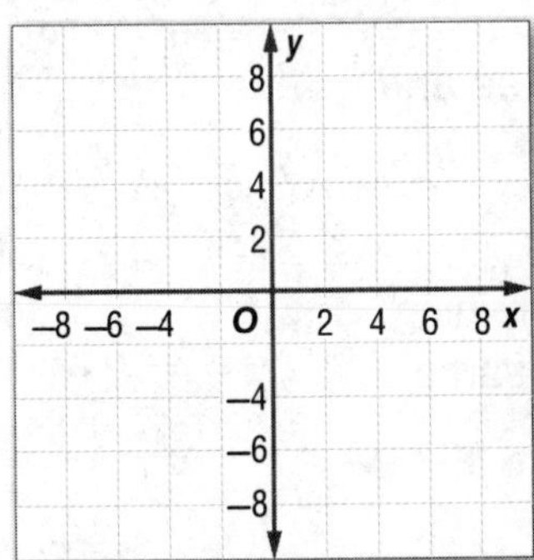

6. $K(3, -4)$, $L(3, 2)$ ________

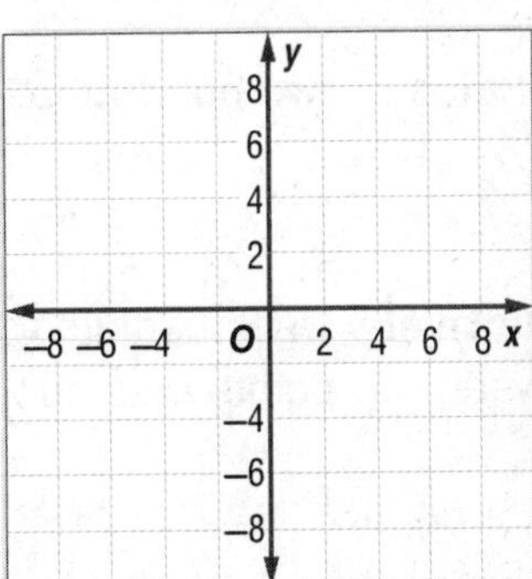

7. $M(5, 1)$, $N(-1, 1)$ ________

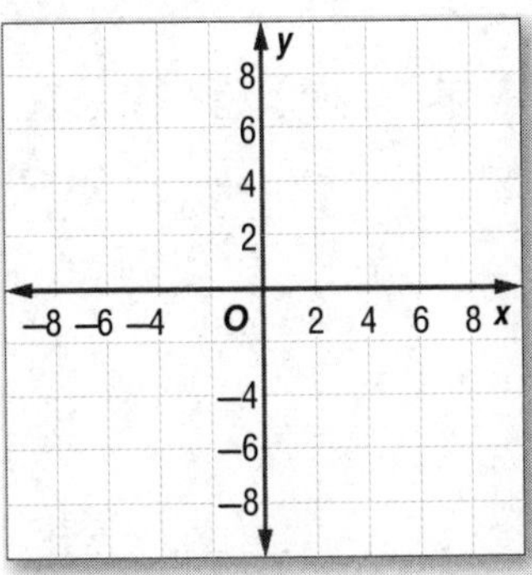

8. $O\left(5\frac{1}{2}, 6\right)$, $P\left(5\frac{1}{2}, 2\right)$ ________

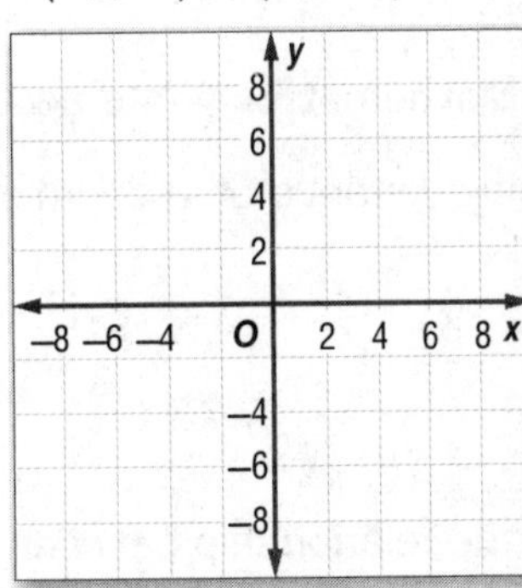

Analyze

With a partner to complete the table below. Use your answers from Exercises 3–6. The first one is done for you.

	Exercise	Coordinates Used	Horizontal or Vertical Line?	Same or Different Quadrant?	Line Length
	2	2 and −2	horizontal	different	4 units
9.	3	and			
10.	4	and			
11.	5	and			
12.	6	and			

13. Compare your answers from Exercises 11 and 12. What is the relationship between the coordinates used and the length of each line?

14. Name the coordinates of two points that have the same *x*-coordinates and are 8 units apart.

15. CCSS **Reason Inductively** Use absolute value to write a rule for determining the distance between two points on a coordinate plane that have the same *x*-coordinate.

Reflect

16. CCSS **Model with Mathematics** Write and solve a real-world problem that involves determining distance on a coordinate plane.

17. Inquiry WHAT is the relationship between coordinates and distance?

21ST CENTURY CAREER in Art

Scientific Illustrator

If you are artistic and have a strong interest in science, you should think about a career as a scientific illustrator. Scientific illustrators combine their artistic abilities with their scientific backgrounds to draw scientifically accurate images. Karen Carr, a wildlife and natural history artist, has artwork in scientific publications, museums, and zoos. To draw animals that are extinct, she examines fossils, talks to scientists, and uses measurements and proportions from scientific literature.

Explore college and careers at ccr.mcgraw-hill.com

Is This the Career for You?

Are you interested in a career as a scientific illustrator? Take some of the following courses in high school.

- Algebra
- Biology
- Geometry
- Life/Figure Drawing
- Physics

Find out how math relates to a career in Art.

You be the Scientfic Illustrator!

Use the information in the table to solve each problem. Write in simplest form.

1. Write the length and height of an Argentinosaurus as decimals. Use bar notation if necessary.

2. How much taller was a Velociraptor than a Microraptor? Write your answer as a decimal. ________

3. Which is greater, the height of the Argentinosaurus or the length of the Camptosaurus? ________

4. How much longer was a Camptosaurus than a Velociraptor? Plot your answer on the number line.

0 1 2 3 4 5 6 7 8 9 10 11 12

5. Compare the heights of all four dinosaurs. Order them from least to greatest.

6. An artist is creating a mural in which a Microraptor is $1\frac{1}{2}$ times the actual size. What is the length of the dinosaur in the mural? ________

Dinosaur Measurements		
Dinosaur	**Length (ft)**	**Height (ft)**
Argentinosaurus	$114\frac{5}{6}$	$24\frac{1}{10}$
Camptosaurus	$16\frac{2}{5}$	$11\frac{4}{5}$
Microraptor	$2\frac{5}{8}$	$\frac{24}{25}$
Velociraptor	$5\frac{9}{10}$	$3\frac{7}{25}$

Career Project

It's time to update your career portfolio! Investigate the education and training requirements for a career as a scientific illustrator.

What are some short-term goals you need to achieve to become a scientific illustrator?

-
-
-
-

Chapter Review

Vocabulary Check

Complete the puzzle by unscrambling the letters below to reveal words from the vocabulary list at the beginning of the chapter.

Q U A D R A N T S :

D I A N N

O R A D N T A S I N A H S

C A O R G A A N T Y H E T F E

X U A X E S I T S D B Y L A N E

Q S E P R R I O E D E T P A X I O U R

Complete each sentence using the vocabulary list at the beginning of the chapter.

1. A ______________ is a number that can be written as a fraction.
2. A number that is less than zero is a ______________.
3. A number that is greater than zero is a ______________.
4. The ______________ of a number is the distance between the number and zero on a number line.
5. The division of a ______________ ends.
6. A decimal whose digits repeat in groups of one or more is a ______________.

Key Concept Check

Use Your FOLDABLES®

Use your Foldable to help review the chapter.

Tape here

Compare and Order Numbers

Examples

Examples

Examples

Got it?

Circle the correct term or number to complete each sentence.

1. The opposite of −4 is (−4, 4).

2. The distance of a number from 0 is its (opposite, absolute value).

3. The value listed first in an ordered pair is the (*x*-coordinate, *y*-coordinate).

4. The absolute value of 17 is (−17, 17).

5. ($1.\overline{25}$, 6.543) is a terminating decimal.

Problem Solving

1. Kirk bought songs for his MP3 player. He needed 6 more songs to have a total of 100. Write an integer to represent how many more songs Kirk needs. (Lesson 1)

2. In a football game, the quarterback was tackled behind the line of scrimmage and lost 7 yards. Represent the loss of 7 yards as an integer. (Lesson 1)

3. Kelsey's bank transactions are shown in the table. A positive number represents a deposit and a negative number represents a withdrawal. What is the absolute value of the transaction in Week 3? (Lesson 2)

Week	Transaction
1	50
2	−15
3	−20
4	30

4. The high temperatures in a city during a 5 day period were −6°, 8°, −2°, 6°, and 11°. Place the temperatures in order from least to greatest. (Lesson 3)

5. **Use Math Tools** Refer to the diagram. Which building is located at (−2, −4)? (Lesson 6)

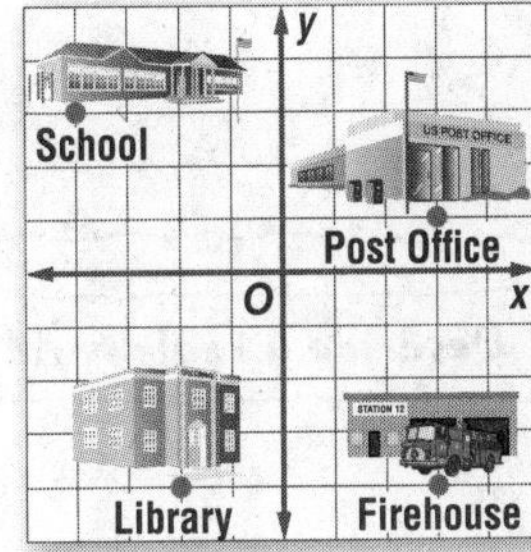

6. **Be Precise** Farah made 28 out of 84 shots on a goal in a recent hockey season. Write her shots made out of shots attempted as a decimal. (Lesson 4)

7. The heights of the lifeguard chairs are $66\frac{1}{3}$ inches and $72\frac{5}{8}$ inches. One section of the lake has a depth of $\frac{203}{4}$ inches, and another section has a depth of $\frac{109}{2}$ inches. Represent each height and depth using a positive or negative number. Then order the numbers from least to greatest. (Lesson 5)

Reflect

Answering the Essential Question

Use what you learned about integers and the coordinate plane to complete the graphic organizer.

Essential Question

HOW are integers and absolute value used in real-world situations?

Vocabulary	Definition
integer	
absolute value	

Describe a real-world situation that can be represented by the absolute value of 27.

Describe a real-world situation that can be represented by the absolute value of −16?

Answer the Essential Question. HOW are integers and absolute value used in real-world situations?

UNIT PROJECT

Get Out the Map! If you could go anywhere in the world, where would you go? This is your chance to explore someplace new. In this project you will:

- **Collaborate** with your classmates as you investigate a new place you would like to travel.
- **Share** the results of your research in a creative way.
- **Reflect** on how mathematical ideas can be represented.

By the end of this project, you will have a better understanding of how to use a map to plan the perfect adventure!

Collaborate

Go Online **Work with your group to research and complete each activity. You will use your results in the Share section on the following page.**

1. Think of someplace new that you would like to travel. Investigate the population of the area you choose as well as any interesting geographical features. Make a list of area attractions that you'd like to visit, such as museums, amusement parks, historical sites, national parks, and so on.

2. Research information about the climate for the area of your location in the time of year you are planning to travel. Important information could include the monthly rainfall, the average daily high and low temperatures, humidity levels, and wind speeds. Create a visual display to share your results.

3. Find lodging in the location you chose. Then, using maps and descriptions available on the Internet, plot three or four area attractions on a coordinate plane. Label each point.

4. Create a budget for each day of the trip showing the cost of the travel, lodging, daily activities, and food. Calculate the cost for your entire family, not just yourself.

5. Use an online map or GPS device to determine the actual distances you will need to travel to get from where you are staying to any attraction you plan to visit. Then find the total distance you will travel during the entire trip.

Share

With your group, decide on a way to present what you have learned from each of the activities about the planning the perfect adventure. Some suggestions are listed below, but you can also think of other creative ways to present your information. Remember to show how you used math to complete each of the activities in this project!

- Create a travel brochure for your current location. Your objective is to increase the tourists for your town. The brochure should include each of the following: a detailed map, recommended restaurants, area attractions, and fun facts.
- Write a journal entry from the perspective of an early explorer who, 1,000 years ago, traveled to the location you chose. Then describe how technology makes planning, budgeting, and navigating at the same location much easier today.

Check out the note on the right to connect this project with other subjects.

with Science

Use the Internet to research what technology was used by early explorers to navigate through unknown territories. Some questions to consider are:

- What tools were used to help explorers travel in the right direction?
- What constellations were used by the explorers and how were they used?

Reflect

6.  **Answer the Essential Question** How can mathematical ideas be represented?

   a. How did you use what you learned about computing with multi-digit numbers, and multiplying and dividing fractions to represent mathematical ideas in this project?

   ____

   ____

   ____

   b. How did you use what you learned about integers and the coordinate plane to represent mathematical ideas in this project?

   ____

   ____

   ____

Glossary/Glosario

Go online for the eGlossary. 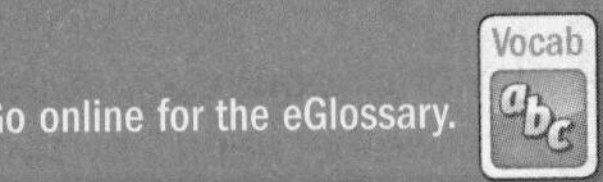

The eGlossary contains words and definitions in the following 13 languages:

Arabic	Cantonese	Hmong	Spanish	Urdu
Bengali	English	Korean	Tagalog	Vietnamese
Brazilian Portuguese	Haitian Creole	Russian		

English / Español

Aa

absolute value The distance between a number and zero on a number line.

valor absoluto Distancia entre un número y cero en la recta numérica.

acute angle An angle with a measure greater than 0° and less than 90°.

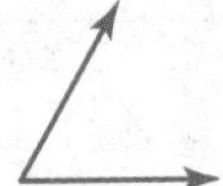

ángulo agudo Ángulo que mide más de 0° y menos de 90°.

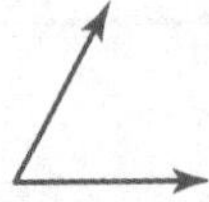

acute triangle A triangle having three acute angles.

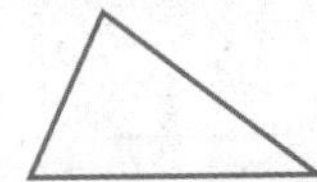

triángulo acutángulo Triángulo con tres ángulos agudos.

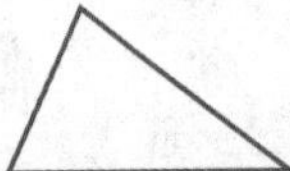

Addition Property of Equality If you add the same number to each side of an equation, the two sides remain equal.

propiedad de adición de la igualdad Si sumas el mismo número a ambos lados de una ecuación, los dos lados permanecen iguales.

algebra A mathematical language of symbols, including variables.

álgebra Lenguaje matemático que usa símbolos, incluyendo variables.

algebraic expression A combination of variables, numbers, and at least one operation.

expresión algebraica Combinación de variables, números y, por lo menos, una operación.

analyze To use observations to describe and compare data.

analizar Usar observaciones para describir y comparar datos.

angle Two rays with a common endpoint form an angle. The rays and vertex are used to name the angle.

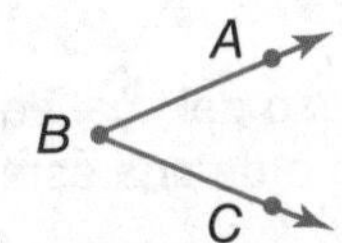

$\angle ABC$, $\angle CBA$, or $\angle B$

ángulo Dos rayos con un extremo común forman un ángulo. Los rayos y el vértice se usan para nombrar el ángulo.

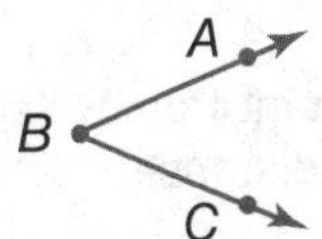

$\angle ABC$, $\angle CBA$ o $\angle B$

arithmetic sequence A sequence in which the difference between any two consecutive terms is the same.

sucesión aritmética Sucesión en la cual la diferencia entre dos términos consecutivos es constante.

Associative Property The way in which numbers are grouped does not change the sum or product.

propiedad asociativa La forma en que se agrupan tres números al sumarlos o multiplicarlos no altera su suma o producto.

average The sum of two or more quantities divided by the number of quantities; the mean.

promedio La suma de dos o más cantidades dividida entre el número de cantidades; la media.

Bb

bar notation A bar placed over digits that repeat to indicate a number pattern that repeats indefinitely.

notación de barra Barra que se coloca sobre los dígitos que se repiten para indicar el número de patrones que se repiten indefinidamente.

base Any side of a parallelogram.

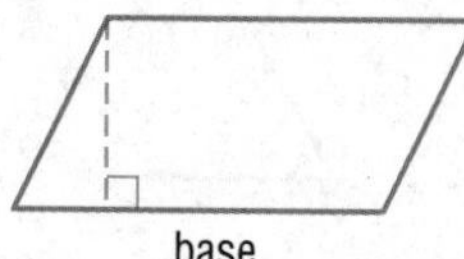

base Cualquier lado de un paralelogramo.

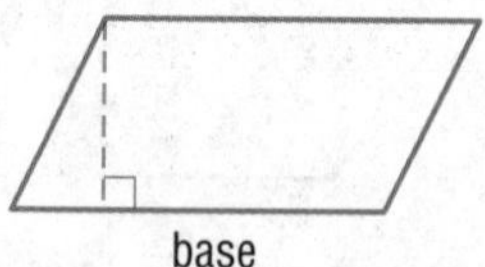

base One of the two parallel congruent faces of a prism.

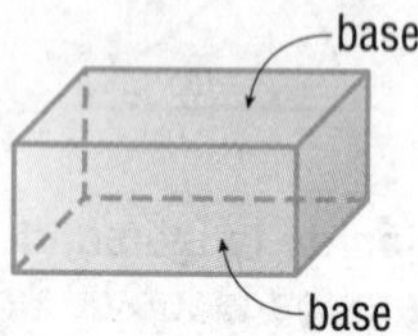

base Una de las dos caras paralelas congruentes de un prisma.

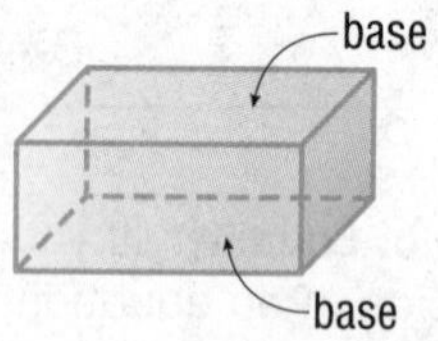

base In a power, the number used as a factor. In 10^3, the base is 10. That is, $10^3 = 10 \times 10 \times 10$.

base En una potencia, el número usado como factor. En 10^3, la base es 10. Es decir, $10^3 = 10 \times 10 \times 10$.

box plot A diagram that is constructed using five values.

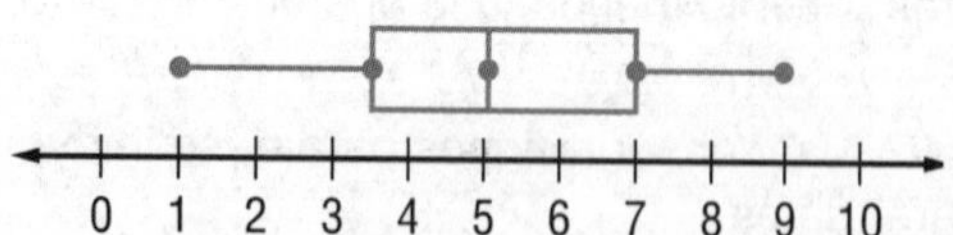

diagrama de caja Diagrama que se construye usando cinco valores.

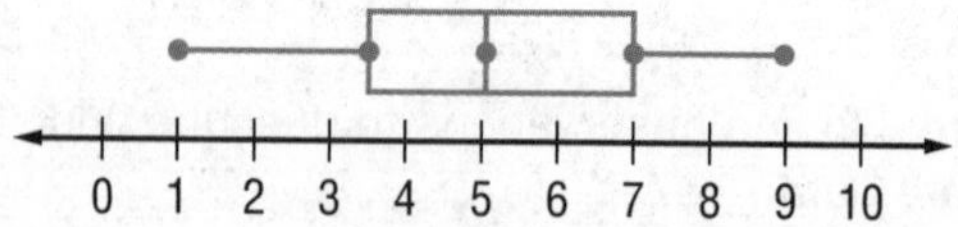

Cc

center The given point from which all points on a circle are the same distance.

centro Un punto dado del cual equidistan todos los puntos de un círculo o de una esfera.

circle The set of all points in a plane that are the same distance from a given point called the center.

círculo Conjunto de todos los puntos en un plano que equidistan de un punto dado llamado centro.

circle graph A graph that shows data as parts of a whole. In a circle graph, the percents add up to 100.

gráfica circular Gráfica que muestra los datos como partes de un todo. En una gráfica circular los porcentajes suman 100.

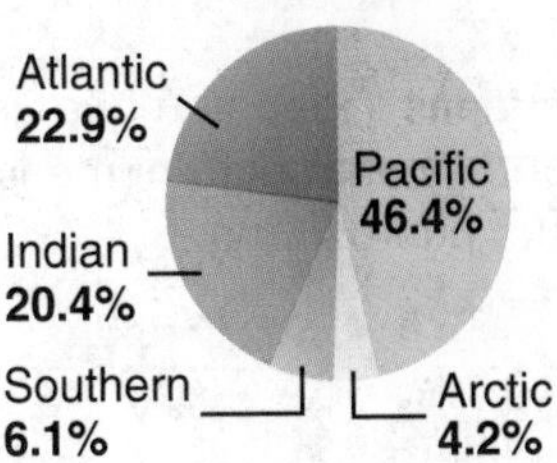

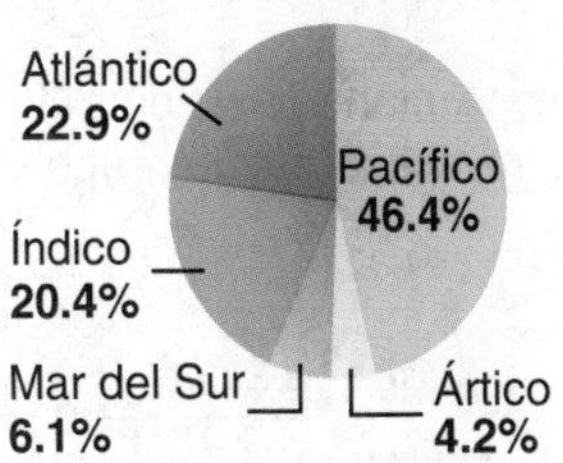

circumference The distance around a circle.

circunferencia La distancia alrededor de un círculo.

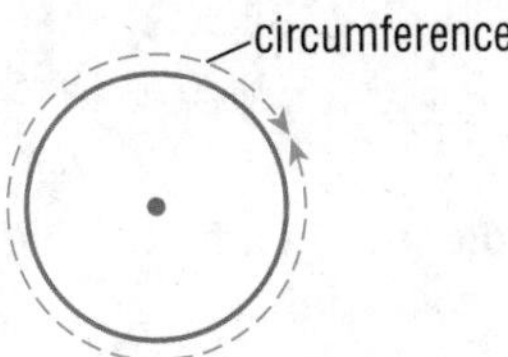

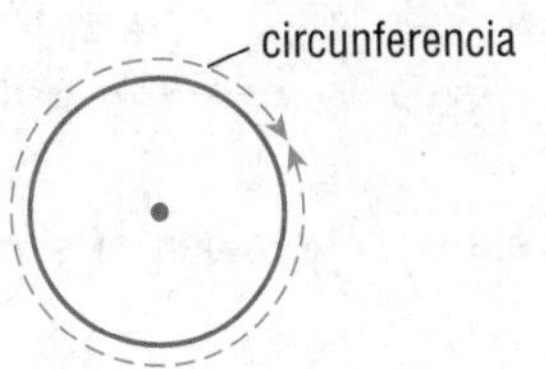

cluster Data that are grouped closely together.

agrupamiento Conjunto de datos que se agrupan.

coefficient The numerical factor of a term that contains a variable.

coeficiente El factor numérico de un término que contiene una variable.

Commutative Property The order in which numbers are added or multiplied does not change the sum or product.

propiedad commutativa La forma en que se suman o multiplican dos números no altera su suma o producto.

compatible numbers Numbers that are easy to use to perform computations mentally.

números compatibles Números que son fáciles de usar para realizar computations mentales.

complementary angles Two angles are complementary if the sum of their measures is 90°.

ángulos complementarios Dos ángulos son complementarios si la suma de sus medidas es 90°.

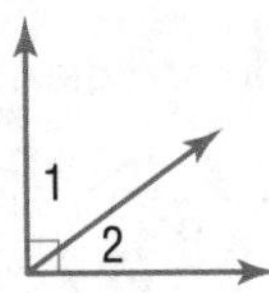

$\angle 1$ and $\angle 2$ are complementary angles.

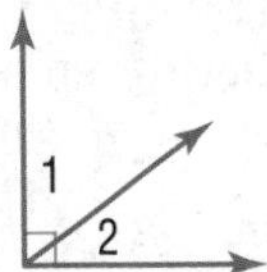

$\angle 1$ y $\angle 2$ son complementarios.

composite figure A figure made of triangles, quadrilaterals, semicircles, and other two-dimensional figures.

figura compuesta Figura formada por triángulos, cuadriláteros, semicírculos y otras figuras bidimensionales.

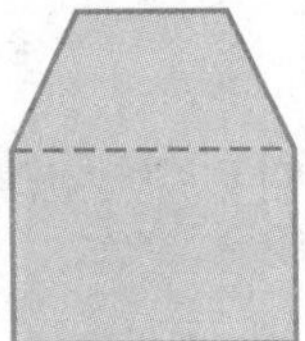

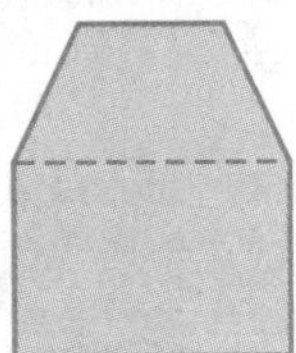

congruent Having the same measure.

congruente Ques tienen la misma medida.

congruent figures Figures that have the same size and same shape; corresponding sides and angles have equal measures.

figuras congruentes Figuras que tienen el mismo tamaño y la misma forma; los lados y los ángulos correspondientes con igual medida.

constant A term without a variable.

constante Un término sin una variable.

coordinate plane A plane in which a horizontal number line and a vertical number line intersect at their zero points.

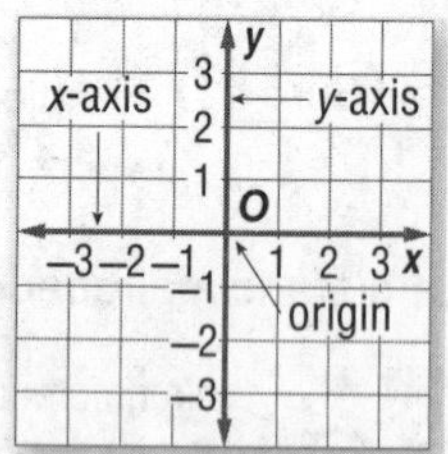

plano de coordenadas Plano en que una recta numérica horizontal y una recta numérica vertical se intersecan en sus puntos cero.

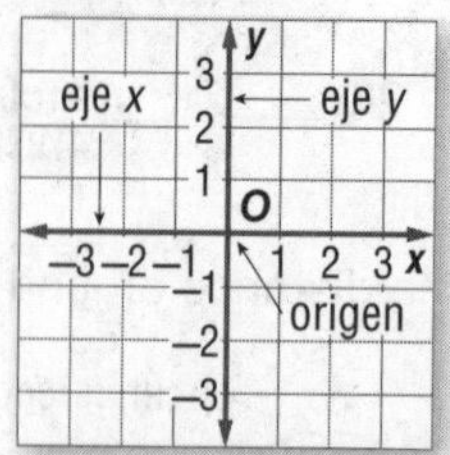

corresponding sides The sides of similar figures that "match."

lados correspondientes Lados de figuras semejantes que coinciden.

cubic units Used to measure volume. Tells the number of cubes of a given size it will take to fill a three-dimensional figure.

3 cubic units

unidades cúbicas Se usan para medir el volumen. Indican el número de cubos de cierto tamaño que se necesitan para llenar una figura tridimensional.

3 unidades cúbicas

Dd

data Information, often numerical, which is gathered for statistical purposes.

datos Información, con frecuencia numérica, que se recoge con fines estadísticos.

decagon A polygon having ten sides.

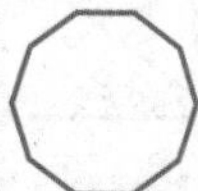

decágono Un polígono con diez lados.

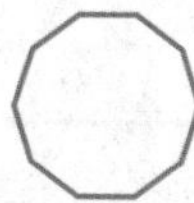

defining the variable Choosing a variable and deciding what the variable represents.

definir la variable Elegir una variable y decidir lo que representa.

dependent variable The variable in a relation with a value that depends on the value of the independent variable.

variable dependiente La variable en una relación cuyo valor depende del valor de la variable independiente.

diameter The distance across a circle through its center.

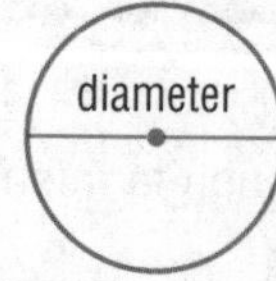

diámetro La distancia a través de un círculo pasando por el centro.

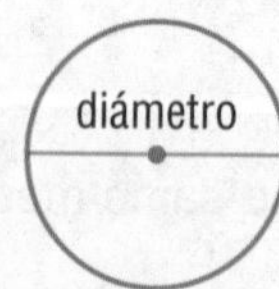

dimensional analysis The process of including units of measurement when you compute.

análisis dimensional Proceso que incluye las unidades de medida al hacer cálculos.

distribution The arrangement of data values.

distribución El arreglo de valores de datos.

Distributive Property To multiply a sum by a number, multiply each addend by the number outside the parentheses.

propiedad distributiva Para multiplicar una suma por un número, multiplica cada sumando por el número fuera de los paréntesis.

Division Property of Equality If you divide each side of an equation by the same nonzero number, the two sides remain equal.

propiedad de igualdad de la división Si divides ambos lados de una ecuación entre el mismo número no nulo, los lados permanecen iguales.

dot plot A diagram that shows the frequency of data on a number line. Also known as a line plot.

diagrama de puntos Diagrama que muestra la frecuencia de los datos sobre una recta numérica.

equals sign A symbol of equality, =.

signo de igualdad Símbolo que indica igualdad, =.

equation A mathematical sentence showing two expressions are equal. An equation contains an equals sign, =.

ecuación Enunciado matemático que muestra que dos expresiones son iguales. Una ecuación contiene el signo de igualdad, =.

equilateral triangle A triangle having three congruent sides.

triángulo equilátero Triángulo con tres lados congruentes.

equivalent expressions Expressions that have the same value.

expresiones equivalentes Expresiones que poseen el mismo valor, sin importer los valores de la(s) variable(s).

equivalent ratios Ratios that express the same relationship between two quantities.

razones equivalentes Razones que expresan la misma relación entre dos cantidades.

evaluate To find the value of an algebraic expression by replacing variables with numbers.

evaluar Calcular el valor de una expresión sustituyendo las variables por número.

exponent In a power, the number that tells how many times the base is used as a factor. In 5^3, the exponent is 3. That is, $5^3 = 5 \times 5 \times 5$.

exponente En una potencia, el número que indica las veces que la base se usa como factor. En 5^3, el exponente es 3. Es decir, $5^3 = 5 \times 5 \times 5$.

face A flat surface.

cara Una superficie plana.

factor the expression The process of writing numeric or algebraic expressions as a product of their factors.

factorizar la expresión El proceso de escribir expresiones numéricas o algebraicas como el producto de sus factores.

first quartile For a data set with median M, the first quartile is the median of the data values less than M.

primer cuartil Para un conjunto de datos con la mediana M, el primer cuartil es la mediana de los valores menores que M.

formula An equation that shows the relationship among certain quantities.

fórmula Ecuación que muestra la relación entre ciertas cantidades.

fraction A number that represents part of a whole or part of a set.

$\frac{1}{2}, \frac{1}{3}, \frac{1}{4}, \frac{3}{4}$

fracción Número que representa parte de un todo o parte de un conjunto.

$\frac{1}{2}, \frac{1}{3}, \frac{1}{4}, \frac{3}{4}$

frequency distribution How many pieces of data are in each interval.

distribución de frecuencias Cantidad de datos asociada con cada intervalo.

frequency table A table that shows the number of pieces of data that fall within the given intervals.

tabla de frecuencias Tabla que muestra el número de datos en cada intervalo.

function A relationship that assigns exactly one output value to one input value.

función Relación que asigna exactamente un valor de salida a un valor de entrada.

function rule An expression that describes the relationship between each input and output.

regla de funciones Expresión que describe la relación entre cada valor de entrada y de salida.

function table A table organizing the input, rule, and output of a function.

tabla de funciones Tabla que organiza las entradas, la regla y las salidas de una función.

gap An empty space or interval in a set of data.

laguna Espacio o intervalo vacío en un conjunto de datos.

geometric sequence A sequence in which each term is found by multiplying the previous term by the same number.

sucesión geométrica Sucesión en la cual cada término después del primero se determina multiplicando el término anterior por el mismo número.

graph To place a dot at a point named by an ordered pair.

gráfica Colocar una marca puntual en el punto que corresponde a un par ordenado.

Greatest Common Factor (GCF) The greatest of the common factors of two or more numbers.

The greatest common factor of 12, 18, and 30 is 6.

máximo común divisor (MCD) El mayor de los factores comunes de dos o más números.

El máximo común divisor de 12, 18 y 30 es 6.

height The shortest distance from the base of a parallelogram to its opposite side.

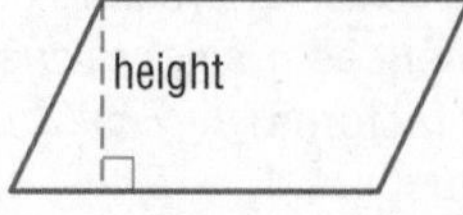

altura La distancia más corta desde la base de un paralelogramo hasta su lado opuesto.

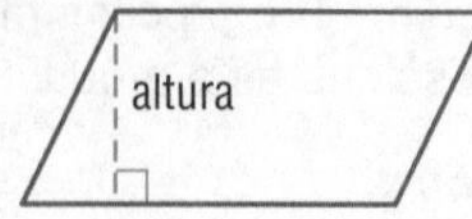

heptagon A polygon having seven sides.

heptágono Polígono con siete lados.

hexagon A polygon having six sides.

histogram A type of bar graph used to display numerical data that have been organized into equal intervals.

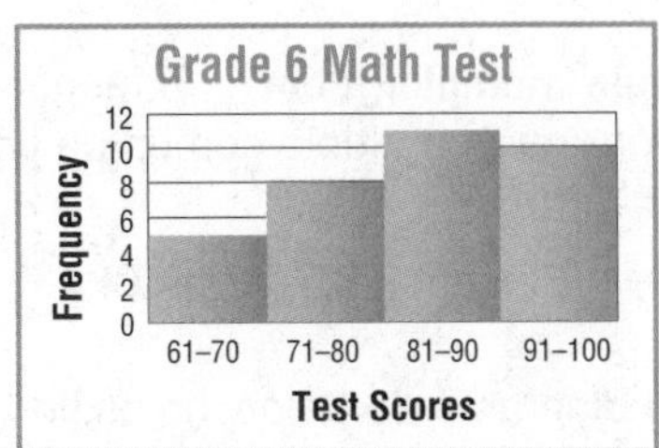

Identity Properties Properties that state that the sum of any number and 0 equals the number and that the product of any number and 1 equals the number.

independent variable The variable in a function with a value that is subject to choice.

inequality A mathematical sentence indicating that two quantities are not equal.

integer Any number from the set {... −4, −3, −2, −1, 0, 1, 2, 3, 4 ...} where ... means *continues without end.*

interquartile range A measure of variation in a set of numerical data, the interquartile range is the distance between the first and third quartiles of the data set.

intersecting lines *Lin*es that meet or cross at a common *point.*

interval The difference between successive values on a scale.

inverse operations Operations which *undo* each other. For example, addition and subtraction are inverse operations.

isosceles triangle A triangle having at least two congruent sides.

hexágono Polígono con seis lados.

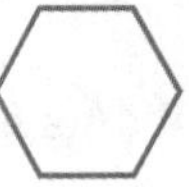

histograma Tipo de gráfica de barras que se usa para exhibir datos que se han organizado en intervalos iguales.

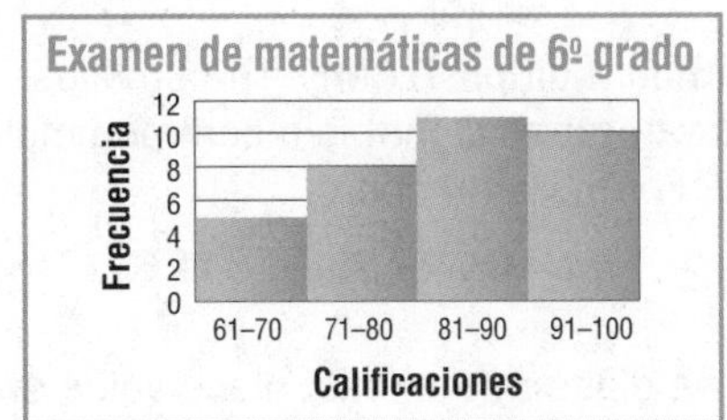

propiedades de identidad Propiedades que establecen que la suma de cualquier número y 0 es igual al número y que el producto de cualquier número y 1 es iqual al número.

variable independiente Variable en una función cuyo valor está sujeto a elección.

desigualdad Enunciado matemático que indica que dos cantidades no son iguales.

entero Cualquier número del conjunto {... −4, −3, −2, −1, 0, 1, 2, 3, 4 ...} donde ... significa que *continúa sin fin.*

rango intercuartil El rango intercuartil, una medida de la variación en un conjunto de datos numéricos, es la distancia entre el primer y el tercer cuartil del conjunto de datos.

rectas secantes *Rectas* que se intersectan o se cruzan en un *punto* común.

intervalo La diferencia entre valores sucesivos de una escala.

operaciones inversas Operaciones que se *anulan* mutuamente. La adición y la sustracción son operaciones inversas.

triángulo isósceles Triángulo que tiene por lo menos dos lados congruentes.

lateral face Any face that is not a base.

cara lateral Cualquier superficie plana que no sea la base.

least common denominator (LCD) The least common multiple of the denominators of two or more fractions.

mínimo común denominador (mcd) El menor múltiplo común de los denominadores de dos o más fracciones.

least common multiple (LCM) The smallest whole number greater than 0 that is a common multiple of each of two or more numbers.

The LCM of 2 and 3 is 6.

mínimo común múltiplo (mcm) El menor número entero, mayor que 0, múltiplo común de dos o más números.

El mcm de 2 y 3 es 6.

leaves The digits of the least place value of data in a stem-and-leaf plot.

hoja En un diagrama de tallo y hojas, los dígitos del menor valor de posición.

like terms Terms that contain the same variable(s) to the same power.

términos semejantes Términos que contienen la misma variable o variables elevadas a la misma potencia.

line A set of *points* that form a straight path that goes on forever in opposite directions.

recta Conjunto de *puntos* que forman una trayectoria recta sin fin en direcciones oputestas.

linear function A function that forms a line when graphed.

función lineal Función cuya gráfica es una recta.

line graph A graph used to show how a set of data changes over a period of time.

gráfica lineal Gráfica que se use para mostrar cómo cambian los valores durange un período de tiempo.

line of symmetry A line that divides a figure into two halves that are reflections of each other.

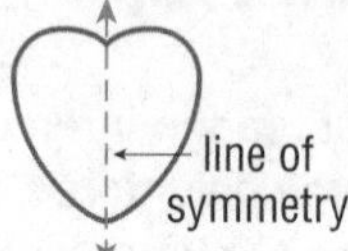

eje de simetría Recta que divide una figura en dos mitades especulares.

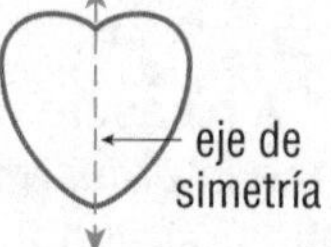

line plot A diagram that shows the frequency of data on a number line. Also known as a dot plot.

esquema lineal Diagrama que muestra la frecuencia de los datos sobre una recta numérica.

line segment A part of a *line* that connects two points.

segmento de recta Parte de una *recta* que conecta dos puntos.

line symmetry Figures that match exactly when folded in half have line symmetry.

simetría lineal Exhiben simetría lineal las figuras que coinciden exactamente al doblarse una sobre otra.

mean The sum of the numbers in a set of data divided by the number of pieces of data.

media La suma de los números en un conjunto de datos dividida entre el número total de datos.

mean absolute deviation A measure of variation in a set of numerical data, computed by adding the distances between each data value and the mean, then dividing by the number of data values.

desviación media absoluta Una medida de variación en un conjunto de datos numéricos que se calcula sumando las distancias entre el valor de cada dato y la media, y luego dividiendo entre el número de valores.

measures of center Numbers that are used to describe the center of a set of data. These measures include the mean, median, and mode.

medidas del centro Numéros que se usan para describir el centro de un conjunto de datos. Estas medidas incluyen la media, la mediana y la moda.

measures of variation A measure used to describe the distribution of data.

medidas de variación Medida usada para describir la distribución de los datos.

median A measure of center in a set of numerical data. The median of a list of values is the value appearing at the center of a sorted version of the list—or the mean of the two central values, if the list contains an even number of values.

mediana Una medida del centro en un conjunto de datos numéricos. La mediana de una lista de valores es el valor que aparece en el centro de una versión ordenada de la lista, o la media de los dos valores centrales si la lista contiene un número par de valores.

mode The number(s) or item(s) that appear most often in a set of data.

moda Número(s) de un conjunto de datos que aparece(n) más frecuentemente.

Multiplication Property of Equality If you multiply each side of an equation by the same nonzero number, the two sides remain equal.

propiedad de multiplicación de la igualdad Si multiplicas ambos lados de una ecuación por el mismo número no nulo, lo lados permanecen iguales.

negative integer A number that is less than zero. It is written with a − sign.

entero negativo Número que es menor que cero y se escribe con el signo −.

net A two-dimensional figure that can be used to build a three-dimensional figure.

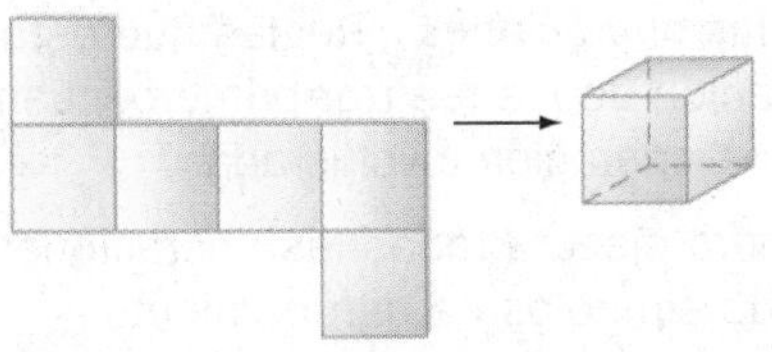

red Figura bidimensional que sirve para hacer una figura tridimensional.

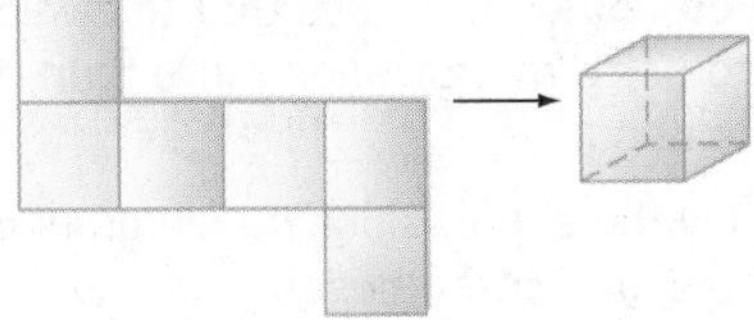

nonagon A polygon having nine sides.

enágono Polígono que tiene nueve lados.

numerical expression A combination of numbers and operations.

expresión numérica Una combinación de números y operaciones.

Oo

obtuse angle Any angle that measures greater than 90° but less than 180°.

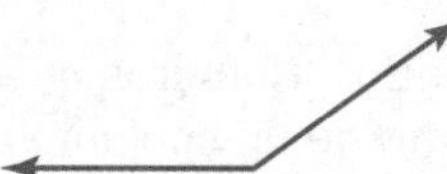

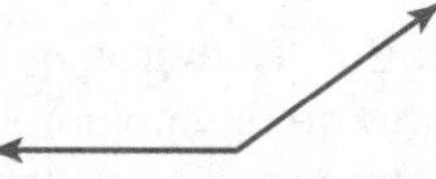

obtuse triangle A triangle having one obtuse angle.

octagon A polygon having eight sides.

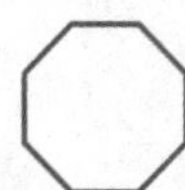

opposites Two integers are opposites if they are represented on the number line by points that are the same distance from zero, but on opposite sides of zero. The sum of two opposites is zero.

ordered pair A pair of numbers used to locate a point on the coordinate plane. The ordered pair is written in the form (x-coordinate, y-coordinate).

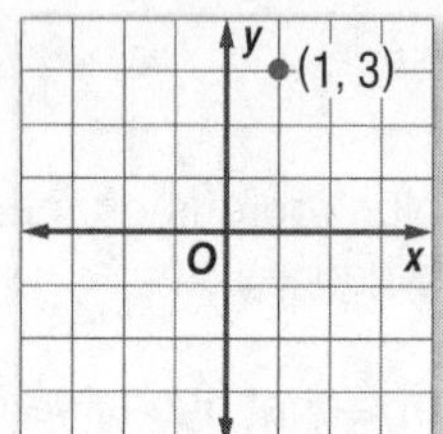

order of operations The rules that tell which operation to perform first when more than one operation is used.

1. Simplify the expressions inside grouping symbols, like parentheses.
2. Find the value of all powers.
3. Multiply and divide in order from left to right.
4. Add and subtract in order from left to right.

ángulo obtuso Cualquier ángulo que mide más de 90° pero menos de 180°.

triángulo obtusángulo Triángulo que tiene un ángulo obtuso.

octágono Polígono que tiene ocho lados.

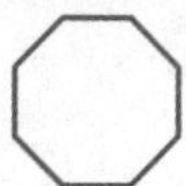

opuestos Dos enteros son opuestos si, en la recta numérica, están representados por puntos que equidistan de cero, pero en direcciones opuestas. La suma de dos opuestos es cero.

par ordenado Par de números que se utiliza para ubicar un punto en un plano de coordenadas. Se escribe de la forma (coordenada x, coordenada y).

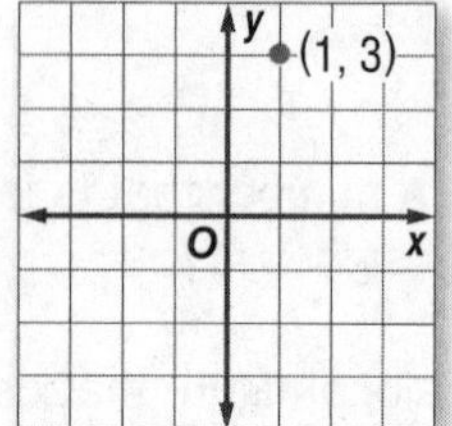

orden de las operaciones Reglas que establecen cuál operación debes realizar primero, cuando hay más de una operación involucrada.

1. Primero ejecuta todas las operaciones dentro de los símbolos de agrupamiento.
2. Evalúa todas las potencias.
3. Multiplica y divide en orden de izquierda a derecha.
4. Suma y resta en orden de izquierda a derecha.

origin The point of intersection of the *x*-axis and *y*-axis on a coordinate plane.

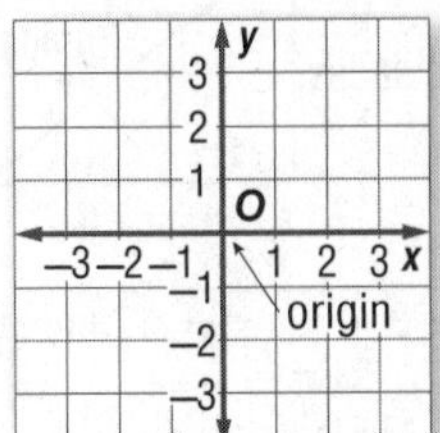

origen Punto de intersección de los ejes axiales en un plano de coordenadas.

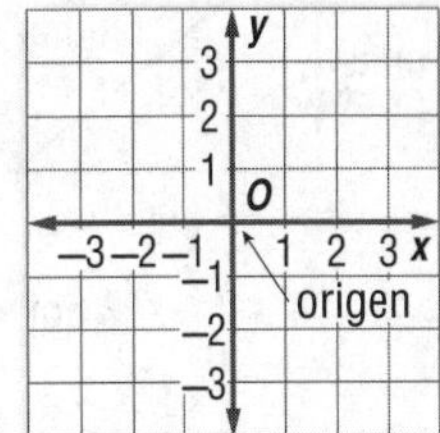

outlier A value that is much higher or much lower than the other values in a set of data.

valor atípico Dato que se encuentra muy separado de los otros valores en un conjunto de datos.

parallel lines Lines in a plane that never intersect.

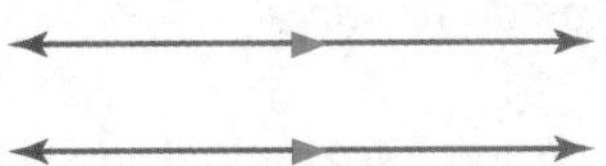

rectas paralelas Rectas en un plano que nunca se intersecan.

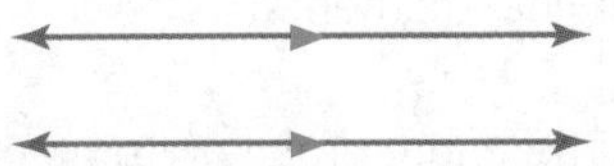

parallelogram A quadrilateral with opposite sides parallel and opposite sides congruent.

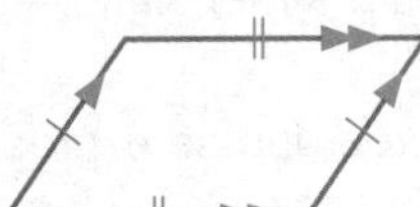

paralelogramo Cuadrilátero cuyos lados opuestos son paralelos y congruentes.

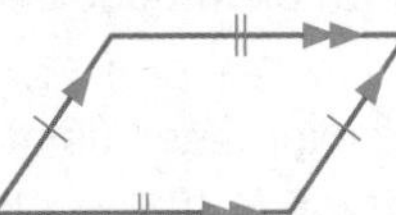

peak The most frequently occurring value in a line plot.

pico El valor que ocurre con más frecuencia en un diagrama de puntos.

pentagon A polygon having five sides.

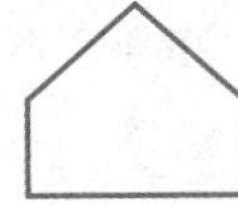

pentágono Polígono que tiene cinco lados.

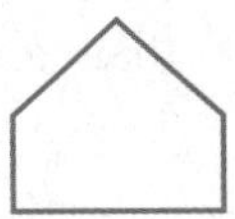

percent A ratio that compares a number to 100.

por ciento Razón en que se compara un número a 100.

percent proportion One ratio or fraction that compares part of a quantity to the whole quantity. The other ratio is the equivalent percent written as a fraction with a denominator of 100.

$$\frac{\text{part}}{\text{whole}} = \frac{\text{percent}}{100}$$

proporción porcentual Razón o fracción que compara parte de una cantidad a toda la cantidad. La otra razón es el porcentaje equivalente escrito como fracción con 100 de denominador.

$$\frac{\text{parte}}{\text{todo}} = \frac{\text{porcentaje}}{100}$$

perfect square Numbers with square roots that are whole numbers. 25 is a perfect square because the square root of 25 is 5.

cuadrados perfectos Números cuya raíz cuadrada es un número entero. 25 es un cuadrado perfecto porque la raíz cuadrada de 25 es 5.

perimeter The distance around a figure.

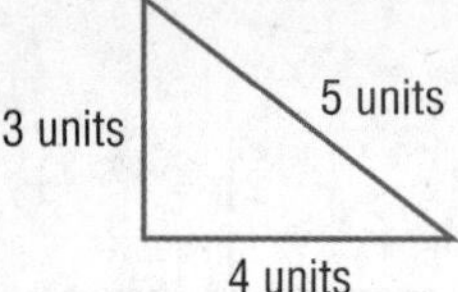

$P = 3 + 4 + 5 = 12$ units

perímetro La distancia alrededor de una figura.

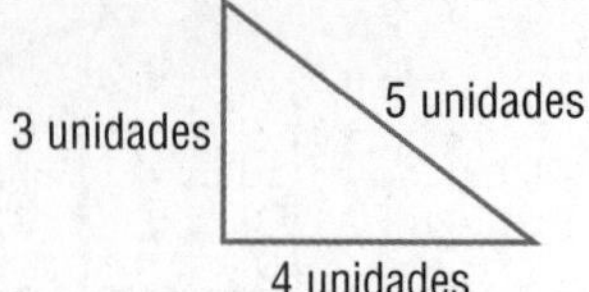

$P = 3 + 4 + 5 = 12$ unidades

pi The ratio of the circumference of a circle to its diameter. The Greek letter π represents this number. The value of pi is always 3.1415926....

pi Razón de la circunferencia de un círculo al diámetro del mismo. La letra griega π representa este número. El valor de pi es siempre 3.1415926....

plane A flat surface that goes on forever in all directions.

plano Superficie plana que se extiende infinitamente en todas direcciones.

point An exact location in space that is represented by a dot.

punto Ubicación exacta en el espacio que se representa con un marca puntual.

polygon A simple closed figure formed by three or more straight line segments.

polígono Figura cerrada simple formada por tres o más segmentos de recta.

population The entire group of items or individuals from which the samples under consideration are taken.

población El grupo total de individuos o de artículos del cual se toman las muestras bajo estudio.

positive integer A number that is greater than zero. It can be written with or without a + sign.

entero positivo Número que es mayor que cero y se puede escribir con o sin el signo +.

powers Numbers expressed using exponents. The power 3^2 is read *three to the second power,* or *three squared.*

potencias Números que se expresan usando exponentes. La potencia 3^2 se lee *tres a la segunda potencia* o *tres al cuadrado.*

prism A three-dimensional figure with at least three rectangular lateral faces and top and bottom faces parallel.

prisma Figura tridimensional que tiene por lo menos tres caras laterales rectangulares y caras paralelas superior e inferior.

properties Statements that are true for any number.

propiedades Enunciados que son verdaderos para cualquier número.

proportion An equation stating that two ratios or rates are equivalent.

proporción Ecuación que indica que dos razones o tasas son equivalentes.

pyramid A three-dimensional figure with at least three triangular sides that meet at a common vertex and only one base that is a polygon.

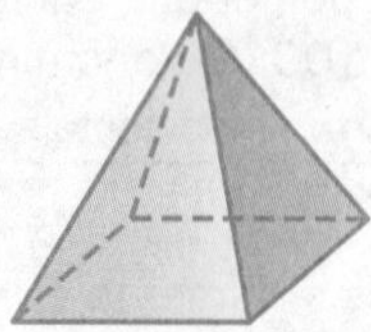

pirámide Una figura de tres dimensiones con que es en un un polígono y tres o mas caras triangulares que se encuentran en un vértice común.

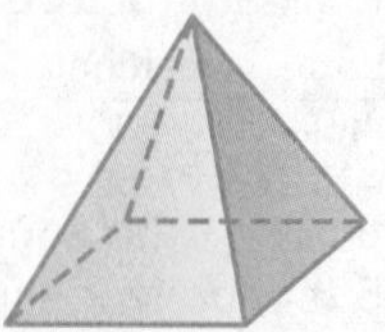

Qq

quadrants The four regions in a coordinate plane separated by the *x*-axis and *y*-axis.

cuadrantes Las cuatro regiones de un plano de coordenadas separadas por el eje *x* y el eje *y*.

quadrilateral A closed figure having four sides and four angles.

cuadrilátero Figura cerrada que tiene cuatro lados y cuatro ángulos.

quartiles Values that divide a data set into four equal parts.

cuartiles Valores que dividen un conjunto de datos en cuatro partes iguales.

Rr

radical sign The symbol used to indicate a nonnegative square root, $\sqrt{\ }$.

signo radical Símbolo que se usa para indicar una raíz cuadrada no negativa, $\sqrt{\ }$.

radius The distance from the center to any point on the circle.

radio Distancia desde el centro de un círculo hasta cualquier punto del mismo.

range The difference between the greatest number and the least number in a set of data.

rango La diferencia entre el número mayor y el número menor en un conjunto de datos.

rate A ratio comparing two quantities with different kinds of units.

tasa Razón que compara dos cantidades que tienen diferentes tipos de unidades.

rate of change A rate that describes how one quantity changes in relation to another. A rate of change is usually expressed as a unit rate.

tasa de cambio Tasa que describe cómo cambia una cantidad con respecto a otra. Por lo general, se expresa como tasa unitaria.

ratio A comparison of two quantities by division. The ratio of 2 to 3 can be stated as 2 out of 3, 2 to 3, 2 : 3, or $\frac{2}{3}$.

razón Comparación de dos cantidades mediante división. La razón de 2 a 3 puede escribirse como 2 de cada 3, 2 a 3, 2 : 3 ó $\frac{2}{3}$.

rational number A number that can be written as a fraction.

número racional Número que se puede expresar como fracción.

ratio table A table with columns filled with pairs of numbers that have the same ratio.

tabla de razones Tabla cuyas columnas contienen pares de números que tienen una misma razón.

ray A line that has one endpoint and goes on forever in only one direction.

rayo Recta con un extremo y la cual se extiende infinitamente en una sola dirección.

reciprocals Any two numbers that have a product of 1. Since $\frac{5}{6} \times \frac{6}{5} = 1$, $\frac{5}{6}$ and $\frac{6}{5}$ are reciprocals.

recíproco Cualquier par de números cuyo producto es 1. Como $\frac{5}{6} \times \frac{6}{5} = 1$, $\frac{5}{6}$ y $\frac{6}{5}$ son recíprocos.

rectangle A parallelogram having four right angles.

rectángulo Paralelogramo con cuatro ángulos rectos.

rectangular prism A prism that has rectangular bases.

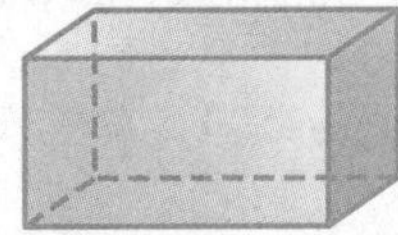

prisma rectangular Una prisma que tiene bases rectangulares.

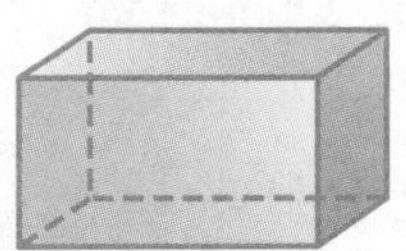

reflection The mirror image produced by flipping a figure over a line.

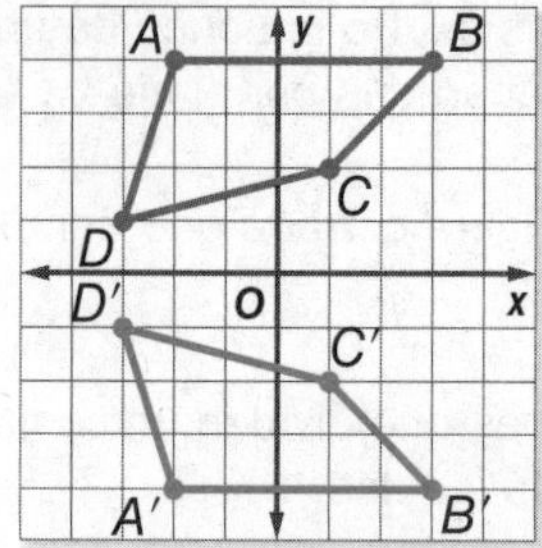

reflexión Transformación en la cual una figura se voltea sobre una recta. También se conoce como simetría de espejo.

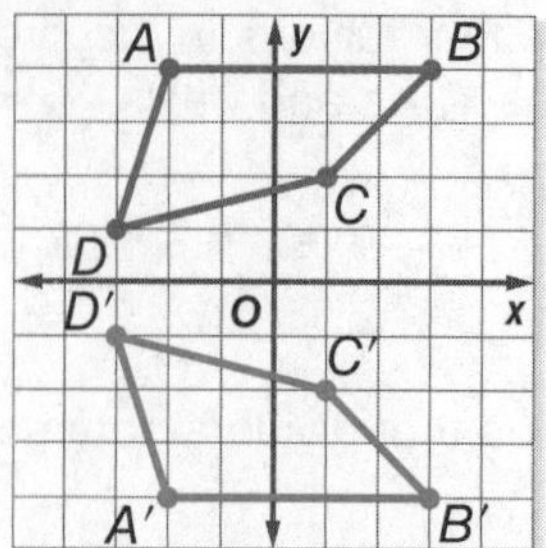

relation A set of ordered pairs such as (1, 3), (2, 4), and (3, 5). A relation can also be shown in a table or a graph.

relación Conjunto de pares ordenados como (1, 3), (2, 4) y (3, 5). Una relación también se puede mostrar en una tabla o una gráfica.

repeating decimal The decimal form of a rational number.

decimal periódico La forma decimal de un número racional.

rhombus A parallelogram having four congruent sides.

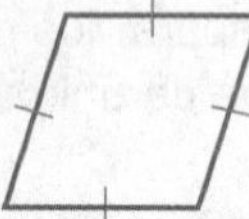

rombo Paralelogramo que tiene cuatro lados.

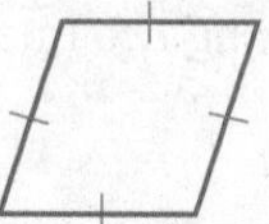

right angle An angle that measures exactly 90°.

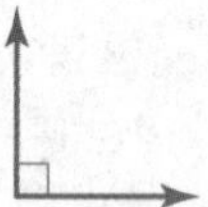

ángulo recto Ángulo que mide exactamente 90°.

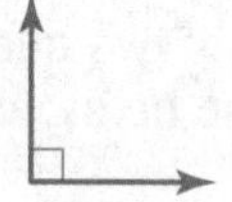

right triangle A triangle having one right angle.

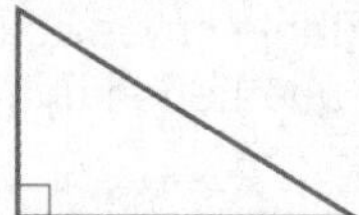

triángulo rectángulo Triángulo que tiene un ángulo recto.

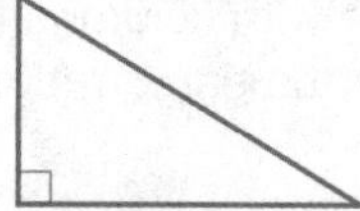

Ss

sample A randomly selected group chosen for the purpose of collecting data.

muestra Grupo escogido al azar o aleatoriamente que se usa con el propósito de recoger datos.

scale The set of all possible values of a given measurement, including the least and greatest numbers in the set, separated by the intervals used.

escala Conjunto de todos los valores posibles de una medida dada, incluyendo el número menor y el mayor del conjunto, separados por los intervalos usados.

scale The scale gives the ratio that compares the measurements of a drawing or model to the measurements of the real object.

escala Razón que compara las medidas de un dibujo o modelo a las medidas del objeto real.

scale drawing A drawing that is used to represent objects that are too large or too small to be drawn at actual size.

dibujo a escala Dibujo que se usa para representar objetos que son demasiado grandes o demasiado pequeños como para dibujarlos de tamaño natural.

scalene triangle A triangle having no congruent sides.

triángulo escaleno Triángulo sin lados congruentes.

scaling To multiply or divide two related quantities by the same number.

homotecia Multiplicar o dividir dos cantidades relacionadas entre un mismo número.

sequence A list of numbers in a specific order, such as 0, 1, 2, 3, or 2, 4, 6, 8.

sucesión Lista de números en un orden específico como, por ejemplo, 0, 1, 2, 3 ó 2, 4, 6, 8.

similar figures Figures that have the same shape but not necessarily the same size.

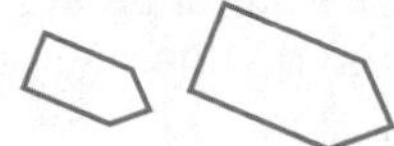

figuras semejantes Figuras que tienen la misma forma, pero no necesariamente el mismo tamaño.

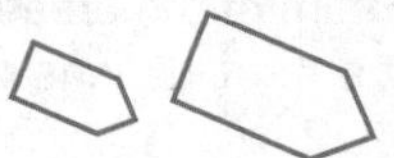

slant height The height of each lateral face.

altura oblicua Altura de cada cara lateral.

solution The value of a variable that makes an equation true. The solution of $12 = x + 7$ is 5.

solución Valor de la variable de una ecuación que hace verdadera la ecuación. La solución de $12 = x + 7$ es 5.

solve To replace a variable with a value that results in a true sentence.

resolver Reemplazar una variable con un valor que resulte en un enunciado verdadero.

square A rectangle having four right angles and four congruent sides.

cuadrado Rectángulo con cuatro ángulos rectos y cuatro lados congruentes.

square root The factors multiplied to form perfect squares.

raíz cuadrada Factores multiplicados para formar cuadrados perfectos.

statistical question A question that anticipates and accounts for a variety of answers.

cuestión estadística Una pregunta que se anticipa y da cuenta de una variedad de respuestas.

statistics Collecting, organizing, and interpreting data.

estadística Recopilar, ordenar e interpretar datos.

stem-and-leaf plot A system where data are organized from least to greatest. The digits of the least place value usually form the leaves, and the next place-value digits form the stems.

Stem	Leaf
1	2 4 5
2	
3	1 2 3 3 9
4	0 4 6 7

4 | 7 = 47

diagrama de tallo y hojas Sistema donde los datos se organizan de menor a mayor. Por lo general, los dígitos de los valores de posición menores forman las hojas y los valores de posición más altos forman los tallos.

Tallo	Hojas
1	2 4 5
2	
3	1 2 3 3 9
4	0 4 6 7

4 | 7 = 47

stems The digits of the greatest place value of data in a stem-and-leaf plot.

tallo Los dígitos del mayor valor de posición de los datos en un diagrama de tallo y hojas.

straight angle An angle that measures exactly 180°.

ángulo llano Ángulo que mide exactamente 180°.

Subtraction Property of Equality If you subtract the same number from each side of an equation, the two sides remain equal.

propiedad de sustracción de la igualdad Si sustraes el mismo número de ambos lados de una ecuación, los dos lados permanecen iguales.

supplementary angles Two angles are supplementary if the sum of their measures is 180°.

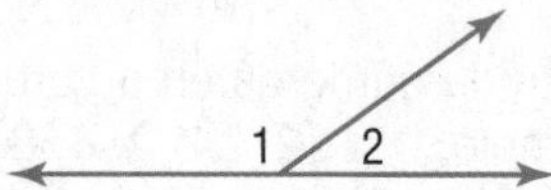

∠1 and ∠2 are supplementary angles.

ángulos suplementarios Dos ángulos son suplementarios si la suma de sus medidas es 180°.

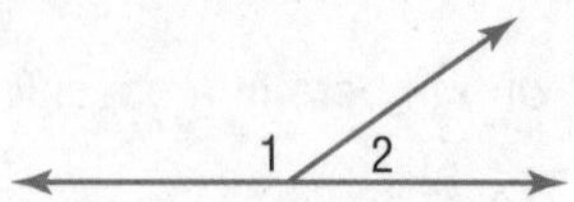

∠1 y ∠2 son suplementarios.

surface area The sum of the areas of all the surfaces (faces) of a three-dimensional figure.
$S.A. = 2\ell h + 2\ell w + 2hw$

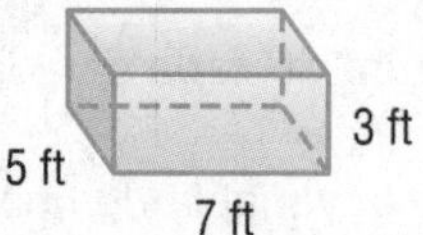

$S.A. = 2(7 \times 3) + 2(7 \times 5) + 2(3 \times 5)$
$= 142$ square feet

área de superficie La suma de las áreas de todas las superficies (caras) de una figura tridimensional.
$S.A. = 2\ell h + 2\ell w + 2hw$

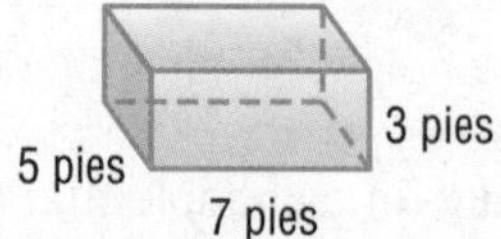

$S.A. = 2(7 \times 3) + 2(7 \times 5) + 2(3 \times 5)$
$= 142$ pies cuadrados

survey A question or set of questions designed to collect data about a specific group of people, or population.

encuesta Pregunta o conjunto de preguntas diseñadas para recoger datos sobre un grupo específico de personas o población.

symmetric distribution Data that are evenly distributed.

distribución simétrica Datos que están distribuidos.

Tt

term Each number in a sequence.

término Cada uno de los números de una sucesión.

term Each part of an algebraic expression separated by a plus or minus sign.

término Cada parte de un expresión algebraica separada por un signo más o un signo menos.

terminating decimal A decimal is called terminating if its repeating digit is 0.

decimal finito Un decimal se llama finito si el dígito que se repite es 0.

third quartile For a data set with median M, the third quartile is the median of the data values greater than M.

tercer cuartil Para un conjunto de datos con la mediana M, el tercer cuartil es la mediana de los valores mayores que M.

three-dimensional figure A figure with length, width, and height.

figura tridimensional Una figura que tiene largo, ancho y alto.

trapezoid A quadrilateral with one pair of parallel sides.

trapecio Cuadrilátero con un único par de lados paralelos.

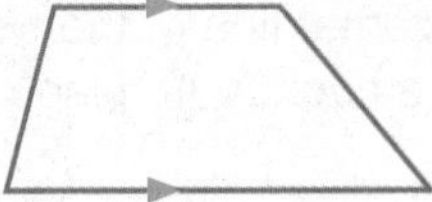

triangle A figure with three sides and three angles.

triángulo Figura con tres lados y tres ángulos.

triangular prism A prism that has triangular bases.

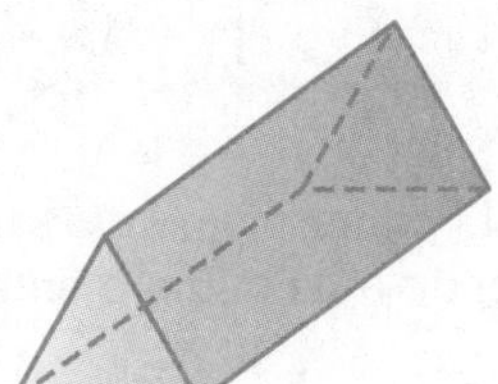

prisma triangular Prisma con bases triangulares.

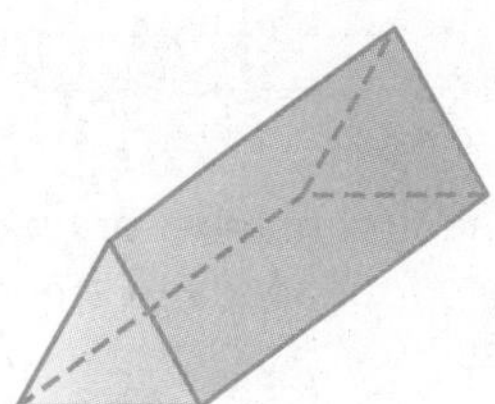

unit price The cost per unit.

precio unitario El costo por cada unidad.

unit rate A rate that is simplified so that it has a denominator of 1.

tasa unitaria Tasa simplificada para que tenga un denominador igual a 1.

unit ratio A unit rate where the denominator is one unit.

razón unitaria Tasa unitaria en que el denominador es la unidad.

variable A symbol, usually a letter, used to represent a number.

variable Un símbolo, por lo general, una letra, que se usa para representar un número.

vertex The point where three or more faces intersect.

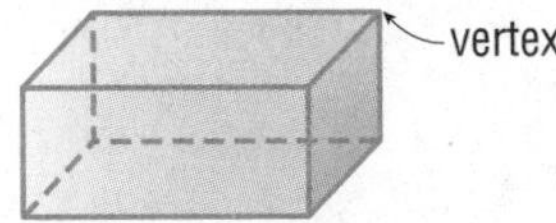

vértice El punto en que se intersecan dos o más caras del prisma.

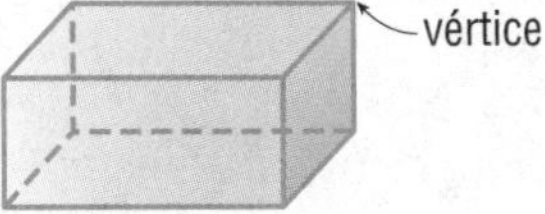

volume The amount of space inside a three-dimensional figure. Volume is measured in cubic units.

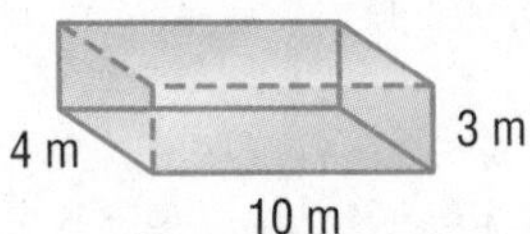

$V = 10 \times 4 \times 3 = 120$ cubic meters

volumen Cantidad de espacio dentro de una figura tridimensional. El volumen se mide en unidades cúbicas.

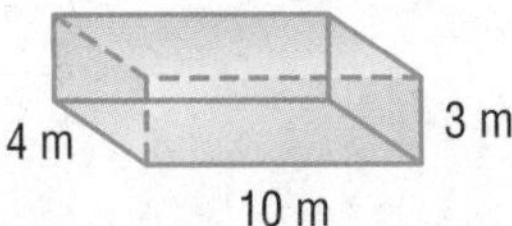

$V = 10 \times 4 \times 3 = 120$ metros cúbicos

x-axis The horizontal line of the two perpendicular number lines in a coordinate plane.

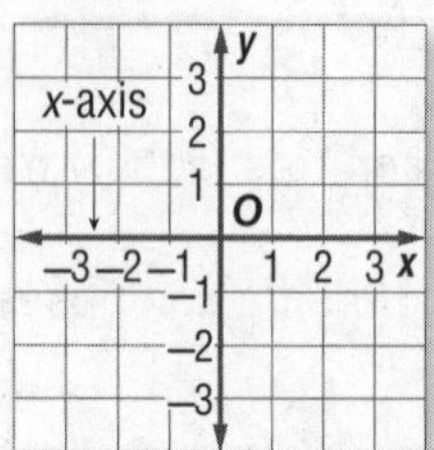

eje *x* La recta horizontal de las dos rectas numéricas perpendiculares en un plano de coordenadas.

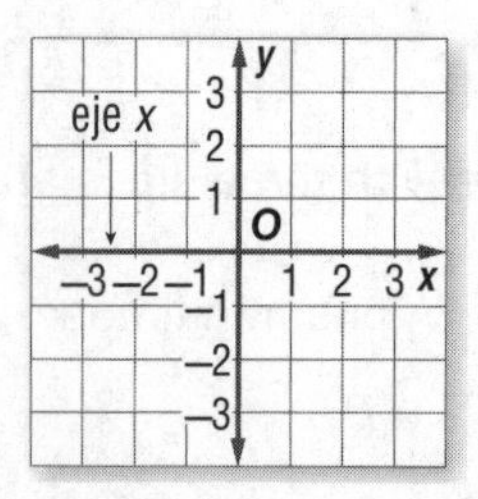

x-coordinate The first number of an ordered pair. The *x*-coordinate corresponds to a number on the *x*-axis.

coordenada *x* El primer número de un par ordenado, el cual corresponde a un número en el eje *x*.

y-axis The vertical line of the two perpendicular number lines in a coordinate plane.

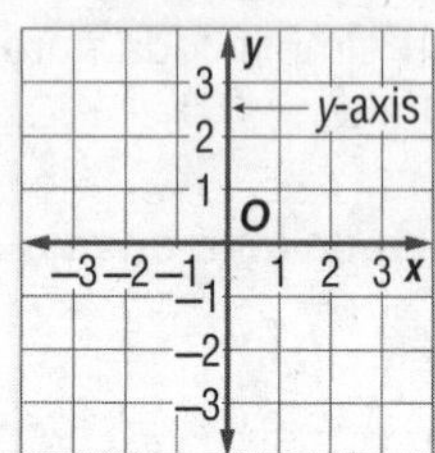

eje *y* La recta vertical de las dos rectas numéricas perpendiculares en un plano de coordenadas.

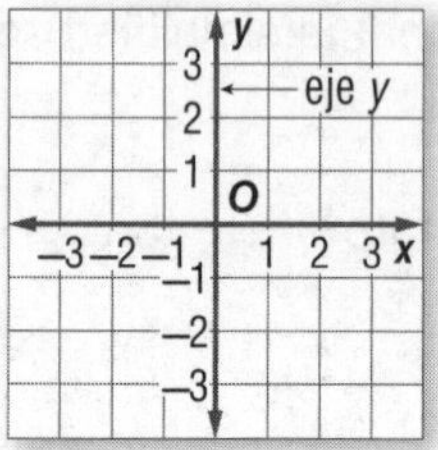

y-coordinate The second number of an ordered pair. The *y*-coordinate corresponds to a number on the *y*-axis.

coordenada *y* El segundo número de un par ordenado, el cual corresponde a un número en el eje *y*.

Chapter 1 Ratios and Rates

Page 6 Chapter 1 Are You Ready?

1. 29 **3.** 6 **5.** $\frac{1}{4}$ **7.** $\frac{13}{25}$

Pages 11–12 Lesson 1-1 Independent Practice

1. 2 **3.** 7 **5.** 30 **7** 60 **9** 9 pansies **11.** 30 days **13.** Sample answer: A gardener has 27 daisies and 36 marigolds. An equal number of each of flower is planted in each row. What is the greatest number of marigolds in each row? 9 marigolds **15.** Yes; Sample answer: when 1 is the only common factor of two numbers, then the GCF is one. The GCF of 5 and 7 is one because they are both prime numbers.

Pages 13–14 Lesson 1-1 Extra Practice

17. 5 **19.** 6 **21.** 15 **23.** 30 **25.** 4 baskets **27.** C **29.** 24 inches **31.** $\frac{3}{5}$ **33.** $\frac{8}{10}$

Pages 23–24 Lesson 1-2 Independent Practice

1. $\frac{2}{1}$; For every 2 flutes, there is 1 drum. **3.** $\frac{2}{5}$; For every 2 boys, there are 5 girls in the class. **5.** 12, 21 **7a.** $\frac{5}{16}$, 5 to 16, or 5:16; The Rangers have made 5 Stanley Cup Finals appearances for every 16 appearances made by the Canadiens. **7b.** $\frac{23}{17}$, 23 to 17, 23:17; The Maple Leafs have made 23 Stanley Cup appearances to every 17 Bruins' appearances. **9.** 1,440; The ratios are 1:2, 1:3, 1:4, and 1:5.

Pages 25–26 Lesson 1-2 Extra Practice

11. $\frac{1}{4}$; for every 1 triangle there are 4 rectangles. **13.** $\frac{1}{3}$; For every 1 puppy, there are 3 kittens available for adoption. **15.** $\frac{2}{7}$, 2 to 7, or 2:7; Two out of every 7 food items donated were cans of fruit. **17.** $\frac{1}{3}$; 1:3; or 1 to 3; Sample answer: If 6 students own a cell phone, 24 − 6 or 18 do not. The ratio is $\frac{6}{18}$ or $\frac{1}{3}$. **19.** G **21.** B **23.** 4 **25.** 195 miles **27.** 15 girls

Pages 35–36 Lesson 1-3 Independent Practice

1. $\frac{12 \text{ oz}}{1 \text{ steak}}$ **3.** $\frac{5.1 \text{ gal}}{1 \text{ container}}$ **5.** Divide the time by the number of laps. Evans drove the fastest at 2.3 minutes per lap. **7.** $4 per mile **9a.** 268 miles **9b.** about 2 h **11.** A unit rate has a denominator of 1. $\frac{\$108}{6 \text{ weeks}} = \frac{\$18}{1 \text{ week}}$ **13.** B

Pages 37–38 Lesson 1-3 Extra Practice

15. 4 tulips per minute **17.** 15 miles per hour **19.** $63 per ticket **21a.** 10.4 m per s **21b.** 9.3 m per s **21c.** 10.3 m per s **23.** C **25.** first kind: $1.67 for one box; second kind: $1.50 for one box **27.** $\frac{1}{5}$ **29.** $\frac{6}{25}$ **31.** 53 miles per hour

Pages 43–44 Lesson 1-4 Independent Practice

1

Number of Pies	5	10	20
Pounds of Apples	2	4	8

8 pounds

3

American Dollars	270	27	9
Mexican Pesos	3,000	300	100

$9

5a.

People Served	24
Liters of Soda	4
Pints of Sherbet	2
Cups of Ice	6

5b. 2 L soda, 1 pt sherbet, 3 c ice; 6 L soda, 3 pt sherbet, 9 c ice **5c.** 3 L soda, 1.5 pt sherbet, 4.5 c ice; Since 18 is half of 36, half the recipe that serves 36 people will serve 18 people. 6 L ÷ 2 = 3 L, 3 pt ÷ 2 = 1.5 pt, and 9 c ÷ 2 = 4.5 c. **7.** Larger Quantity: ×; larger batches, Smaller Quantity: ÷; unit rate

9.

Girls	10	5	15
Boys	8	4	12

No; if 5 girls and 5 boys are added, there would be 15 girls and 13 boys in the class. Using the ratio table, you can see that there should be 12 boys for 15 girls.

Pages 45–46 Lesson 1-4 Extra Practice

11.

Number of Adults	1	2	3	4
Number of Students	7	14	21	28

4 adults

13.

Ounces of Nectar	16	2	12
Number of Birds Fed	80	10	60

60 birds

15. 160 mi **17.** B **19.** 8; Sample answer: There are 60 minutes in an hour. 15 × 4 = 60, so 2 × 4 = 8. **21.** 7; 3

Pages 51–52 Lesson 1-5 Independent Practice

1. 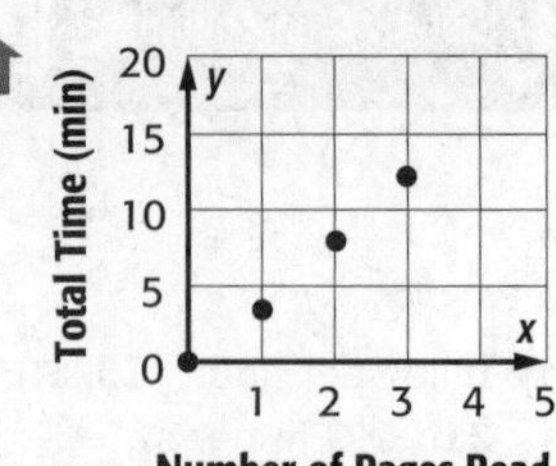

3.

Ken's Home Supply		
Fencing (ft), x	Cost ($), y	(x, y)
1	5	(1, 5)
2	10	(2, 10)
3	15	(3, 15)
4	20	(4, 20)

Wayne's Warehouse		
Fencing (ft), x	Cost ($), y	(x, y)
1	6	(1, 6)
2	12	(2, 12)
3	18	(3, 18)
4	24	(4, 24)

5. Sample answer: As the number of feet of fencing increases, the cost at Wayne's Warehouse increases at a faster rate than the cost at Ken's Home Supply. The cost at Wayne's Warehouse is shown on the graph as a steeper line. **7.** Sample answer: Lauren earns $7 an hour tutoring. Make a table showing the relationship between the number of hours she tutors and the amount of money she earns.

9. 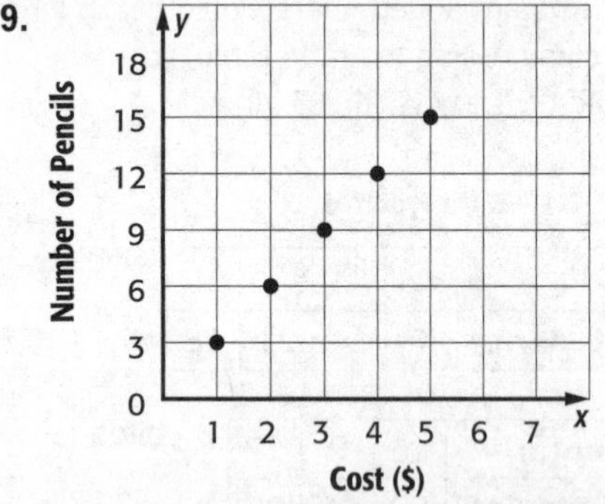

Sample answer: The points at (1, 3), (2, 6), (3, 9), and (5, 15) represent a ratio equivalent to 1:3. The ratio 4:12 is equivalent to 1:3. So, the cost of 12 pencils is $4.

Pages 53–54 Lesson 1-5 Extra Practice

11. 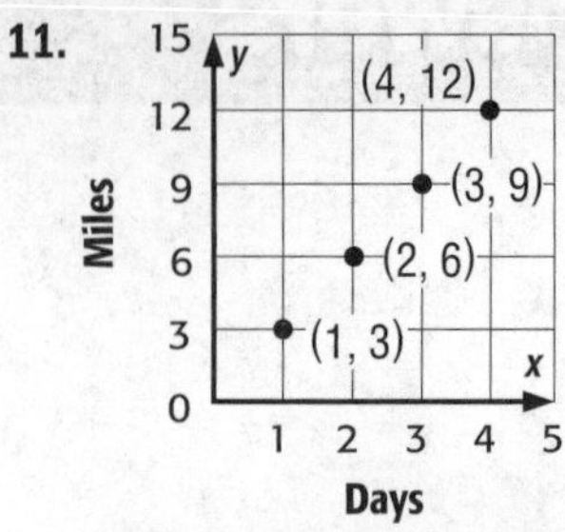

13.

Tiger Exhibit		
Animals, x	Employees, y	(x, y)
1	2	(1, 2)
2	4	(2, 4)
3	6	(3, 6)
4	8	(4, 8)

Elephant Exhibit		
Animals, x	Employees, y	(x, y)
1	4	(1, 4)
2	8	(2, 8)
3	12	(3, 12)
4	16	(4, 16)

15. Sample answer: The number of employees for the elephant exhibit increases at a faster rate than the number of employees for the tiger exhibit. The line representing the elephant exhibit is a steeper line. **17.** $90 **19.** $\frac{1}{5}$ **21.** 6 students

Page 57 Problem-Solving Investigation The Four-Step Plan

Case 3. 6,482 steps **Case 5.** $890

Pages 63–64 Lesson 1-6 Independent Practice

1. No; Since the unit rates, $\frac{\$0.50}{1 \text{ bagel}}$ and $\frac{\$0.38}{1 \text{ bagel}}$, are not the same, the rates are not equivalent. **3.** Yes; Since $\frac{3 \text{ h} \times 3}{\$12 \times 3} = \frac{9 \text{ h}}{\$36}$, the fractions are equivalent; $\frac{3 \text{ h}}{\$12} = \frac{9 \text{ h}}{\$36}$. **5.** No; Sample answer: since $\frac{8 \text{ pairs}}{\$12} \neq \frac{3 \text{ pairs}}{\$6}$, the ratios are not equivalent. **7.** No; Sample answer: Kiera did $\frac{6 \text{ problems}}{30 \text{ minutes}}$ or $\frac{1 \text{ problem}}{5 \text{ minutes}}$, and Heath did $\frac{18 \text{ problems}}{40 \text{ minutes}}$ or $\frac{9 \text{ problems}}{20 \text{ minutes}}$. So, the ratios are not equivalent. **9a.** Yes; Sample answer: The cross products 5×9 and 3×15 are both 45. **9b.** No; Sample answer: The cross product 7×5 or 35, is not equal to 2×21 or 42. **11.** B

Pages 65–66 Lesson 1-6 Extra Practice

13. Yes; Since the unit rates are the same, $\frac{32 \text{ words}}{1 \text{ minute}}$, the rates are equivalent; $\frac{96 \text{ words}}{3 \text{ minutes}} = \frac{160 \text{ words}}{5 \text{ minutes}}$.

15. No; Since $\frac{16 \text{ students}}{28 \text{ students}} \neq \frac{240 \text{ students}}{560 \text{ students}}$, the ratios are not equivalent. **17.** yes; The length to width ratio for the model and sofa form equivalent fractions. **19.** Yes; sample answer: $\frac{\$35}{5 \text{ weeks}} = \frac{\$7}{1 \text{ week}}$ and $\frac{\$56}{56 \text{ days}} = \frac{\$56}{8 \text{ weeks}}$ or $\frac{\$7}{1 \text{ week}}$. **21.** H **23.** 150 **25.** 126 **27.** $4

Pages 75–76 Lesson 1-7 Independent Practice

1. 90 cookies **3.** 840 gal **5.** 60 students
9. Elisa did not set up the equivalent ratios in the correct order. She should have set it up as $\frac{1}{12} = \frac{■}{276}$. There are 23 teachers at the preschool. **11.** 15 people

Pages 77–78 Lesson 1-7 Extra Practice

13. 54 teenagers **15.** 3 baseballs **17.** 28 DVDs **19.** C **21.** 27 **23.** $\frac{1}{7}$ **25.** $\frac{1}{6}$ **27.** 32 cars

Page 81 Chapter Review Vocabulary Check

Across
3. ordered pair **9.** graph **11.** *y* coordinate
Down
1. *y* axis **5.** scaling **7.** ratio table

Page 82 Chapter Review Key Concept Check

1. d **3.** a **5.** b

Page 83 Chapter Review Problem Solving

1. $\frac{1}{3}$, 1 to 3, or 1:3; one out of 3 DVDs Amos owns is an action DVD.

3.

Number of Trucks	3	24	48
Number of Vehicles	8	64	128

48 trucks

5. yes; $\frac{4}{90} = \frac{2}{45}$ and $\frac{2}{45} = \frac{2}{45}$ **7.** 72 students

Chapter 2 Fractions, Decimals, and Percents

Page 88 Chapter 2 Are You Ready?

1. 4 **3.** 18 **5.** 60 **7.** 3 rotations

Pages 93–94 Lesson 2-1 Independent Practice

1. $\frac{1}{2}$ **3.** $\frac{33}{100}$ **5.** 0.385 **7.** 0.16 **9.** Mercury: 87.96; Venus: 224.7; Mars: 686.98 **11a.** meat: $\frac{7}{20}$; vegetables: $\frac{3}{20}$; sauce: $\frac{1}{20}$; bread: $\frac{1}{20}$ **11b.** $\frac{1}{5}$ lb **11c.** $\frac{3}{5}$ lb **13.** Sample answer: $\frac{1}{5}$ in. and $\frac{7}{20}$ in. **15.** Always; a decimal that ends in the thousandths place can have a denominator of 1,000. Since 1,000 is divisible by 2 and 5, the denominator of every such terminating decimal is divisible by 2 and 5. **17.** C

Pages 95–96 Lesson 2-1 Extra Practice

19. $\frac{13}{20}$ **21.** $9\frac{7}{20}$ **23.** 0.622 **25.** 14.6 **27.** 23.375 **29.** 0.6 **31.** D **33.** $\frac{4}{5}$ **35.** 1; 5 **37.** 18; 25

39.

Multiplication Problem	Product
36 × 100	3,600
36 × 10	360
36 × 1	36
36 × 0.1	3.6
36 × 0.01	0.36

Pages 105–106 Lesson 2-2 Independent Practice

1. $\frac{1}{50}$ **3.** $\frac{17}{20}$ **5.** 20% **7.** 35% **9.** $\frac{7}{25}$ **11.** $\frac{9}{50}$
13. Do not prefer: 80%, prefer: 20%; the sum of the percents is 100. **15.** Sample answer: $\frac{11}{20} = \frac{55}{100}$ or 55%, $\frac{3}{5} = \frac{60}{100}$ or 60%, $\frac{7}{10} = \frac{70}{100}$ or 70% **17.** $\frac{8}{45}$; The other numbers are equivalent to $\frac{9}{20}$. **19.** D

Pages 107–108 Lesson 2-2 Extra Practice

21. $\frac{47}{100}$ **23.** $\frac{22}{25}$ **25.** 84% **27.** 72% **29.** 95% **31a.** 44% **31b.** 16% **31c.** 60% **31d.** 40% **33.** H **35.** 68.5 **37.** 325.5 **39.** $2.76

Pages 113–114 Lesson 2-3 Independent Practice

1. 0.35 **3.** 0.31 **5.** 22% **7.** 10% **9.** 0.04 **11.** 12%
13. C: $59.50, A: $70, B: $87.50 **15.** 0.88, 0.90, 0.92
17. Sample answer: Since $\frac{3}{4}$ is equal to 0.75, write $43\frac{3}{4}\%$ as 43.75%. Then change 43.75% to the decimal 0.4375.
19. D

Pages 115–116 Lesson 2-3 Extra Practice

21. 0.03 **23.** 0.11 **25.** 62% **27.** 87% **29.** 0.65 **31.** 82% **33.** D **35.** 0.25 **37.** = **39.** >
41. Aliah's brother

Pages 121–122 Lesson 2-4 Independent Practice

1. 3.5; $3\frac{1}{2}$ **3.** 0.0015; $\frac{3}{2{,}000}$ **5.** 250% **7.** 420% **9.** 850% **11.** 0.9% **13.** 140% **15.** 0.003; $\frac{3}{1{,}000}$; 3 out of every 1,000 people are Japanese. **17a.** 0.0005 **17b.** sulfur **19.** 30 mph **21.** B

Pages 123–124 Lesson 2-4 Extra Practice

23. 4; 4 **25.** 0.0004; $\frac{1}{2{,}500}$ **27.** 3,500% **29.** 0.77% **31.** 9.8% **33.** 0.0012 **35.** 125% **37.** 133% **39.** D
41. <
43.

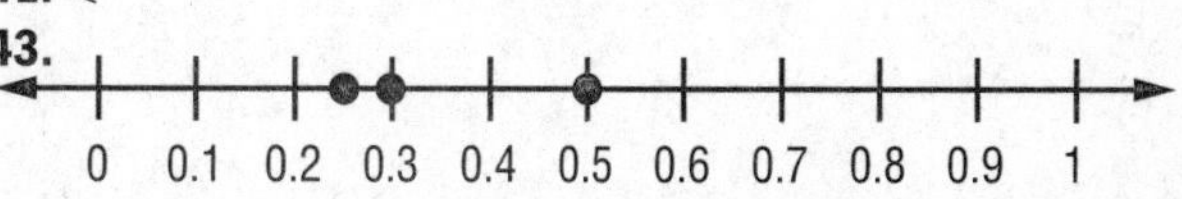

Page 127 Problem-Solving Investigation Solve a Simpler Problem

Case 3. 2 books **Case 5.** 21 bracelets

Pages 133–134 Lesson 2-5 Independent Practice

1. < **3.** < **5.** $\frac{1}{4}, \frac{1}{2}, \frac{2}{3}, \frac{5}{6}$ **7.** Alex; 0.35 < 0.40 **9a.** 0.20, 0.25, 0.20 **9b.** Two are the same; 0.25 is the greatest score. **11.** 8%, 17%, 0.2, $\frac{11}{20}$ **13.** $\frac{3}{9}, \frac{3}{8}$, and $\frac{3}{7}$; Because the numerators are the same, the larger the denominator, the smaller the fraction. **15.** A

Pages 135–136 Lesson 2-5 Extra Practice

17. = **19.** > **21.** $\frac{1}{2}, \frac{9}{16}, \frac{5}{8}, \frac{3}{4}$ **23.** 3%, $\frac{2}{50}$, 0.08 **25.** $\frac{1}{2}$, 0.55, $\frac{5}{7}$ **27.** B **29.** $\frac{1}{8}, \frac{3}{16}, \frac{11}{32}, \frac{3}{4}$ **31.** 4.29 **33.** $\frac{37}{50}$

Pages 141–142 Lesson 2-6 Independent Practice

1. $\frac{1}{2}$ of \$120 is \$60. **3.** $\frac{2}{5}$ of 15 is 6. **5.** 24 + 24 + 24 = 72 **7.** $\frac{3}{4}$ of 20 yr is 15 yr. **9a.** No; 40% is $\frac{2}{5}$. $\frac{2}{5}$ of 15 is 6. He needs 7 baskets to win a prize. **9b.** about 50% **11.** Sample answer 71 − 37 = 34 missed shots and $\frac{34}{71}$ is about $\frac{35}{70}$ or $\frac{1}{2}$. Since $\frac{1}{2}$ = 50%, he missed about 50% of his shots. **13.** about 75% **15.** More; Rachel rounded \$32 down to \$30, so the actual amount she will save will be more than \$12. **17.** Sample answer: First, round 42% to 40%, and \$122 to \$125. Next, rewrite 40% as $\frac{2}{5}$. Then find $\frac{1}{5}$ of \$125. Finally, multiply this result by 2 to find $\frac{2}{5}$ of \$125.

Pages 143–144 Lesson 2-6 Extra Practice

19. $\frac{1}{2}$ of 60 is 30. **21.** $\frac{1}{4}$ of 120 is 30. **23.** 19 + 19 + 19 = 57 **25.** 61 + 61 + 61 + 61 = 244 **27.** $\frac{3}{4}$ of 8 h is 6 h. **29.** about 423; 47 × 9 = 423 **31.** about 75% **33.** B **35.** G **37.** 0.07 **39.** 0.15 **41.** 0.06 **43.** 22.5 ft^2

Pages 151–152 Lesson 2-7 Independent Practice

1. 46 **3.** 92 **5.** 9 **7.** 336 **9.** \$8.40 **11.** No; 70% of 30 is 21, not 24. 80% of 30 is 24. **13.** Sample answers given for examples. Percent: equal to; 100%; less than; 25%; greater than; 125%. Fraction: equal to; $\frac{3}{3}$; less than; $\frac{1}{3}$; greater than; $\frac{4}{3}$ **15.** Yes; 16% of 40 is 6.4 and 40% of 16 is 6.4 **17.** C

Pages 153–154 Lesson 2-7 Extra Practice

19. 16.5 **21.** 161 **23.** 159.84 **25.** 0.45 **27.** 18 ounces of bleach **29.** 398.24 **31.** 212.85 **33.** C **35.** Sample answer: about 135 cars; Round 28% to 30%; 30% of 450 is 135. **37.** 90 **39.** 7.5 **41.** 4.8

43.

Trip	Part of Club
Carnegie Museum of Art	0.32
Fallingwater	0.20
Westemoreland Museum of American Art	0.48

Pages 159–160 Lesson 2-8 Independent Practice

1. 70

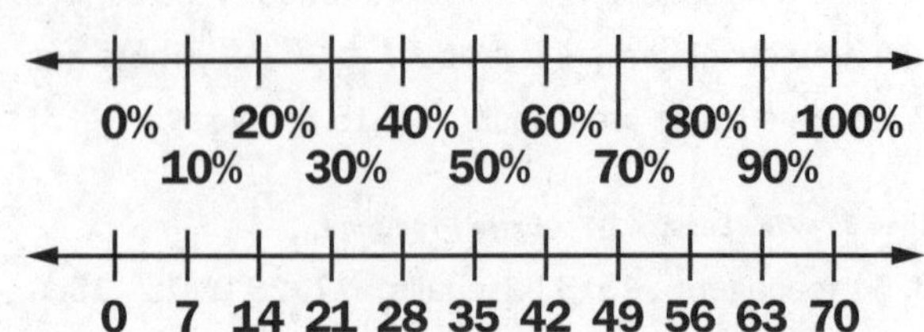

3. $\frac{22}{■} = \frac{44}{100}$; 50 **5.** \$50 **7.** 15 cups **9.** 20 cups **11.** Sample answer: $\frac{■}{25} = \frac{84}{100}$; 21 **13.** 18 karats; 24 is the whole and 75 is the percent, so $\frac{18}{24} = \frac{75}{100}$. **15.** B

Pages 161–162 Lesson 2-8 Extra Practice

17. 400

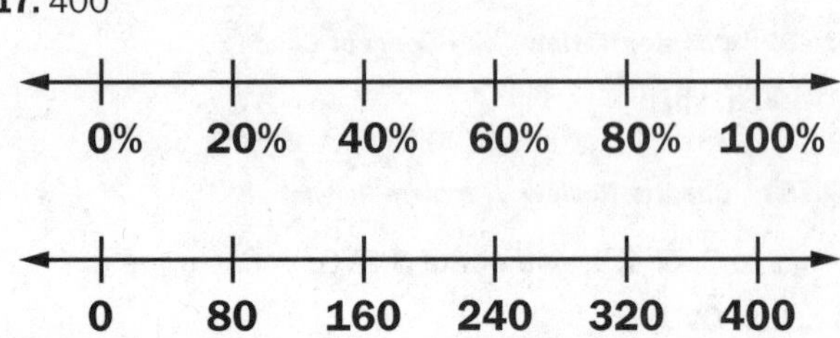

19. $\frac{270}{■} = \frac{90}{100}$; 300 **21.** 24 **23.** 300 students **25.** 240 ounces **27.** H **29.** 300 cans **31.** 49 **33.** 123 **35.** 16 **37.** 4.5 miles

Page 165 Chapter Review Vocabulary Check

1. percent **3.** percent proportion **5.** least common denominator

Page 166 Chapter Review Key Concept Check

1. not correct;

$$\frac{4}{5} \rightarrow 5\overline{)4.0} = 0.8, \quad 4.0 - 40 = 0$$

3. not correct;

$$120\% = \frac{120}{100} = \frac{6}{5} = 1\frac{1}{5}$$

Page 167 Chapter Review Problem Solving

1. savings: $\frac{2}{5}$; charity: $\frac{3}{20}$; shopping: $\frac{9}{20}$ **3.** $\frac{4}{5}$ **5.** Miguel **7.** 15 in.; Sample answer: $\frac{1}{4}$ of 60 is 15.

Chapter 3 Compute with Multi-Digit Numbers

Page 176 Chapter 3 Are You Ready?

1. 300 **3.** 1,078 **5.** 49 **7.** 4 million albums

Pages 181–182 Lesson 3-1 Independent Practice

1. 16.7 **3** 103.01 **5.** 80.02 **7** 1.73 s **9.** $24.85 **11.** Luis did not annex a zero before subtracting. 8.9 − 3.72 = 5.18 **13.** Sample answer: When you add the whole numbers the sum is 20. The sum of the decimals will be added on, which will make the sum greater than 20. **15.** 6; A zero is annexed to subtract from 8.5. So, you will subtract 10 − 4 to find the value in the hundredths place.

Pages 183–184 Lesson 3-1 Extra Practice

17. 7.9 **19.** 39.99 **21.** 14.82 **23.** $207.85 **25.** 34.15 degrees Celsius **27.** B **29.** 166.85 miles **31.** 34 **33.** Thursday

Pages 189–190 Lesson 3-2 Independent Practice

1. 30; 10 × 3 = 30 **3** 160; 20 × 8 = 160 **5.** 240; 30 × 8 = 240 **7.** about 80 million tons **9** More; her wage and hours worked were rounded up, so the actual total is less than the estimate. **11a.** Raj needs to save $132 more. **11b.** No; $6 × 20 = $120 **13.** about $12 **15.** B

Pages 191–192 Lesson 3-2 Extra Practice

17. 30 × 80 = 2,400 **19.** about 90 **21.** yes **23.** no; 32 **25.** Calories: 400; Vitamin C: 400 mg; carbohydrates: 100 g; calcium: 80 mg **27.** yes; 4 × $3.69 ≈ 4 × $4 = $16 **29.** B **31.** 345 **33.** $4.37

Pages 197–198 Lesson 3-3 Independent Practice

1. 8.4 **3.** 2.6 **5.** 0.06 **7** $215.27 **9** 134.6°F; Sample answer: 1,346 × 10 = 13,460. Since 13.46 has 2 decimal places, 13.46 × 10 = 134.60. **11.** Sample answer: 32 mm; 20 × 1.95 = 39; 4 × 1.75 = 7; 39 − 7 = 32 **13.** Sample answer: First evaluate 1.17 × 100 to be 117. Then, multiply 117 by 5.4 to get the answer of 631.8. Or first evaluate 5.4 × 100 to be 540. Then multiply 540 by 1.17 to get the answer of 631.8. Or first evaluate 5.4 × 10 to be 54 and 1.17 × 10 to be 11.7. Then multiply 54 by 11.7 to get the answer of 631.8. **15.** No; zero represents the number of hundredths and should be counted.

Pages 199–200 Lesson 3-3 Extra Practice

17. 8.5 **19.** 19.2 **21.** 0.084 **23.** 2.24 g **25.** 93.5 in^2 **27.** 7.44 miles **29.** 4,515 square inches **31.** 6 **33.** 13 **35.** 5; 3; 35

Pages 205–206 Lesson 3-4 Independent Practice

1. 0.28 **3** 1.092 **5.** 167.0067 **7** 84.474 ft; 46.93 × 1.8 ≈ 45 × 2 = 90; 84.474 ≈ 90 **9** $5.76; Each price is about $1. He bought about 6 pounds of fruit. 6 × 1 = 6 ≈ $5.76 **11.** 1.03515

13.

×	2	0.2	0.02	0.002
3	6	0.6	0.06	0.006
0.3	0.6	0.06	0.006	0.0006
0.03	0.06	0.006	0.0006	0.00006
0.003	0.006	0.0006	0.00006	0.000006

Sample answer: The factor 0.002 has three decimal places and the factor 0.003 has three decimal places. So, the product will have six decimal places. **15.** 32.013341...; Sample answer: 3.9853 × 8.032856 rounds to 4 × 8 = 32, so the answer must be about 32. **17.** greater than 0.4; It is being multiplied by a decimal greater than 1. **19.** D

Pages 207–208 Lesson 3-4 Extra Practice

21. 2.48 **23.** 16.128 **25.** 0.02255 **29a.** Junnie **29b.** 0.62 mile farther **31.** G **33.** 7.275 mi **35.** 12 **37.** 7 boxes

Page 213 Problem-Solving Investigation Look for a Pattern

Case 3. $47.70 **Case 5.** 30.5, 39, 48.5; Add 3.5; then add 4.5; then add 5.5, and so on.

Pages 219–220 Lesson 3-5 Independent Practice

1. 29 **3.** 15 **5** 130 R30 **7.** 170 **9** 60 mi **11.** 175 names **13.** 144 cups **15.** Sample answer: Mary saved $2,400 in 12 months. What was the average amount she saved each month?; $200 **17.** No; Sample answer: if the remainder equals the divisor, then the quotient should be increased by 1. **19.** B

Pages 221–222 Lesson 3-5 Extra Practice

21. 57 R3 **23.** 10 R21 **25.** 166 R24 **27.** 224 R1 **29.** 845 cups **31.** B **33.** H **35.** 60 **37.** 700 **39.** 10 vans

Pages 227–228 Lesson 3-6 Independent Practice

1. 33 ÷ 3 = 11 **3.** 36 ÷ 12 = 3 **5** Sample answer: about 3 **7** about 6 gal; 53 ÷ 8.5 ≈ 54 ÷ 9 = 6 **9.** 1; 2; 4; 8 **11.** Sample answer: 160.23 ÷ 6.54 **13.** B

Pages 229–230 Lesson 3-6 Extra Practice

15. 46 ÷ 23 = 2 **17.** about 7 inches; 45.9 ÷ 7 ≈ 49 ÷ 7 = 7 **19** $474.72 ÷ 12 ≈ $480 ÷ 12 = $40 **21a.** 5; 100 × 5 = 500 **21b.** 10 **23.** about 12 children **25.** 490 ÷ 70 = 7 **27.** 0.147 **29.** 7.3456; 0.73456; 0.073456; Sample answer: Divide by ten to move the decimal point one place value to the left.

Selected Answers

Pages 235–236 Lesson 3-7 Independent Practice

1. 13.1 **3.** 23.7 **5.** 1.2 **7.** \$770.56 **9.** 22.8 ft; Area of a rectangle is length times width, so divide the area by the length to find the width. $752.4 \div 33 = 22.8$ **11.** Brand B; The cost of each bottled water for Brand B is about \$0.44. For Brand A the cost is about \$0.58 and for Brand C the cost is about \$0.46. So Brand B has the best cost per bottle. **13.** Amanda placed the decimal point in the wrong place of the quotient. $11.2 \div 14 = 0.8$ **15.** Since $40 \div 20 = 2$, the answer is about 2.

Pages 237–238 Lesson 3-7 Extra Practice

17. 18.4 **19.** 1.6 **21.** 1.9 **23.** \$3.75 **25.** Dominoes; The cost of each domino set is about \$0.66. The cost of each peg game is about \$0.83, and the cost of each mini football is about \$0.75. So, the domino set has the best cost per toy. **27.** \$62.46 **29.** 83 **31.** 31.1 **33.** 8.8 kg

Pages 243–244 Lesson 3-8 Independent Practice

1. 3.6 **3.** 250 **5.** 450 **7.** 20 steps **9a.** 24 h **9b.** 21.12 h **11** **a.** 2.2 times **b.** 3.8 times **13.** $49 \div 7$; the quotient is 7 and all of the other problems have a quotient of 0.7.

Pages 245–246 Lesson 3-8 Extra Practice

15. 0.2 **17.** 0.0492 **19.** 420 **21.** 6 pieces **23.** about 4.4 times **25.** D **27.** I **29.** $<$ **31.** $>$ **33.** $\frac{7}{12}$ of her free time

Page 249 Chapter Review Vocabulary Check

Across

1. compatible numbers **5.** decimal

Down

3. multidigit **5.** dividend

Page 249 Chapter Review Key Concept Check

Across

1. 483 **3.** 178 **5.** 21 **9.** 4930 **13.** 203

Down

1. 40 **3.** 108 **5.** 239 **7.** 72 **9.** 463 **11.** 880

Page 251 Chapter Review Problem Solving

1. 60 mph $\times$ 3 h = 180 mi **3.** 73.08 ft^2 **5.** 186,000 **7.** about 4.5 feet; 5.75×0.8 is about 6×0.8 or 4.8, which is closest to 4.5

Chapter 4 Multiply and Divide Fractions

Page 256 Chapter 4 Are You Ready?

1. 12 **3.** 6 **5.** $4\frac{11}{21}$ **7.** $6\frac{7}{8}$ in.

Pages 261–262 Lesson 4-1 Independent Practice

1. $\frac{1}{4} \times 20 = 5$

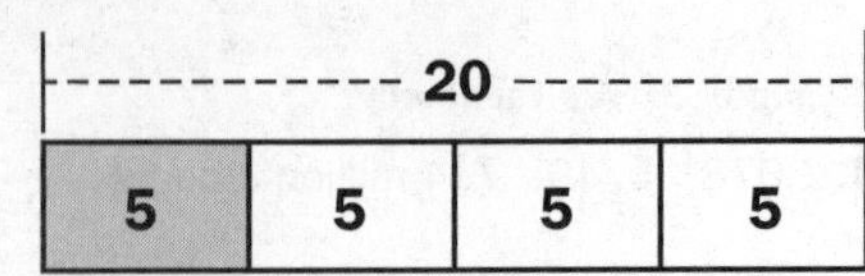

3. $1 \times 0 = 0$ **5.** $12 \times \frac{1}{4} = 3$ pizzas **7a.** about 20 lb **7b.** about \$75 **9.** Sample answer: $10 \times 3 = 30$ in^2 **11.** Sample answer: $\frac{5}{9}$; $8\frac{1}{2}$ is about $\frac{5}{9} \times 9 = 5$. **13.** C

Pages 263–264 Lesson 4-1 Extra Practice

15. 14 **17.** $\frac{2}{9} \times 90 = 20$ **19.** $\frac{1}{2} \times 4 = 2$ **21.** $8 \times 1 = 8$ cm^2 **23.** 23 movies **25.** B **27.** C **29.** $\frac{1}{2}$ **31.** 0

33.

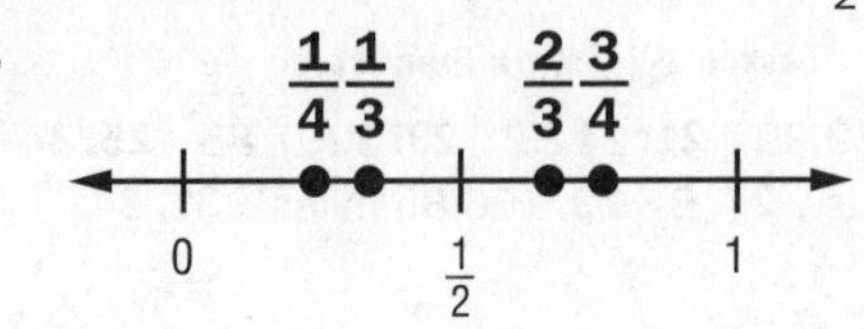

35. 120 square feet

Pages 269–270 Lesson 4-2 Independent Practice

1. 15 **3.** 2 **5.** $\frac{22}{5}$ or $4\frac{2}{5}$ **7.** $2\frac{2}{5}$ in. **9.** neither; $\frac{4}{5} \times 30 = 24$ and $\frac{2}{3} \times 36 = 24$. So, $24 = 24$. **11.** seventh **13.** He multiplied by $\frac{8}{8}$ instead of multiplying by $\frac{8}{1}$. $\frac{3}{4} \times \frac{8}{1} = \frac{24}{4}$ or 6. **15.** C

Pages 271–272 Lesson 4-2 Extra Practice

17. 6 **19.** $\frac{7}{5}$ or $1\frac{2}{5}$ **21.** $\frac{25}{2}$ or $12\frac{1}{2}$ **23.** 146 days **25.** B **27.** 6 yd **29.** 286 **31.** 153 **33.** 3 inches

Pages 277–278 Lesson 4-3 Independent Practice

1. $\frac{2}{15}$ **3.** $2\frac{2}{3}$ **5.** $\frac{1}{6}$ **7.** $\frac{3}{8}$ **9.** $\frac{3}{10}$

11a. Sample answer: Olivia withdrew $\frac{3}{4}$ of her savings. She used $\frac{1}{5}$ of what was left to buy a book. If she had \$100 in savings, how much did she spend on the book?

11b.

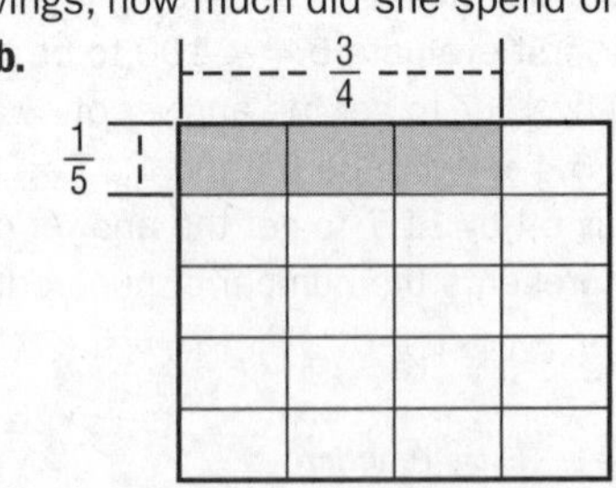

11c. Sample answer: Multiply $\frac{1}{5} \times \frac{1}{4}$. Multiply the product, $\frac{1}{20}$ by \$100. She spent \$5 on a book. **13.** Sample answer: $a = \frac{3}{8}$ and $b = \frac{5}{7}$; $a = \frac{5}{8}$ and $b = \frac{3}{7}$; $a = \frac{5}{14}$ and $b = \frac{3}{4}$ **15.** C

Pages 279–280 Lesson 4-3 Extra Practice

17. $\frac{6}{35}$ **19.** $4\frac{1}{8}$ **21.** $\frac{1}{3}$ **23.** Nyemi: 138; Luke: 69; Natalie: 23 **25.** $\frac{1}{2}$ c; $\frac{3}{4} \times \frac{2}{3} = \frac{3 \times 2}{4 \times 3} = \frac{1}{2}$ **27.** H **29.** 648 **31.** 1,320 **33.** A; The product must be less than 5 because the factors are 5 and a fraction less than 1.

Pages 285–286 Lesson 4-4 Independent Practice

1. $1\frac{1}{6}$ **3.** $2\frac{27}{32}$ **5.** 9 **7.** $9\frac{1}{4}$ mi **9.** $3\frac{3}{8}$ c
11a. about $69\frac{27}{40}$ million mi **11b.** about $139\frac{7}{20}$ million mi **11c.** about $487\frac{29}{40}$ million mi **11d.** about $882\frac{11}{20}$ million mi **13.** $1\frac{1}{24}$ **15.** B; the product must be greater than $\frac{2}{3}$ and less than $2\frac{1}{2}$.

Pages 287–288 Lesson 4-4 Extra Practice

17. $2\frac{1}{8}$ **19.** $\frac{17}{20}$ **21.** $12\frac{3}{4}$ **23.** $1,259\frac{1}{4}$ in^2
25. 31 inches **27.** D **29.** H **31.** 4 **33.** 8 **35.** 3
37.

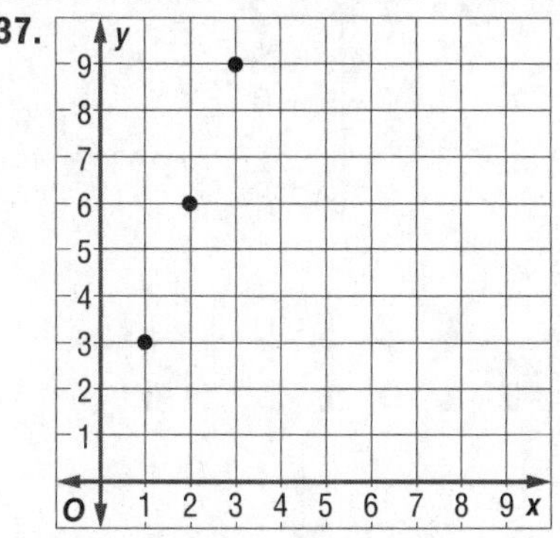

Pages 293–294 Lesson 4-5 Independent Practice

1. 6 **3.** 52 **5.** $2\frac{1}{2}$ **7.** 1,500 lb **9.** $4\frac{4}{5}$ oz
11a. The *x*-value represents the number of quarts and the *y*-value represents the equivalent number of gallons. **11b.** Sample answer: The point on the line whose *y*-value is equal to 2.5 is (10, 2.5), so 10 qt = 2.5 gal. **13.** 16 in. is equivalent to 1 ft 4 in.; $1\frac{1}{2}$ ft is equivalent to 1 ft 6 in.; So, 16 in. < $1\frac{1}{2}$ ft. **15.** Sample answer: 5 pt; 80 fl oz

Pages 295–296 Lesson 4-5 Extra Practice

17. $4\frac{1}{2}$ **19.** 24 **21.** $6\frac{1}{2}$ **23.** 3,520 ft **25.** No; 15 in. + $4\frac{1}{2}$ in. + $6\frac{3}{4}$ in. = $26\frac{1}{4}$ in.; $2\frac{1}{2}$ ft = 30 in.; So, $26\frac{1}{4}$ in. < 30 in. **27.** $\frac{1}{2}$ **29.** D **31.** D **33.** 39 **35.** 7.5 **37.** $41\frac{2}{3}$ feet

Page 299 Problem-Solving Investigation Draw a Diagram

Case 3.

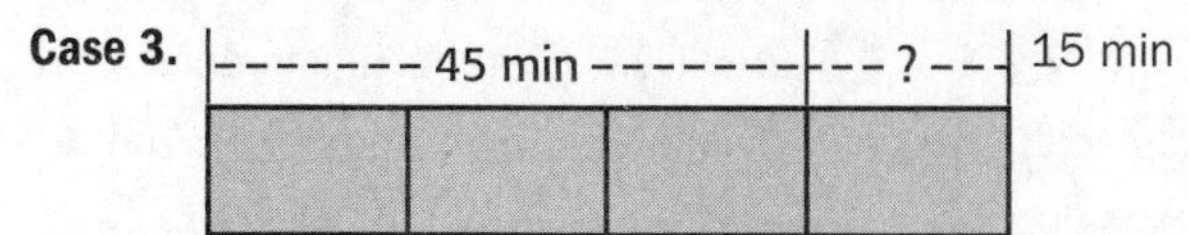

Case 5.

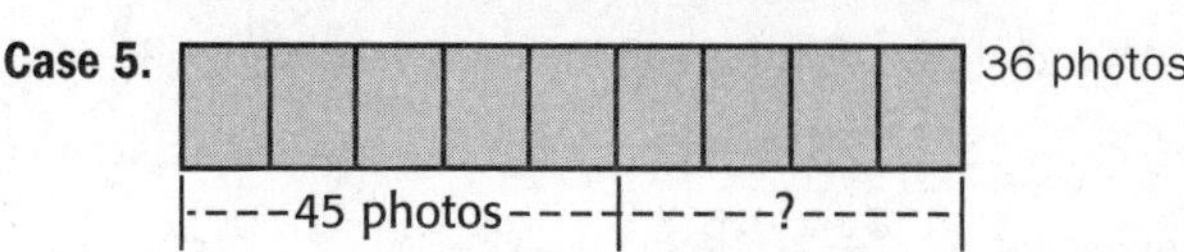

Pages 309–310 Lesson 4-6 Independent Practice

1. $\frac{5}{3}$ **3.** 1 **5.** $6\frac{2}{3}$ **7.** 10 **9.** $4\frac{1}{2}$ **11.** 110 horses
13. 6 activities; $4 \div \frac{2}{3} = 4 \times \frac{3}{2} = \frac{12}{2} = 6$ **15.** Daniella did not multiply by the reciprocal of 4, which is $\frac{1}{4}$. $\frac{8}{9} \div 4 = \frac{8}{9} \times \frac{1}{4} = \frac{8}{36}$ or $\frac{2}{9}$ **17.** D

Pages 311–312 Lesson 4-6 Extra Practice

19. $\frac{9}{7}$ **21.** $3\frac{1}{3}$ **23.** $3\frac{3}{5}$ **25.** $7\frac{1}{5}$ **27.** 8 dinners **29.** A
31. 16 bags; He can only fill another $\frac{2}{3}$ of a bag, so he can't make 17 bags. **33.** 12 **35.** 15 **37.** 16
39. 84 people

ages 321–322 Lesson 4-7 Independent Practice

1. $\frac{1}{4}$ **3.** $\frac{1}{12}$ **5.** $\frac{1}{24}$ **7.** Sample answer: David has $\frac{5}{6}$ foot of tape. He uses $\frac{1}{12}$ foot of tape to hang each photo on the bulletin board. How many photos can he hang on the bulletin board? 10 photos

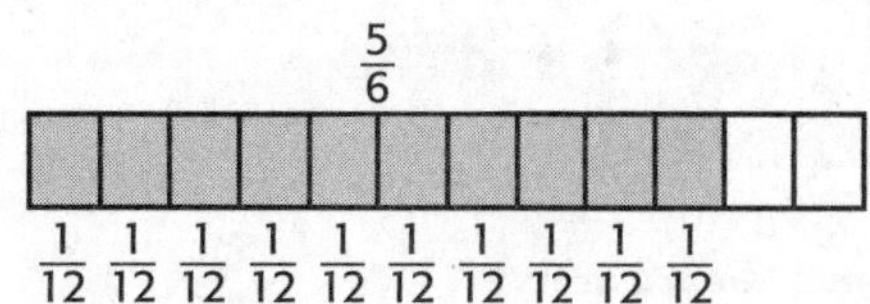

9. $\frac{3}{4} \div \frac{3}{8} = 2$; 2 T-shirts
11.

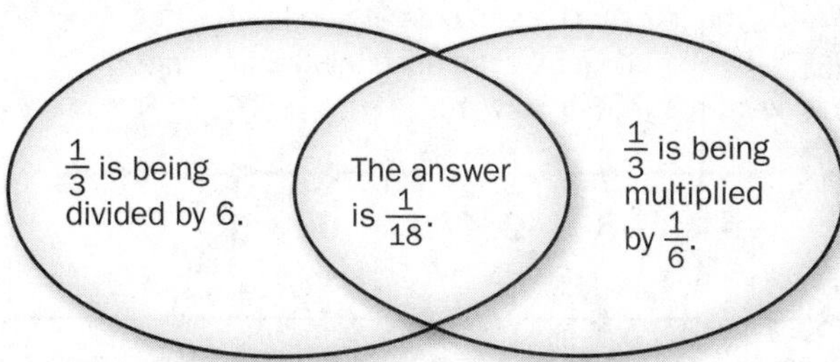

13. greater than 1; the dividend is greater than the divisor; less than 1; the dividend is less than the divisor **15.** D

Pages 323–324 Lesson 4-7 Extra Practice

17. $\frac{1}{20}$ **19.** $\frac{1}{7}$ **21.** $\frac{5}{12}$ **23.** $\frac{1}{10} \div 4 = \frac{1}{40}$; $\frac{1}{40}$ kilometer
25. $\frac{7}{8} \div \frac{1}{24} = 21$; 21 cycles **27.** D **29.** I **31.** 4 **33.** 4
35. 4 **37.** 2 feet

Pages 329–330 Lesson 4-8 Independent Practice

1. $\frac{5}{12}$ **3.** $\frac{2}{3}$ **5.** 28 slices **7.** fraction; improper; reciprocal; fractions **9.** less than; Sample answer: The expression $5\frac{1}{6} \div 3\frac{5}{8}$ represents $5\frac{1}{6}$ being divided into a greater number of parts than the expression $5\frac{1}{6} \div 2\frac{2}{5}$. If $5\frac{1}{6}$ is divided into a greater number of parts, each part will be smaller. So, $5\frac{1}{6} \div 3\frac{5}{8} < 5\frac{1}{6} \div 2\frac{2}{5}$.

Pages 331–332 Lesson 4-8 Extra Practice

11. $2\frac{3}{4}$ **13.** $2\frac{2}{3}$ **15.** 2 **17.** 2 **19.** $2\frac{1}{10}$ **21.** $6\frac{3}{8} \div \frac{3}{8}$; 17 bags **23.** D **25.** 20 books; Twenty books weigh 25 pounds and twenty-one books weigh $26\frac{1}{4}$ pounds, so the bag can hold 20 books. **27.** $\frac{6}{7}$ **29.** $\frac{1}{8}$ **31.** $\frac{1}{2}$ **33.** 6 groups

Page 335 Chapter Review Vocabulary Check

1. mixed number **3.** reciprocal **5.** denominator **7.** unit ratio **9.** Compatible numbers

Page 336 Chapter Review Key Concept Check

1. not correct; $13 \times \frac{1}{3} = \frac{13}{5}$ or $4\frac{1}{3}$ **3.** correct **5.** correct

Page 337 Chapter Review Problem Solving

1. 120 in^2 **3.** $\frac{7}{16}$ **5.** $\frac{1}{20}$ **7.** 8,800 yd **9.** $1\frac{2}{5}$ gallons

Chapter 5 Integers and the Coordinate Plane

Page 342 Chapter 5 Are You Ready?

1. = **3.** < **5.** > **7.** peanuts

Pages 349–350 Lesson 5-1 Independent Practice

1. −3; The integer 0 represents at sea level.
3. −5; The integer 0 represents neither moving backward nor moving forward.

5.

7.

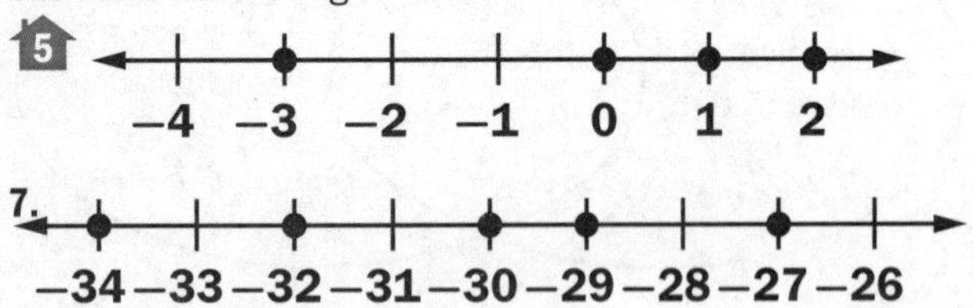

9. Sample answers are given.

Positive Integer	Negative Integer
• gain • above • earn • +	• lose • below • spend • −

11. Negative; Sample answer: A drop of 15° would result in a temperature of 0°F. Since the drop of 20° is greater than 15°, the temperature is below zero and will be represented by a negative integer. **13.** D

Pages 351–352 Lesson 5-1 Extra Practice

15. −25; The integer 0 represents neither spending nor earning.

17.

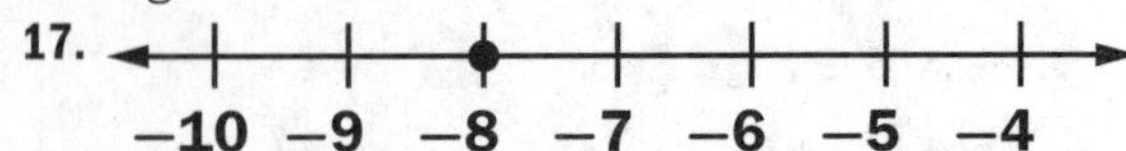

19. **21.**

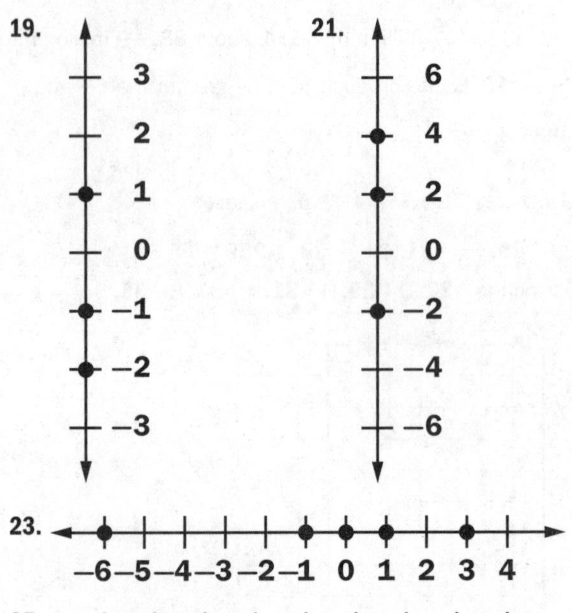

23.

25.

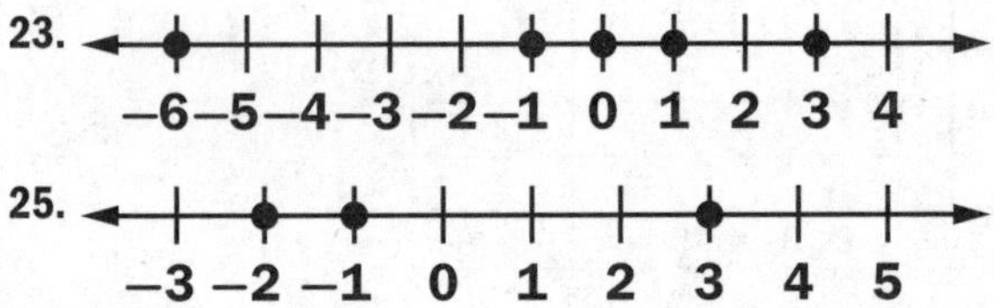

27. The integer 0 represents neither owing money nor having money. **29.** < **31.** > **33.** < **35.** 104 raffle tickets

Pages 359–360 Lesson 5-2 Independent Practice

1. −6 **3.** 0 **5.** −9 **7.** 14 **9.** 21 **11.** 4 feet **13.** 70°F **15.** 14 **17.** Absolute value cannot be a negative number. So, the absolute value of −14 is 14, not −14. **19.** Never; distance cannot be negative. **21.** Absolute value is distance and distance cannot be negative. **23.** C

Pages 361–362 Lesson 5-2 Extra Practice

25. −15 **27.** 9 **29.** −8 **31.** 0 **33.** 15 **35.** 11 **37.** 9 yards **39.** −10 **41.** B **43.** D **45.** Acetone **47.** > **49.** $8\frac{3}{4}$ tsp

Pages 367–368 Lesson 5-3 Independent Practice

1. > **3.** > **5.** −9 < 26; The temperature in Flagstaff, Arizona, was warmer. **7.** −79, −55, 18, 44, 101, 143 **9a.** Sun **9b.** Sun, 100-Watt Bulb, Full Moon, Venus, Andromeda Galaxy, Alpha Centauri **9c.** −27 **11.** Sample answer: Elise owes her brother $15. Jacob has $7. Elise has less money than Jacob. **13.** $-\frac{12}{4}, -\frac{1}{2}, \frac{1}{6}, \frac{7}{8}$, and $\frac{5}{2}$

Pages 369–370 Lesson 5-3 Extra Practice

15. $<$ **17.** $<$ **19.** $-20 > -25$; Michael owes less money than Yvonne. **21.** −221, −89, −71, −10, 54, 63

23a. *B C A D*

−5 −4 −3 −2 −1 0 1 2 3 4 5 6 7

23b. D; Since $-4 < -3 < 2 < 6$, player D had the most strokes over par. **25.** F **27.** 0.75 **29.** 0.15 **31.** Mitchell

Page 373 Problem-Solving Investigation Work Backward

Case 3. 1,220 meters **Case 5.** 3

Pages 383–384 Lesson 5-4 Independent Practice

1 $0.4\overline{6}$ **3.** $-0.\overline{6}$ **5.** $3.3\overline{409}$ **7.** $0.\overline{34}$ **9.** $-\frac{9}{10}$
11. $-3\frac{4}{5}$ **13.** $\frac{4}{13}$ **15** $0.\overline{7}$ **17a.** 43 **17b.** $\frac{24}{43}$; 0.558
19. $\frac{17}{36}$ is not a terminating decimal since decimals are based on powers of 10 and 36 is not a factor of any power of 10. **21.** B

Pages 385–386 Lesson 5-4 Extra Practice

23. $0.\overline{27}$ **25.** −0.7 **27.** $-1.\overline{80}$ **29.** $-\frac{3}{20}$ **31.** $-12\frac{27}{50}$
33. B **35.** $-\frac{5}{4}$ **37.** $>$ **39.** $>$ **41.** $>$
43. $47.394 > 47.362$

Pages 391–392 Lesson 5-5 Independent Practice

1. $>$ **3** $=$ **5** $-2\frac{3}{4}, -2.\overline{2}, 2.8, 3\frac{1}{8}$ **7.** $-4\frac{1}{2}, -2\frac{3}{8}$, 1.35, 5.6 **9** −10.8, −9.7, 9.0, 11.4 **11.** always; The greater a number is, the farther away from zero. Therefore, its opposite will also be farther from zero. **13.** The first decimal is a terminating decimal, so its thousandths place is zero. The second decimal has a repeating digit of 3 so its thousandths place is 3. $-0.330 > -0.\overline{333}$ **15.** D

Pages 393–394 Lesson 5-5 Extra Practice

17. $>$ **19.** $>$ **21.** $<$ **23.** $1\frac{1}{5}$, 1.25, $1.2\overline{5}$, $1\frac{3}{4}$ **25.** $<$
27. $>$ **29.** A **31.** D **33.** −6.25, −2.45, 5.50, 7.80
35–41.

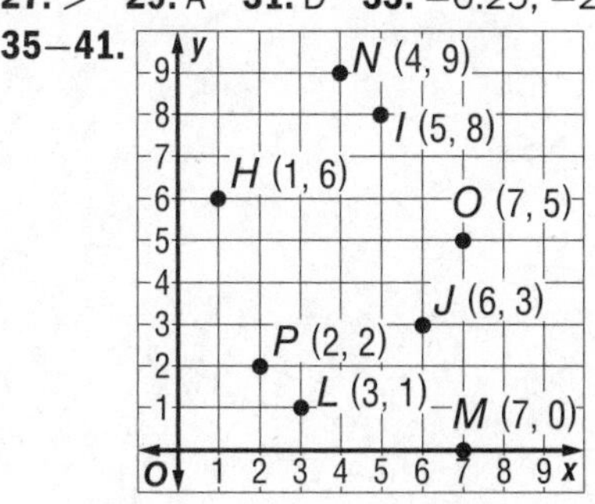

Pages 399–400 Lesson 5-6 Independent Practice

1. (2, 2); I **3.** (−4, 2); II **5** (5, 0); none **7** *Z*; II **9.** *A*; IV **11.** *N*; none **13a.** The Clock **13b.** the Wonder Wheel; (2, −4) **13c.** the Big Coaster; (−3, 1) **13d.** (−1, −2) **15.** Quadrants I and III; Sample answer: In Quadrant I, both coordinates are positive and in Quadrant III, both coordinates are negative. **17.** Sample answer: The first coordinate tells the location in relation to the *y*-axis. The second coordinate tells the location in relation to the *x*-axis. Any point is defined by only one ordered pair.

Pages 401–402 Lesson 5-6 Extra Practice

19. (1, 3); I **21.** (−2, 1); II **23.** (−4, −5); III **25.** *L*; II **27.** *S*; IV **29.** *B*; III **31a.** (4, 2) **31b.** (4, 4) **31c.** (−1, −4) **33.** G
35.

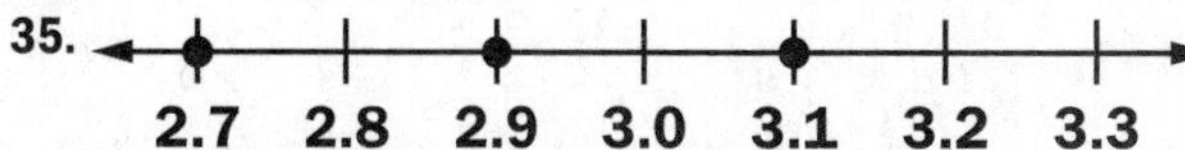

37. 19 magazines

Pages 407–408 Lesson 5-7 Independent Practice

1–8.

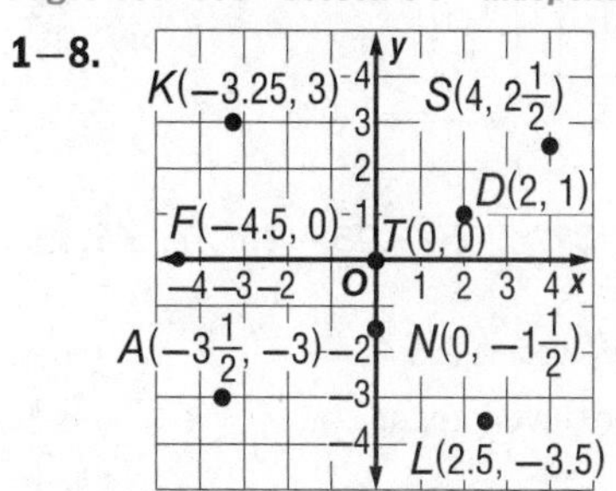

9–11.

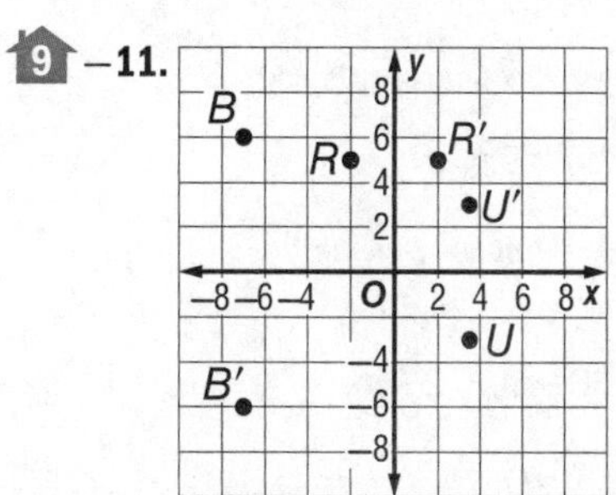

13. (4.25, −1.75)
15

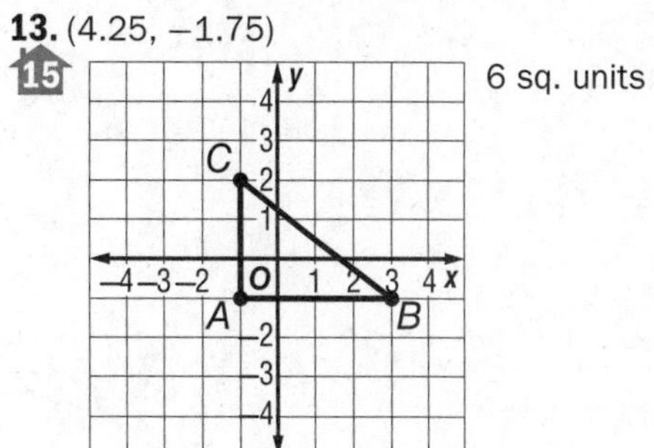

6 sq. units

17. Sample answer: (7, 2), (−5, 2) **19.** always; The *y*-coordinate will be the opposite of the original following the reflection across the *x*-axis. The *x*-coordinate will be the opposite of the original following the reflection across the *y*-axis.

Pages 409–410 Lesson 5-7 Extra Practice

21–28.

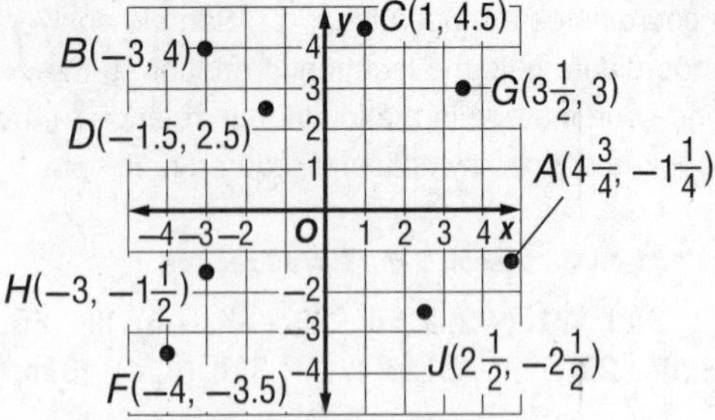

29–31.

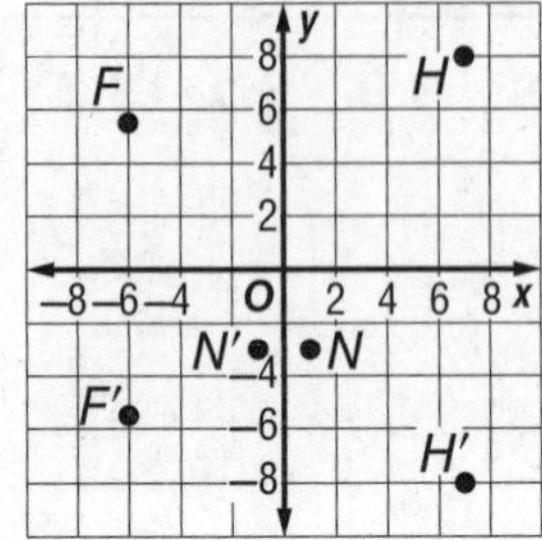

33. (4.75, 2.25) **35.** D **37.** (2.5, −3.25) **39.** 27 **41.** 25

Page 417 Chapter Review Vocabulary Check

1. rational number **3.** positive integer
5. terminating decimal

Page 418 Chapter Review Key Concept Check

1. 4 **3.** x-coordinate **5.** 6.543

Page 419 Chapter Review Problem Solving

1. 6 **3.** 20 **5.** library
7. $-\frac{109}{2}$ in., $-\frac{203}{4}$ in., $66\frac{1}{3}$ in., $72\frac{5}{8}$ in.

Index

Dd

Ee

Nn

Name ____

0 1 2 3 4 5 6 7 8 9

−11 −10 −9 −8 −7 −6 −5 −4 −3 −2 −1 0 1 2 3 4 5 6 7 8 9 10 11

Name ____________________

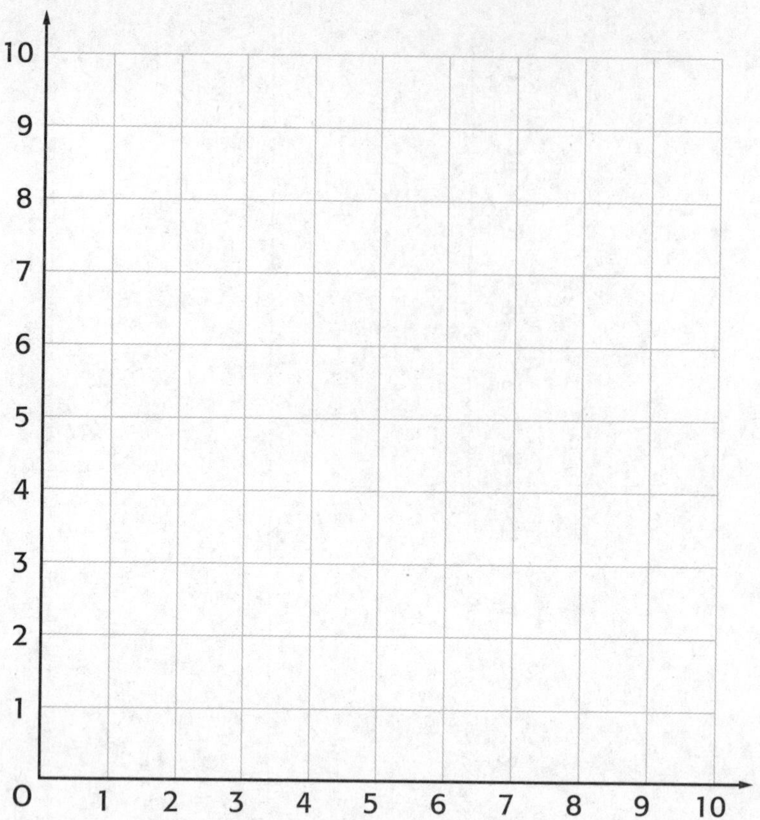

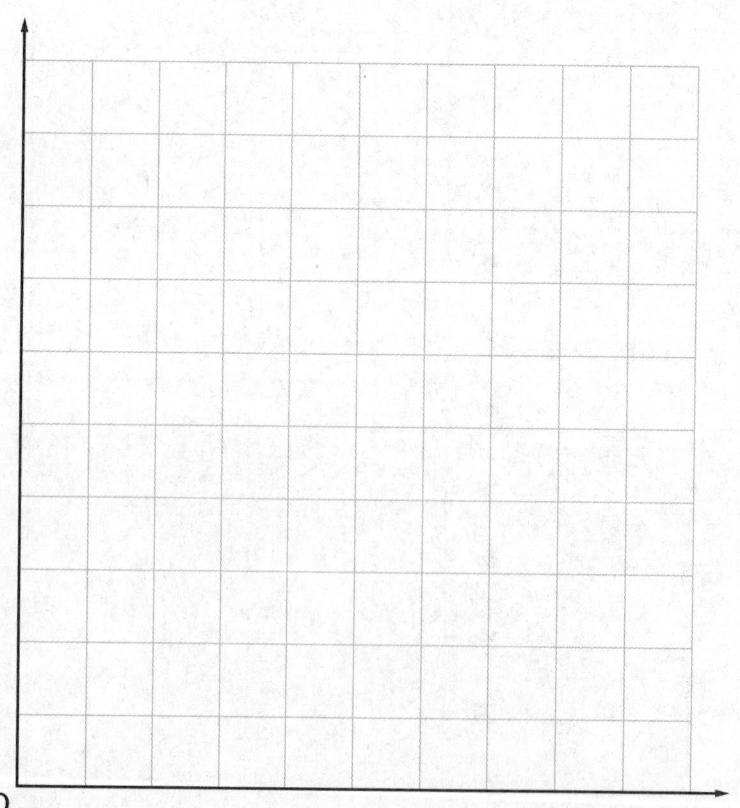

Name ____

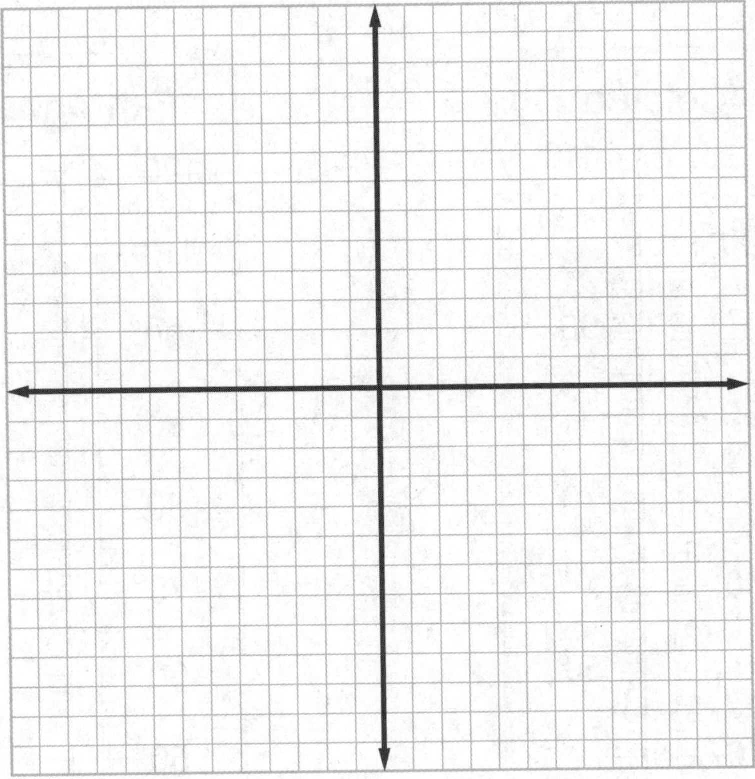

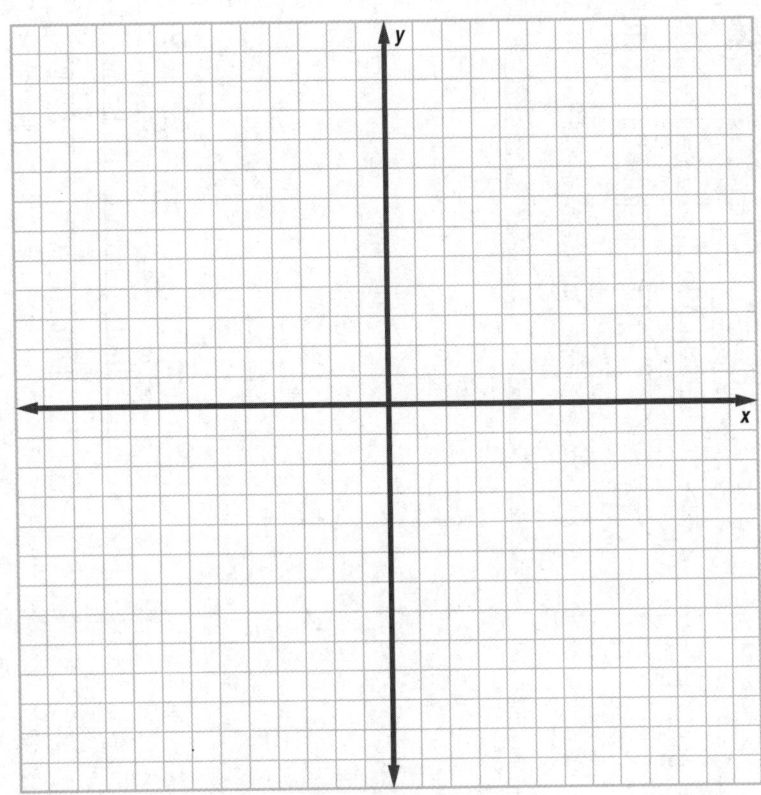

Name ____________________

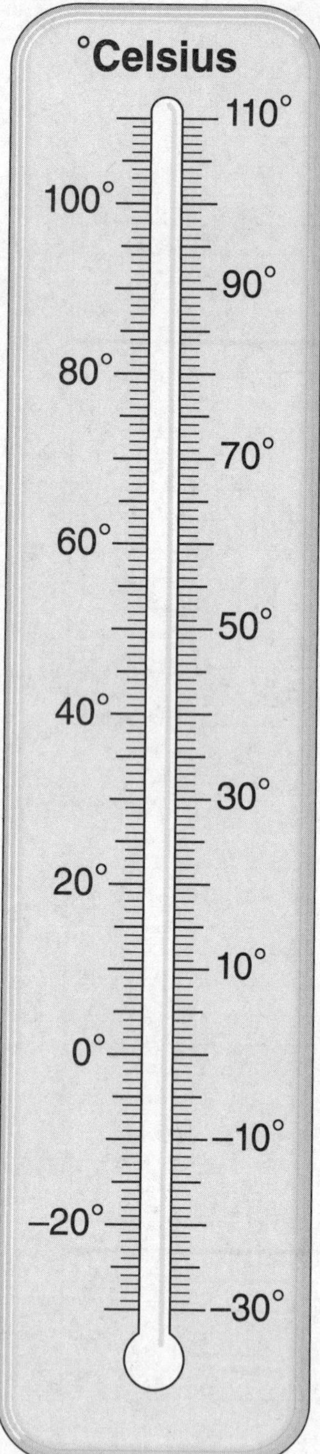
°Celsius
110°
100°
90°
80°
70°
60°
50°
40°
30°
20°
10°
0°
–10°
–20°
–30°

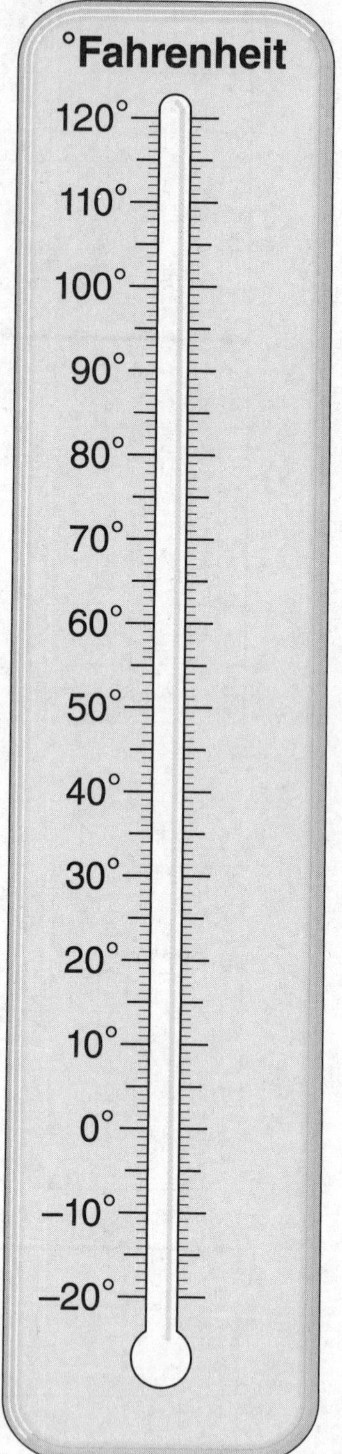
°Fahrenheit
120°
110°
100°
90°
80°
70°
60°
50°
40°
30°
20°
10°
0°
–10°
–20°

What Are Foldables and How Do I Create Them?

Foldables are three-dimensional graphic organizers that help you create study guides for each chapter in your book.

Step 1 Go to the back of your book to find the Foldable for the chapter you are currently studying. Follow the cutting and assembly instructions at the top of the page.

Step 2 Go to the Key Concept Check at the end of the chapter you are currently studying. Match up the tabs and attach your Foldable to this page. Dotted tabs show where to place your Foldable. Striped tabs indicate where to tape the Foldable.

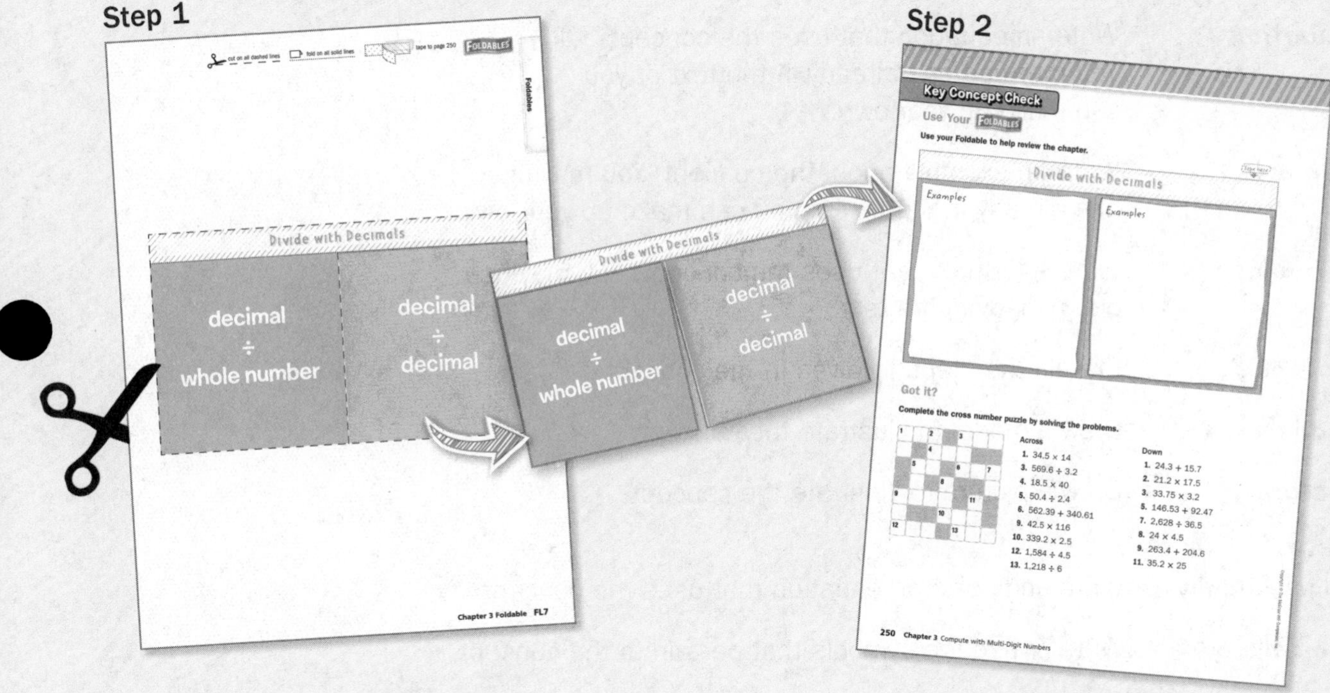

How Will I Know When to Use My Foldable?

When it's time to work on your Foldable, you will see a Foldables logo at the bottom of the **Rate Yourself!** box on the Guided Practice pages. This lets you know that it is time to update it with concepts from that lesson. Once you've completed your Foldable, use it to study for the chapter test.

How Do I Complete My Foldable?

No two Foldables in your book will look alike. However, some will ask you to fill in similar information. Below are some of the instructions you'll see as you complete your Foldable. **HAVE FUN** learning math using Foldables!

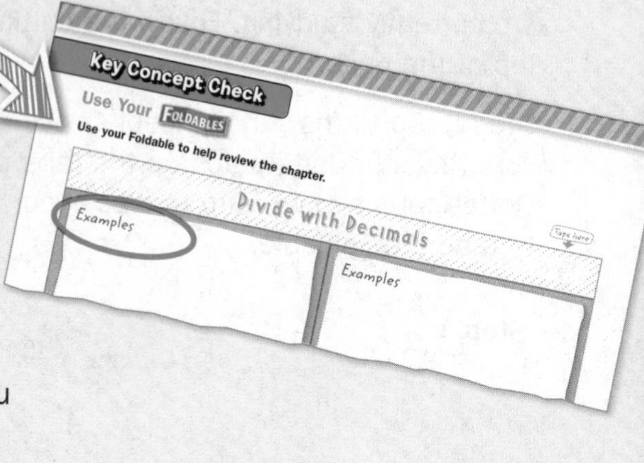

Instructions and what they mean

Best Used to...	Complete the sentence explaining when the concept should be used.
Definition	Write a definition in your own words.
Description	Describe the concept using words.
Equation	Write an equation that uses the concept. You may use one already in the text or you can make up your own.
Example	Write an example about the concept. You may use one already in the text or you can make up your own.
Formulas	Write a formula that uses the concept. You may use one already in the text.
How do I ...?	Explain the steps involved in the concept.
Models	Draw a model to illustrate the concept.
Picture	Draw a picture to illustrate the concept.
Solve Algebraically	Write and solve an equation that uses the concept.
Symbols	Write or use the symbols that pertain to the concept.
Write About It	Write a definition or description in your own words.
Words	Write the words that pertain to the concept.

Meet Foldables Author Dinah Zike

Dinah Zike is known for designing hands-on manipulatives that are used nationally and internationally by teachers and parents. Dinah is an explosion of energy and ideas. Her excitement and joy for learning inspires everyone she touches.

cut on all dashed lines fold on all solid lines tape to page 82

FOLDABLES®

Foldables

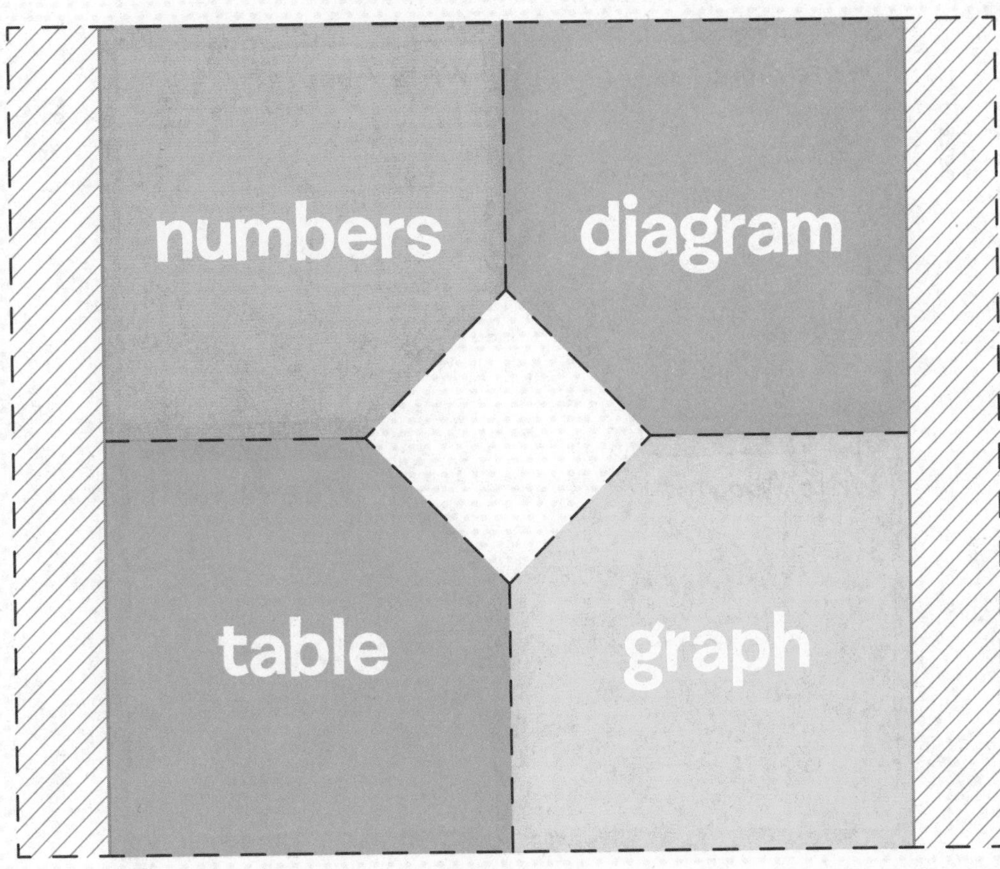

cut on all dashed lines　fold on all solid lines　tape to page 82　FOLDABLES®

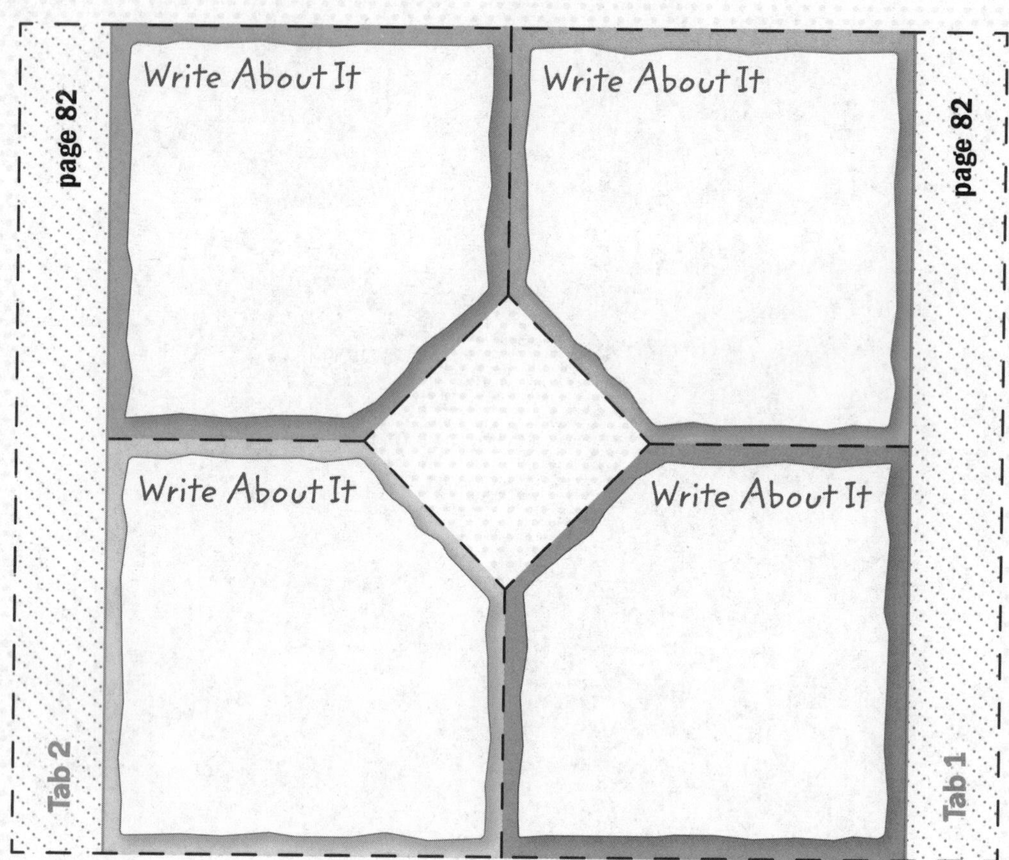

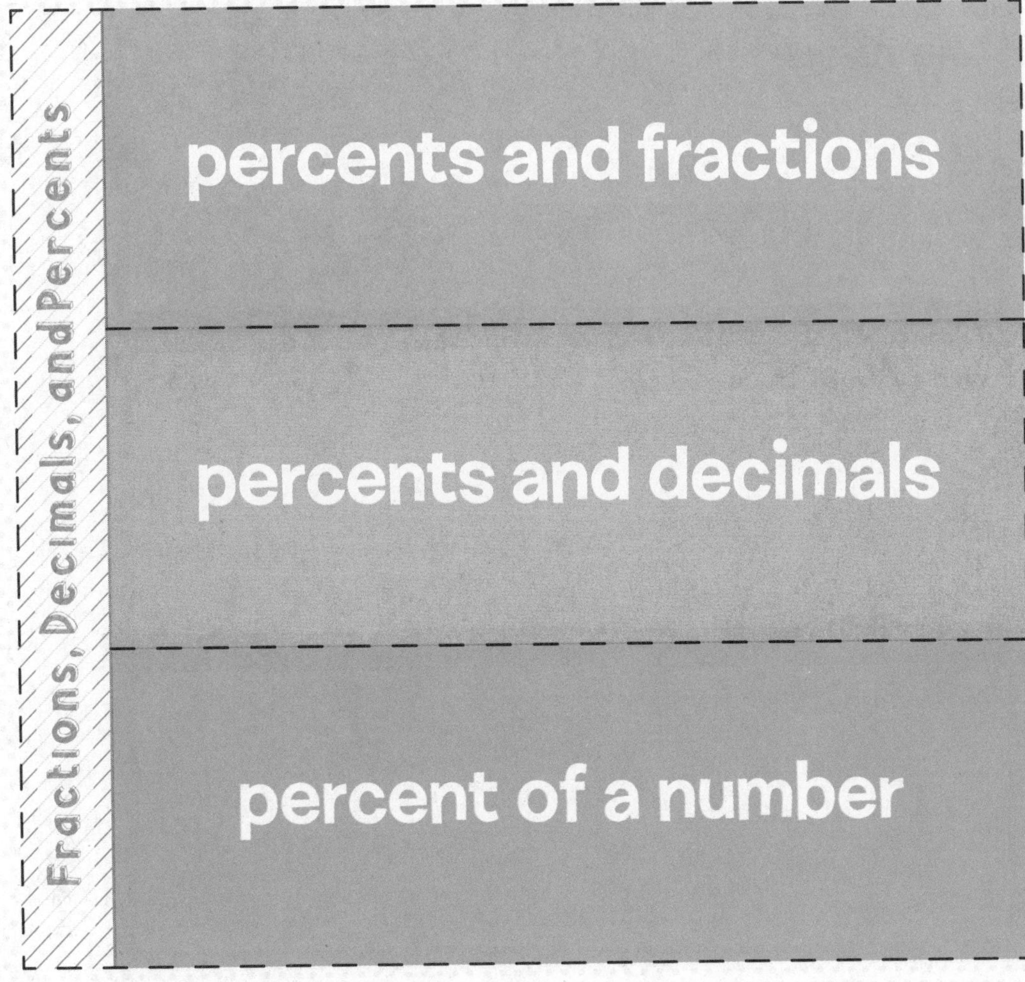
Fractions, Decimals, and Percents
percents and fractions
percents and decimals
percent of a number

cut on all dashed lines

fold on all solid lines

tape to page 166

FOLDABLES®

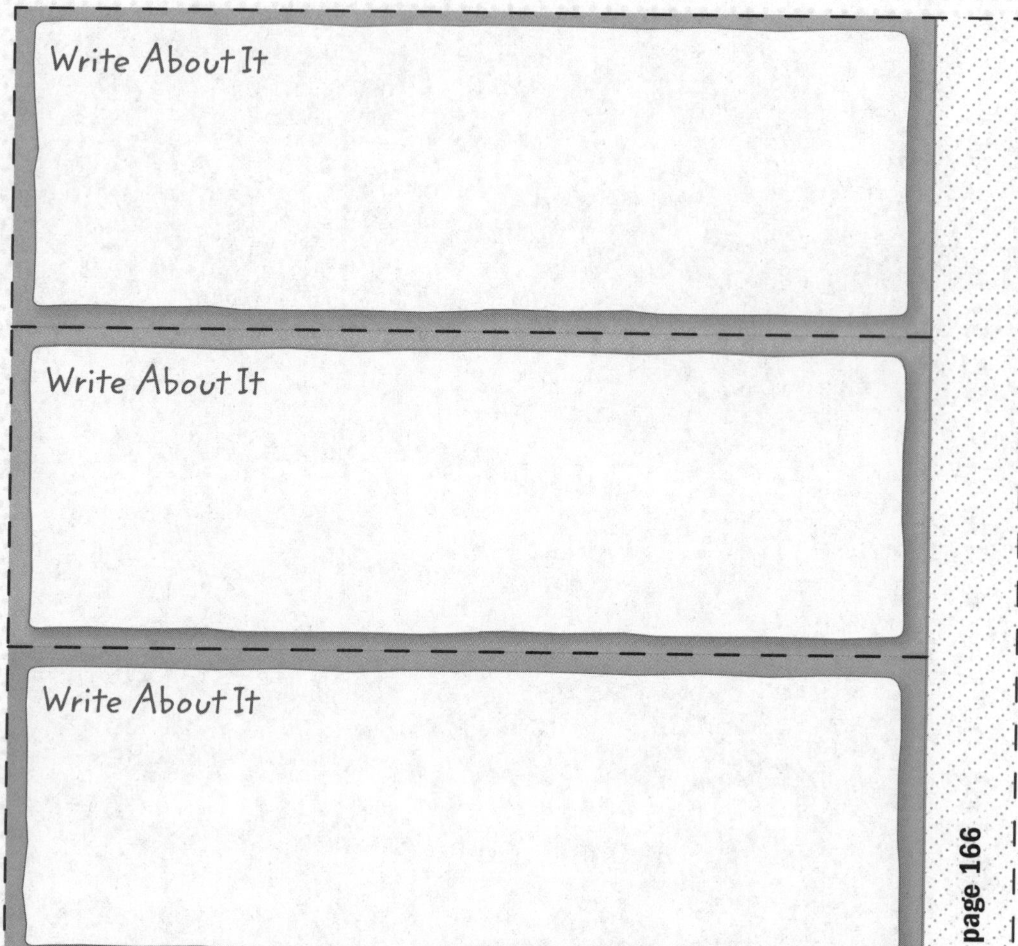

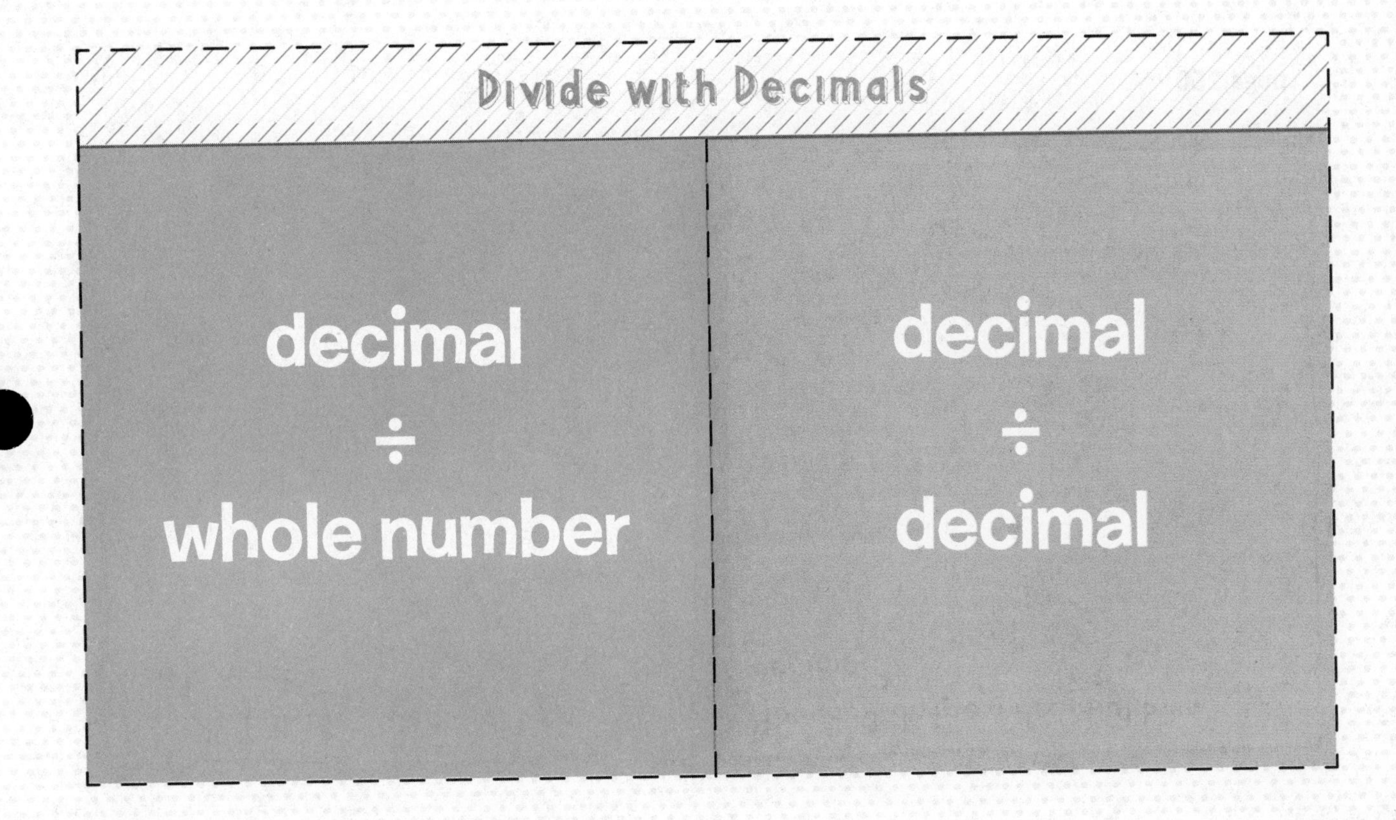

Divide with Decimals

decimal ÷ whole number	decimal ÷ decimal

cut on all dashed lines — fold on all solid lines — tape to page 250

FOLDABLES

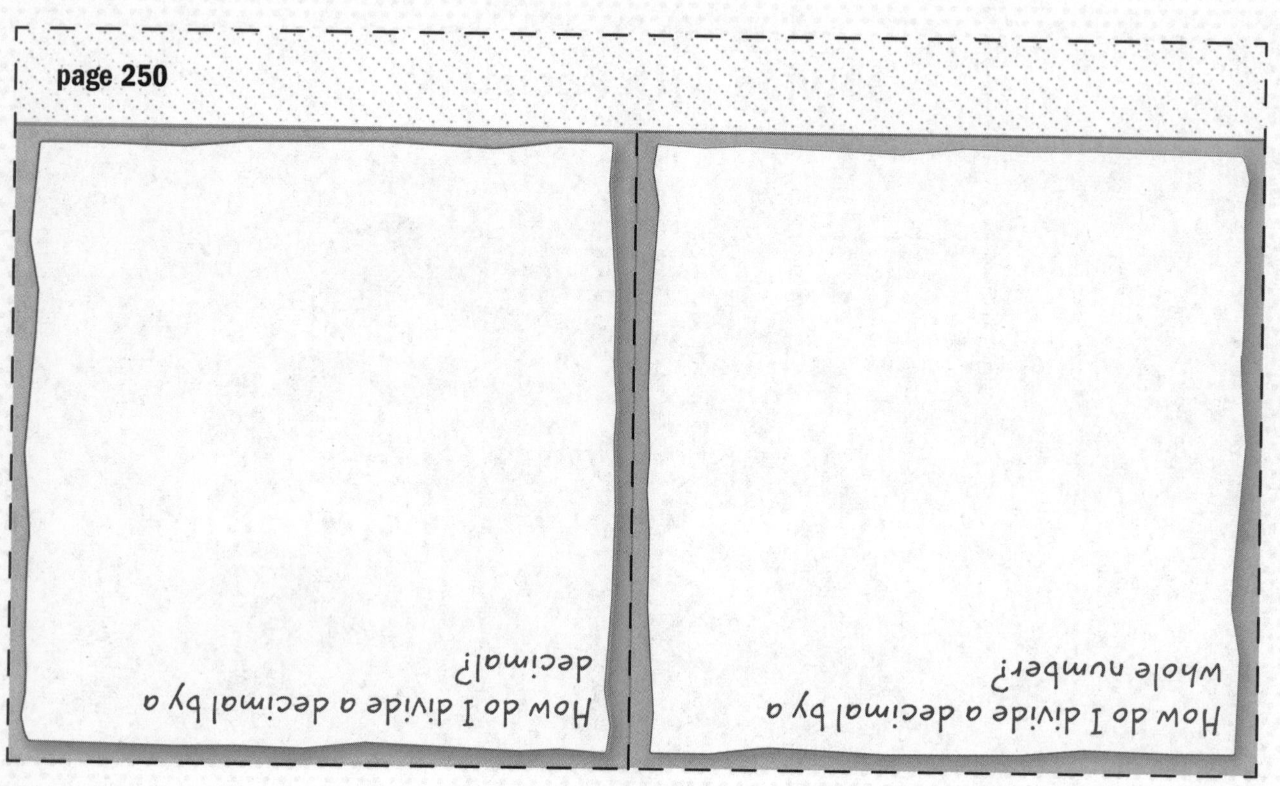

cut on all dashed lines | fold on all solid lines | tape to page 336

FOLDABLES®

Multiply and Divide Fractions

multiply	divide
Example fraction x whole number	Example whole number ÷ fraction
Example fraction x fraction	Example fraction ÷ fraction

cut on all dashed lines

fold on all solid lines

tape to page 336

FOLDABLES®

page 336

Tab 3

How do I divide a whole number by a fraction?

How do I multiply a fraction by a whole number?

page 336

Tab 2

How do I divide a fraction by a fraction?

How do I multiply a fraction by a fraction?

page 336

Tab 1

How do I divide a mixed number by a fraction?

How do I multiply a fraction by a mixed number?

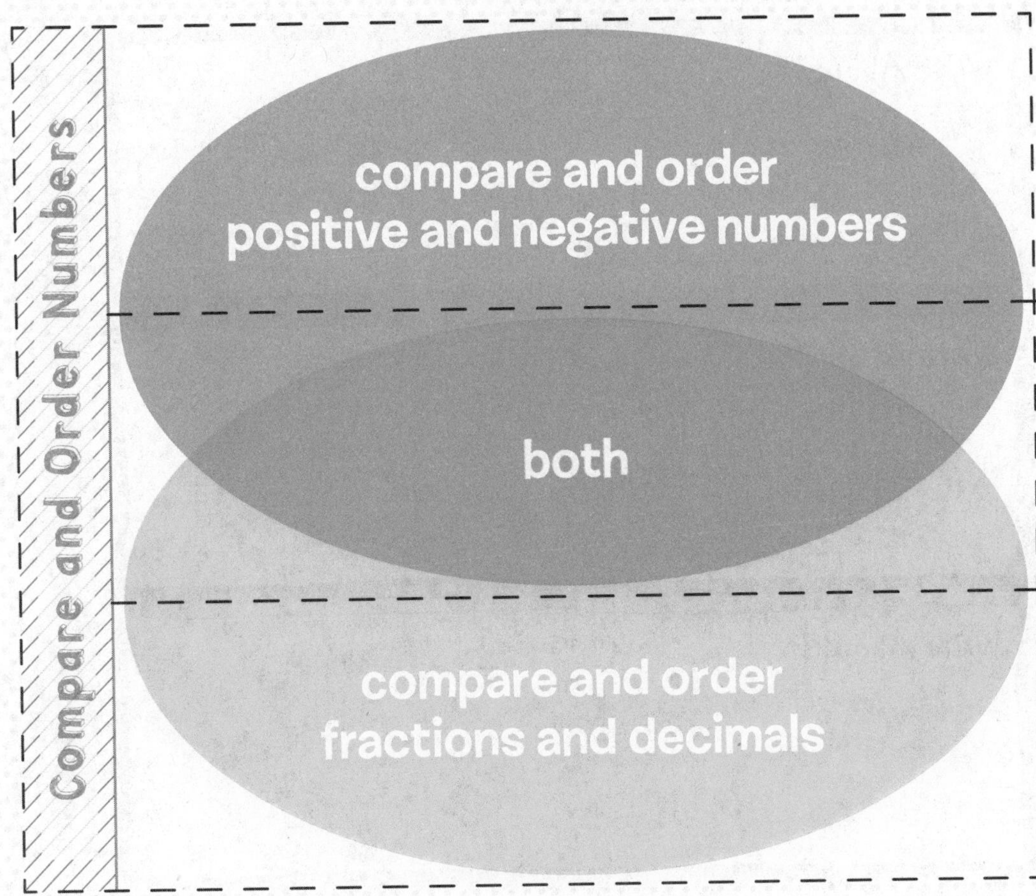
Compare and Order Numbers
compare and order
positive and negative numbers
both
compare and order
fractions and decimals

cut on all dashed lines fold on all solid lines tape to page 418 FOLDABLES®

Write About It

Write About It

Write About It

page 418